The Orders of Nature

The Orders of Nature

Lawrence Cahoone

SUNY
PRESS

Published by State University of New York Press, Albany

Printed in the United States of America

For information, contact State University of New York Press, Albany, NY
www.sunypress.edu

Production by Eileen Nizer
Marketing by Michael Campochiaro

Library of Congress Cataloging-in-Publication Data

Cahoone, Lawrence E., 1954–
The orders of nature / Lawrence Cahoone.
p. cm.
Includes bibliographical references (p.) and index.
ISBN 978-1-4384-4415-4 (hc : alk. paper)– 978-1-4384-4416-1 (pb : alk. paper)
1. Naturalism. I. Title.

B828.2.C34 2012
146—dc23 2011051054

10 9 8 7 6 5 4 3 2 1

We have a way of discussing the world, when we talk of it at various hierarchies, or levels . . . at one end we have the fundamental laws of physics . . . if we go higher up from this, in another level we have properties of substances—like 'refractive index' . . . or 'surface tension'. . . . As we go up in this hierarchy of complexity, we get to things like muscle twitch or nerve impulse . . . Then come things like 'frog' . . . And then . . . we come to words and concepts like 'man,' and 'history.' . . . And going on, we come to things like evil, and beauty, and hope. . . . Which end is nearer to God . . . Beauty and hope, or the fundamental laws? . . . I do not think either end is nearer to God . . . To stand at either end . . . hoping that out in that direction is the complete understanding, is a mistake. . . . The great mass of workers in between, connecting one step to another, are improving all the time our understanding of the world, both from working at the ends and working in the middle, and in that way we are gradually understanding this tremendous world of interconnecting hierarchies.

—Richard Feynman, *The Character of Physical Law*

Contents

PART III. NATURALISTIC SPECULATIONS

for Harry and Rose,

my best guess

and in memory of

Sarah Rose Broeder (1995–2011)

and Paul Baeten (1959–2012)

Acknowledgments

The author's Acknowledgements page of most academic books expresses gratitude to helpers while accepting sole responsibility for error. That usually reads like a pleasant expression of humility. But it must be taken especially seriously in the present case. Many have given help without which this project would have been impossible, but because the whole has evolved over the years under multiple influences—not least the author's—each of them would undoubtedly find something in the finished product objectionable.

My undergraduate teacher, the late Bernard Kaplan, set me on this interdisciplinary path long ago—the path of, as Donald Campbell called it, "incompetence in many fields at once." I owe thanks for instruction or critical readings to Andrea Borghini, Gregg DiGirolamo, Robert Garvey, Kornath Madhavan, Kenneth Mills, Karen Ober, Florence Shepard, Janine Shertzer, Abner Shimony, John Stachel, Robert Ulanowicz, Bruce Weber, and anonymous readers and editorial board members of SUNY Press. I am grateful for help, comments, and encouragement from members of the Society for the Advancement of American Philosophy, especially John Shook, Paul Thompson, and Kathleen Wallace; the Metaphysical Society of America, particularly Wes DeMarco and past president Dan Dahlstrom; and the Center for Process Studies, especially Philip Clayton, John Cobb, and Brian Henning. The support of the Committee on Faculty Scholarship of the College of the Holy Cross, and the encouragement and leadership of former president Michael McFarland, are much appreciated. The influence of Peter Brecher and of Walter Wright has been formative throughout. I owe special gratitude for multiple readings and/or tutorials, personal and virtual, to Elizabeth Baeten, Tian-yu Cao, Robert Cohen, Valerius Geist,

Andrew Hwang, Matthew Koss, Joseph Margolis, Robert Neville, Holmes Rolston, Stanley Salthe, Karsten Stueber, and William Wimsatt. My greatest debt is, as always, to my muse Elizabeth Baeten and to our children, Harrison and Isabel Rose.

I thank the publishers and/or authors of the following works for permission to reproduce figures (which I have sometimes modified as noted): Peter Hobson, *The Cradle of Thought: Exploring the Origins of Thinking*, Pan Macmillan, 2004, p. 107, Figure 2 (my Figure 9.1); and William Wimsatt, "The Ontology of Complex Systems: Levels, Perspectives, and Causal Thickets," *Biology and Society: Reflections on Methodology*, edited by Mohan Matthen and R. X. Ware, *Canadian Journal of Philosophy*, Supplementary Vol. 20, 1994:207–74, p. 230 (my Figure 3.2). Portions of some of my chapters have appeared earlier in: "Local Naturalism," *Contemporary Pragmatism* 6 (2), December 2009; "Reduction, Emergence, and Ordinal Physicalism," *Transactions of the Charles S. Peirce Society,* vol.44, no.1, Winter 2008; "Arguments From Nothing: God and Quantum Cosmology," *Zygon: Journal of Religion and Science* 44 (4) December 2009; as well as some discussions from my *Cultural Revolutions: Reason versus Culture in Philosophy, Politics, and Jihad,* Pennsylvania State University Press, 2005.

Introduction

We live in nature. That is surely a plausible truth, even if a partial one that leaves questions unanswered. Whatever else we are, however else we address other features of human existence, part of the truth about us seems to be captured by that claim. Today a considerable number of thinkers, and citizens, accept it as the primary truth. Many contemporary philosophers think of themselves as naturalists. Some in fact think naturalism so obviously valid as not to need philosophical argument.

But this surface complacency hides disagreement. Naturalism is widely understood to say that everything is in or part of nature, that nothing is supra-natural. That would seem to exclude divinity—to naturalism's credit, for some, but to its discredit, for others. Many identify naturalism with physicalism, the claim that everything is physical, a property of the physical, or determined by the physical. Consequently other philosophers have doubted that such a naturalism can give an adequate account of mind, culture, ethics, freedom, or art, that it "reduces" the most complex human features of reality to the most simple. Those are the traditional objections to naturalism. Just as trenchant today is the arguably "postmodern" objection, evident both in European and Anglo-American philosophy, that general or systematic metaphysics is an anachronistic, failed genre. Philosophizing about the world or reality in general, many think, is a mistake. So naturalism as general metaphysics is as illegitimate as any general metaphysics. This view was fueled by twentieth-century philosophical claims that deny we can have any knowledge characterized by certainty, finality, transcendence, a "privileged perspective," non-trivial self-evident truths, valid first principles, or a view of "the Whole." Following these claims, some naturalists think naturalism is not metaphysical at all, that nature is what we are left with when we *abandon* metaphysics. They hold we can endorse a "natural ontological attitude," accept science as the institution with the most knowledge about things, while dropping any general metaphysical characterization or reading of science's achievements (Fine 1991, Rorty 1991). So while many assume naturalism, or occasionally employ it, few want to explore the meaning and validity of a systematic naturalism.

There certainly is metaphysics in contemporary philosophy, including mainstream analytic or Anglo-American philosophy. Such work is usually concerned to articulate the necessary and sufficient conditions of features of reality, like possibility or entities or dispositions or individuals. But general or systematic metaphysics is less common. Historically, metaphysics has a set of canonical problems, like the existence of God, whether all is matter or mental or spiritual entities also exist, and how to understand phenomena like possibilities, cultural objects, meanings, universals, etc. A systematic metaphysics tries to inquire into many of these things all at once in a coordinated way. It is this kind of general inquiry that has the worst current reputation, seeming the most obvious suspect for an inquiry still seeking an anachronistic view of the Whole, which is impossible, or a view from "nowhere," which is inconceivable, or claiming to incorporate its own meta-language—the language in which the basic terms of the theory are defined—which is illogical. Thus a systematic metaphysical naturalism continues to arouse a variety of negative responses: if it's naturalism, it's not metaphysics; if it's systematic metaphysics, it's not naturalism; if it's both, it epitomizes the errors of traditional philosophy exposed by thinkers with names like Nietzsche, Dewey, and Carnap, not to mention Wittgenstein, Heidegger, Quine, Derrida and Rorty.

My disagreement is rooted not in a defense of tradition *per se*, but in the conviction that the above worries, while partly right, are exaggerated. I think it *is* possible to perform, and reach conclusions in, systematic or general metaphysics without imagining the achievement of certainty, or imperialistically "grounding" or "founding" other kinds of inquiry, or speaking of the Whole, or Simples, or the Foundational, or other sins of the past from which philosophy since the mid-twentieth century has sought repentance. It is possible to forge a metaphysics that aspires not to finality or the end of inquiry but to an adequate, yet corrigible, set of concepts for further inquiry, always vulnerable to our conceptual criticism and best empirical guesses about the world. And it is possible to formulate a naturalism that, while incorporating the physical, is not physicalist, accepts reductive and non-reductive scientific explanations as complimentary, and coheres with the work of multiple sciences, hence is scientific yet pluralistic. The remaining danger in pursuing such a project is that it might be wrong. But that is the danger inherent in inquiry, indeed in all human judgment, so hardly a reason for skipping the party.

There remains the indifferent response to general metaphysics which we could perhaps call the Alfred E. Neuman position: "What? Me worry about the nature of reality?" Today this view seems more common among some philosophers than non-philosophers. Aristotle thought that humans always have a series of unreflective judgments of reality in general, so the choice is between a naïve and a reflective metaphysics. It can be argued that reconsidering naivety is more pressing today than in his time, for we are forced by modern life to employ very different methods or pictures of reality—the scientific, practical, religious, personal, etc.—among which we must negotiate. Some think this diversity shows that the search for a general description is wrong-headed, and prefer a sophisticated superficiality. But that leaves specialized languages immune from critical relation to each other. It is true that we cannot expect a final unity, but between fragmentation and unification there are many degrees of relation and overlapping patchwork. It is hard to disagree with Aristotle that reflection is better.

Our particular journey will require us to trek to the intersection of *two* roads less traveled, one in our conception of metaphysics, and one in understanding nature. The first will hold that we can forge an approach to metaphysics that avoids the problems raised against general metaphysics. The second holds that nature and the natural sciences can be understood in such a way as to be compatible with the existence, irreducible distinctiveness, and causal relevance of meanings, culture, minds, and purposes in a predominantly material and physical universe. In short, naturalism can be scientific yet pluralist. If these claims are valid, then the way is clear to say that such a naturalism is our most likely approximately true general description of reality. That, at any rate, is what I will argue.

Now, beginning an inquiry into ethics or aesthetics or philosophy of science rightly presupposes definitions on which the inquiry is based, boundaries dictating what is *not* to be studied. But a systematic metaphysics claims an open-ended subject matter—what is not at least indirectly an object of its inquiry? This raises the problem of starting point in a particularly acute way. The fear is that, there being no philosophically neutral way to begin, the selection of a starting point, which probably will mean selection of both a preferred type of evidence or subject matter and a preferred way of handling that subject matter, will create an unjustified evidence-filter that biases and skews the enterprise.

For our world has lots in it. It seems partly physical and material. But it includes dimensions or zones of being or phenomena that seem neither physical nor publicly accessible, for example, my feelings and thoughts, as well as layers of interpretation that differ from one public to another, e.g., different cultures, societies, and the historical funding they bring, as well as all specialized forms of culture—science, art, politics, sports, etc. Our lives are multiple. We engage in publicly normed social activities, lose ourselves in private torment or reverie, become ecologists on camping trips, pour out our personal experience in public signs, act as idiosyncratic individuals then as role-performers or group-members, deal with technologies through scientific physicalism before we go to church, pepper economic activities explainable by functionalism with emotive decisions based on charisma and tribal resentments, act like a materialist one moment, a dualist the next, an idealist the third, etc. Edmund Husserl called the experiential *Lebenswelt* or lifeworld "pre-theoretical," but it might be better to call the lifeworld *poly*-theoretical, characterized (as Husserl knew) by the leavings of many specialized inquiries which feed their popularized and technological influences back into the world we share from their offices, libraries, and labs. Thus our everyday lifeworld is narrative and logical and experiential and physical and biological and psychological and historical and cultural and personal and social and semiotic and spiritual. How to begin to analyze that?

Certainly we will start with presuppositions. *Absolute* neutrality, or presupposition-less-ness, is unavailable. No inquiry starts from zero. But there are degrees and varying thicknesses of presuppositions. *Relative* neutrality, neutrality with respect to the choice between some particular X and –X, is entirely feasible. You can write an essay or teach a course that is neutral with respect to the existence of God, the legitimacy of America's invasion of Iraq, or the best flavor of ice cream. In everyday and scientific inquiry people have little trouble recognizing the difference between what they can't be neutral about, because it structures their inquiry—you can't study home runs without believing the game of baseball exists—and those things they can and perhaps should be initially neutral about, since choosing sides would bias the inquiry toward a candidate solution before you even get started. At the same time, our general metaphysics must be able to handle things in a preliminary way without predetermining our conclusions about them. Like any theory, it can be adequately justified only if its evidence can be described independent of itself. Consequently, we need *two* metaphysical languages: one that labels the evidence and sets out our approach to it in the most neutral way available, then a language that, we conclude, best accounts for that evidence. We begin with a language comparatively neutral with respect to competing theories and on that basis argue for a partisan language or theory.

I claim that we best start with a recognition of *pluralism*, the most open, least decisive perspective available. "Pluralism" in metaphysics just means there are indefinitely many things and properties, kinds of things and properties, and kinds of methods or evidence or arguments that may provide truth. Pluralism is the most (not completely) neutral, the least (not utterly) suppositional, the minimally (not un-) biased, language with which to begin. What this means, what kind of pluralism, or what language would qualify as pluralistic, must await later discussion (Chapter 1). But it is crucial to recognize that the fact that pluralism is the proper starting point does *not* mean it is the proper endpoint.

So I am nominating pluralism as our background metaphysical language. Such pluralism makes no *a priori* claim that any philosophical method is out of bounds. It needs saying after a century in which Western philosophy fragmented into schools denying even the relevance of competitors: to start with pluralism is to refuse to accept the general or *a priori* primacy of any one feature of reality, method, or type of evidence. To take normative logic, sense data, material objects, natural language, or social action, organic processes, public experience, or mind, private experience, history, or cultural signs—to prioritize any one of these as *the* context for inquiry is equally legitimate and inadequate. Thus neither ideal language philosophy, positivism, ordinary language philosophy, physicalism, pragmatism, process philosophy, German Idealism, phenomenology, fundamental ontology, hermeneutics, or poststructuralism, hence neither what are called analytic philosophy, American philosophy, or continental philosophy, are right or wrong, or even more right or wrong than the others, *in general*. We cannot presume any of them.

Given a pluralistic language for discriminating what there is to be accounted for, I will then argue that naturalism gives the most robust, comprehensive, and likely true account of it. But this requires the right kind of naturalism. First, such naturalism needs to be *local*, claiming not that all being is natural or part of nature, but that of what does, has or will exist, nature constitutes the most robustly accessible elements. Whether all beings or properties are natural is not decided *a priori*. The task is to see what and how much can be incorporated into, or inferred from, nature. Second, the nature it conceives is *pluralistic*. Metaphysically, we assume nature contains an indefinitely large number of entities, structures, processes, and properties. Just what kinds, how many kinds, and which kinds depend on, are constituted by, or determined by which, is a question to be answered by our best explanatory practices, not *a priori*.

Such a naturalism rejects the dominant bipolar disorder of modern philosophy, the belief that reality is constituted by *at most* two kinds of entities or properties, the physical and the mental, a disorder shared by idealism, dualism, and physicalism or materialism, reductive or nonreductive. That disorder encouraged us to think physics is the only metaphysically interesting natural science, that human mentality is the only part of nature that creates problems for a (physically oriented) metaphysics, that knowledge and mind are solely human possessions, that all the other natural sciences—chemistry, the Earth sciences, biology, engineering—are metaphysically unimportant. This dualism has been repeatedly and recursively applied, multiplying sub-schools (for example, between "scientific" naturalists and "humanistic" naturalists), but always with the same tendencies. It arguably has something to do with the congealing of twentieth-century Western philosophy into two opposed hermetic traditions, analytic and continental philosophy, one (in its metaphysics) tending to focus on highly specialized problems in the interpretation of physics and the possible reduction of mentality, the other rejecting natural science as inhospitable to whatever matters to the human prospect (there being some exceptions on both sides who, as is said, prove the rule). In a broader context, both are manifestations of the conflict of C. P. Snow's the "two cultures" (Snow and Collini, 1993).

In contrast the current naturalism bases itself in multiple sciences, not just physics. It accepts *emergence*, the presence of irreducible properties at levels of complexity, or what is the same thing, the reality and causal relevance of hierarchically arranged complex systems and processes. The idea is not new. It is a re-fashioning of a genre of post-Darwinian naturalistic metaphysics active from 1870–1930, and epitomized by the British Emergentists of the 1920s. Such thinkers—who pre-dated the division of philosophy into analytic and continental schools—accepted the metaphysical relevance of multiple sciences and saw nature as complex and evolving. Largely cast aside by philosophy, their conceptions remained alive in the work of some revisionist theologians and interdisciplinary scientists concerned with emergence and hierarchical systems theory, recently resurrected by scientific work on complex systems.

Such a naturalism must be interdisciplinary. That presents unique challenges, and accounts for some of the strangeness of what follows. For to work across disciplines means adopting a language foreign to most inquirers. The problem is well described by Donald Campbell. He once proposed a "fish-scale model of omniscience," claiming that human knowledge covers reality, to the degree it does, only collectively as a set of overlapping fish-scales of different perspectives and methods. The quotation is lengthy, but worth our time.

> To do the job right . . . I need my fish-scale to overlap to some extent with each of the following areas: physiological psychology . . . psycholinguistics, social psychology . . . sociology . . . sociology of knowledge . . . history of science, social evolution, biological evolution, philosophy of science, and philosophical epistemology. It is obvious, but still needs saying that I am not competent in any one of these areas, not even in the social psychology that

> is my job description. . . . If there is a coherent specialty lying in this interdisciplinary space . . . it may have to be done by marginal scholars who are willing to be incompetent in a number of fields at once. Such bravery is apt to fluctuate wildly. . . . Trial-and-error is essential . . . for conceptual innovation . . . [a] necessary process of sticking one's neck out and making mistakes. . . . knowing is fundamentally a social process, a social interaction of individual gambits . . . requir[ing] corrective critical responses from those whose areas overlap. This is your responsibility, and to exercise it effectively, you have to be willing to keep up the conversation with those who speak your language imperfectly, patiently correcting their misconceptions while still encouraging their efforts. It is at this point where our collective process so often fails, where bold explorers of interdisciplinary space get no response, critical or otherwise . . . and end in paranoid isolation if they persist at all. (Campbell 1988a, pp. 438–39)

There is no avoiding the fact that graduate education—in the sciences as well as philosophy—trains each of us to practice *one* subfield in *one* methodological language, and success, if it comes, typically comes from and within that particular community. Now, as Max Weber made clear, there is no point in romantic bemoaning of specialization. It is inevitable in any modern field of research or practice, and it has produced tremendous advances in knowledge. But it comes at a price, even if the price is usually worth it. The cost is that at any moment a field or subfield has a set of canonical problems and prime candidate theories for addressing them, which dictate the "phase space" of relevance, so that any work that does not advance a candidate theory is regarded as irrelevant. Silence, not condemnation, is the usual response to work that crosses intellectual specializations.

Successful interdisciplinary work entails a multi-phased learning process. The aspiring "interdisciplinarian" starts with a patchwork of competences, and must move from the more to the less familiar. First is the absolutely necessary moment of "charity," in which whatever the other discipline says must be trusted to make sense, and to be logically and rhetorically formulated in the way it must be, even if that is not apparent at first. For the standard the inquirer pursues is that of the practitioner of the *other* discipline; the goal is to understand that field's concepts the way its practitioners understand them, like an insider, evident in both theoretical statements and practical handling. This is probably unattainable, but the interdiciplinarian must take it as the goal nevertheless. Throughout, she remains responsible to criticisms from those disciplinary specialties. Eventually, though, this attitude must be supplemented by another, the willingness to test the other field's claims against one's *own* language for describing the world, the language of an outsider. This is necessary, not only for critique of the former, but even for understanding it, since we never quite understand an intellectual claim until we have to decide whether or not it is true. The result, hopefully, in the third moment, is that one's language has been stretched, and, if luck is with us, shows its mettle not only in meeting tests of experience better but in its capability to describe the limitations of both the other disciplinary language *and* its former self (Simpson 2001).

This dialectic must be provoked repeatedly at increasing levels of complexity of discussion. And it must happen socially. There is knowledge in an inquirer's way of handling material that never makes it into her explicit, published or sometimes even spoken formulations. The hardest task is the social one, for nobody (or at least, no academic) is strong or noble enough repeatedly to stick her neck out in the physical presence of the headsman (typically, not more than once). But neither does a neck massage from an excessively genteel critic do any good. Finding the right interlocutors is a job for any inquirer who has come unstuck in the disciplinary world. While it is probably true that we are condemned to the task of, in Otto Neurath's famous figure, rebuilding our ship while at sea, the interdisciplinary inquirer recognizes that we inquirers sail not *a* ship, but an armada of flotsam, from pieces of driftwood to dugouts to rafts to elegant dinghies and sailboats, each sustaining its tiny crew in a vaguely common direction by its own methods, some on course, some zigzagging, some in the doldrums, some sinking. Part of the job is for each crew to stay afloat. But the other part—our current task—is to find a way to lash our cognitive vehicles together, because in inquiry, which is by nature a public dialogue, only together can reliable progress be made.

The four chapters of Part I introduce the metaphysical perspective. In Chapter 1, I outline a pluralistic approach to metaphysics that combines what I call a local, or non-global, metaphysics with an argument for naturalism. Chapter 2 is an historical overview that explores the relation of the current naturalism to older forms. Chapter 3 presents a defense of explanatory and ontological emergence compatible with scientific reduction, on the way to arguing against physicalism, whether reductive or non-reductive. Nature includes, but is neither defined or determined by, the physical. Chapter 4 suggests a set of basic concepts for analyzing nature understood pluralistically, inspired by hierarchy theory. The result is a metaphysics of five serially dependent and increasingly complex orders of nature: physical, material, biological, mental, and cultural.

The heart of the book, Part II, describes those orders. Chapter 5 is a description of the physical order derived from physics that teases out a tentative ontology of the physical. Chapter 6 discusses astrophysics, Earth science, chemistry, and thermodynamics to show that what we normally call "matter" is a complex, developmental feature of nature dependent on special conditions. Chapter 7 presents a philosophical account of the phenomena of life and the indispensability of teleological (or more precisely, teleonomic) explanation. Chapter 8 gives a theory of mind and mental causation as an animal, not merely human, phenomenon. Chapter 9 ascribes the uniqueness of the human mind to a form of social relating that permits the joint manipulation of signs, hence the ability to recognize and handle meanings, leading to the creation of culture. Chapter 10 sketches a theory of knowledge based in evolutionary epistemology to explain the validity of the scientific knowledge used in this book, and its compatibility with the cultural status of human knowing, in a circular but non-vicious fashion.

Something further must be said about Part II, and by implication, the book as a whole. Its chapters in several cases present basic science, that is, some introduction to modern physics, astronomy, chemistry, biology, neurology, ethology, paleoanthropolgy, etc. This may be tedious for those already familiar with these fields, and for others impatient to reach the philosophical argument. But one of the convictions fueling this book is that the empirical contents of multiple sciences, and the ways we describe them, are metaphysically important. A systematic metaphysics must honor these distinctive languages, for each impacts how the metaphysician should handle not only it, but all the others. For example, the problems of philosophy of mind are transformed once one reviews the definition of the physical with respect to physics, the teleonomic behavior of biotic systems, and basic ethology. The influence runs in the other direction too. What makes living and chemical systems distinct from the physical puts limits on how we conceive the latter. In each chapter I have tried to segregate the scientific expository work into sections and paragraphs, alerting the reader what to expect.

Part III is a speculative coda to the work, extending the argument to two canonical metaphysical issues on the basis of the foregoing. Chapter 11 forges arguments for the existence of a Ground of Nature, or God, as the cause of the universe and the source of its fine-tuned physical constants, in the process discussing recent work in physical cosmology. Chapter 12 seeks a contemporary contribution to the tradition of natural religion, claiming that the Ground—understood as at least partly physical and lacking omniscience and omnipotence—has initiated a nature whose contingent evolution in fact exhibits a direction: the creation of systems of increasing complexity. That direction is not a matter of necessity, but it seems to be an historical fact. This supports a natural claim of value: complexity *qua* complexity is *a* (not *the*) good. At any rate, these final chapters do not supplant, but expand, the pluralistic naturalism of the earlier chapters. For the point of the work as a whole is that a properly understood naturalism is likely true.

Part I

A Kind of Naturalism

1

From Pluralism to Naturalism

This chapter will argue for a systematic naturalistic metaphysics, understood in a particular way. The motivation is to answer, or diminish, traditional objections to naturalism and recent objections to systematic metaphysics. Doing so will distance my naturalism from others, including some held to be obvious, intuitive, or valid without argument as a kind of non-metaphysical default position. For naturalism is a metaphysics and it must be defended against two sorts of criticisms: that naturalism is too narrow, being incompatible with any adequate account of mind, meanings, culture and, relevant only to some, the divine; and that no systematic or general metaphysics, naturalist or physicalist or dualist or idealist, can be justified.

The naturalism described herein will be distinctive. As Section I of this chapter will describe, it understands metaphysics as fallibilist and *a posteriori*, and rejects metaphysical and methodological globalism, the notion that the validity of a metaphysical analysis of a thing or order of things hangs on the valid characterization of the most inclusive order in which it functions. Hence (in Section II) it avoids all talk of the Whole or Foundations, instead adopting a radically pluralist language for the discrimination of any being or evidence whatsoever. That is our background metaphysical language. Naturalism will then (in Section III) be hypothesized as the most robust theory to account for whatever is discriminated within this pluralism. That is, a localist approach to metaphysics allows us to adopt a naturalistic perspective within or on the basis of pluralism, resulting in a pluralistic form of naturalism capable of employing the work of multiple sciences while blunting traditional criticisms. Given all this, I will argue (in Section IV) that *naturalism is at least locally true*. The task will then be to work out such a naturalism in conversation with multiple sciences, showing that important features of reality can be included in it. That will require the rest of the book. The present chapter tries only to outline the project and show that it stands a chance.

I. A Fallibilist and Local Metaphysics

John Herman Randall, a philosopher at Columbia University, argued that metaphysics, on Aristotle's view, is distinguished from other inquiries by its subject-matter, not a special method. It investigates, "the general characters and the ultimate distinctions illustrated and exhibited in each specific and determinate kind of existence and existential subject-matter" (Randall 1958, p. 144). This approach differs, he claimed, from the traditions of metaphysics that have sought the Unity of existence, trying to synthesize all knowledge into a unified system, or the True Being behind all appearances (Randall 1958, pp. 124–33). Following Aristotle, Randall argues metaphysics is the inquiry that seeks the most generic features of the plural kinds of determinate beings that obtain—all existence being at least partly, not completely, determinate—and are studied by all other disciplines. This means what distinguishes metaphysics from other inquiries is its generality, not its method. Philosophy, including metaphysics, is inquiry, continuous with other forms of inquiry from physics to art history. It is only more general.

A century earlier the American philosopher Charles Peirce argued there is no type or line of argument that is infallible or certain or complete; there are only degrees of likelihood, trustworthiness, and confidence. We never know anything with certainty, and we never know everything about anything. We can hope for neither certainty nor completeness in any inquiry. Peirce extended this as far as to include even deductive arguments, for the simple reason that even mathematicians make mistakes. Sometimes these are errors of reasoning, more often ambiguities which accumulate along a chain of arguments, as do perturbations in some physical systems. As a result, Peirce claimed that philosophy ought more to trust a plurality of seemingly reliable and compatible arguments from different sets of premises than a single deductive series of arguments each member of which is dependent for its reliability on the preceding argument.

> Philosophy ought to imitate the successful sciences in its methods . . . to trust rather to the multitude and variety of its arguments than to the conclusiveness of any one. Its reasoning should not form a chain which is no stronger than its weakest link, but a cable whose fibers may be ever so slender, provided they are sufficiently numerous and intimately connected. (Peirce 1992b, p. 29)

This means avoiding, as a basic or global strategy, deductive or axiomatic methods, as well as dialectical method (in which the content of one concept leads to an alternate concept it philosophically implies or presupposes, the combination of which implies a third that overcomes the antithesis between them). It does *not* mean, of course, that deduction and dialectic are never to be used, only that they ought not characterize the overall argumentative structure. Under Peirce's cable metaphor the justification of any claim will be a bundle of more or less independent reasonings toward the claim, what we might call *argumentative pluralism.*

Akin to Peirce's, and Randall's, approach is something that, oddly, seems to go unrecognized in some quarters: that the validity of a metaphysical theory can hang on empirical generalizations which might later be shown to be false by improved empirical methods. In short, metaphysics can be *a posteriori*. An example is Abner Shimony's notion of "experimental metaphysics."[1] Shimony holds, as did most philosophers of the seventeenth century, that metaphysics ought to make sense in terms of the best science of the time. The early modern philosophers, however, attempted to do so by creating an *a priori* philosophy, in which the justification of their chosen ontology was deductive, although their reason for choosing it—in the order of discovery, one might say—was in fact its inferential appropriateness to current science. Shimony, following Peirce, is unafraid to infer, fallibly, from the empirical science to the ontology. Given his work in the conceptual foundations of quantum mechanics, he attempted to glean what must be true of the ontology of the natural world for the science to be as it is, taking into account differences of interpretation, likelihood of theory stability, and guesses at what may come later on. Of course, as Shimony rightly says, "One should not anticipate straightforward and decisive resolution of metaphysical disputes by the outcomes of experiments," since the significance of those outcomes will be highly mediated by other notions and dependent on conceptual analysis, all legitimately evaluated with respect to coherence with explanations of other phenomena (Shimony 1993b, p. 64). Any of our claims, including metaphysical claims, are open to rejection based on their failure adequately to cohere with our other reliable guesses about things.

William Wimsatt, partly inspired by Peirce's cable notion, has recently developed another idea connected to argumentative pluralism, *robustness* (Wimsatt 2007, pp. 42–74). Those phenomena are robust to which we have multiple means of access, whether via multiple sensory modalities, multiple ways of measuring, or multiple independent theoretical inferences. The conviction is that multiplicity of independent sources of measurement, experience, or description, must enhance confidence (which is not to say achieve certainty). Following Donald Campbell's invocation of the importance of coincidence of object boundaries for vision (opacity) and touch (impenetrability), Wimsatt notes that access by multiple sensory modalities is a deeply entrenched human criterion of objectivity (Campbell 1960). One might say empiricists, positivists, and phenomenologists made similar claims, but they gave evidential priority to degree of immediacy rather than relative invariance across inquirers, observational circumstances, or areas of inquiry. Robustness is the Peircean alternative to an idealized immediacy that twentieth-century philosophy showed to be unavailable. Wimsatt suggests robustness is the appropriate argumentative strategy for error-prone beings of finite reasoning capacity, namely, us.

It should be noted in passing that a fallibilist and *a posteriori* metaphysics is entirely compatible with epistemic realism, the claim that our true knowledge is made true, at least in part, by its objects. (A fuller discussion must be postponed to Chapter 10). Certainly the validity, or truth, of our judgments is relative to a host of nested characteristics of the judgment: its natural language, its logic, its conceptual grammar, its perspective, its encompassing theory, etc. A chastened realism can admit all that. Particularly important for what follows, the fact that we aspire to true judgments made true by a relation to their objects does *not* say what kind of objects there are. There is a tendency in the discussion of epistemic realism versus anti-realism (the view that truth is fixed by relations among our judgment) in the philosophy of science to assume that realism must refer to *entities*. Some of the most sophisticated commentators continue to presuppose that realism is tied to entities, and so claim, for example, that since quantum mechanics undermines traditional notions of entity it likewise undermines realism.[2] But surely what reality is like, or what the objects of an inquiry are, are contingent questions that should not be preempted by the definition of knowledge or truth. Structures, relations, processes, interactions, events, states, or properties are no less real and may, given the circumstances, be more explanatorily relevant than entities. Epistemic realism need not presuppose a particular metaphysics.

Now to a key methodological point: metaphysics can be "local." Localism in metaphysics signifies a rejection of methodological globalism. The globalism being rejected is evident throughout the history of philosophy in thinkers as disparate as Plato and Democritus, Hegel and Quine. The rejected view claims that the metaphysical validity of any description or explanation of any being or order of beings necessarily hangs on the relation of that being or order to more inclusive orders of beings, hence transitively to the most inclusive order. Bertrand Russell and others rebelled against F. H. Bradley's idealism for just this reason, that it seemed to imply that the metaphysical connections among the plural orders of things was so tight that nothing valid could be said about a cup or spoon until one knew the role of the cup or spoon in the context of the Whole (although Russell went on to construct what is arguably another version of the same approach). If we reject such globalism, the task of metaphysics is to begin with robust or more reliably accessible and knowable orders of things, and, having described them and their properties and performances, to relate those orders to other orders that are less robustly accessible or more controversial. Metaphysics on this conception is local, it describes one neighborhood, then another, then another, and relations among them. In Wimsatt's term, it proceeds "piecewise."

Notice that this localism is not synonymous with what Husserl called "regional ontology" or Strawson "descriptive metaphysics" (Husserl 1982, Strawson 1990). Those try to describe the nature or the necessary and sufficient conditions of a kind of being or beings, e.g., experienced physical objects or individuals. These are specialized metaphysics of a particular zone of reality. Local metaphysics does describe particular orders but afterward invites us to push outward to other types of orders. It is general but not global. It does not say that the location of an order of beings in more inclusive orders of beings is *irrelevant* to the understanding of the former, or that we ought to renounce the aim of pressing our understanding as far as possible. Rather, it regards the location of an order in more inclusive orders as an ongoing project whose present unavailability does not undermine the validity of local ontologies. For it is the local descriptions against which any broader and more inclusive scheme must be tested. A robust approach to metaphysics does not hold its description of types of being hostage to a description of the most inclusive order. Hence localism concerns itself first of all with those descriptions of beings that remain invariant with respect to differences of global ontology.

Imagine three philosophers sitting at a lunch counter discussing metaphysics, one an eliminative materialist, another a Spinozist, the third a Berkeleyan idealist. The Spinozist drops her spoon and the others lunge to grab it before it falls. The question is: to what degree are their antithetical beliefs about the most inclusive order of being entangled in, hence determinative of, their perceptions, attitudes, actions, and expectations about the spoon, e.g., about what it is, what its use is, what is happening to it, or what ought to be done about it? With respect to local description, the answer seems to be: negligibly little. All three believe that spoons are for eating, hands can grab spoons, friends help friends, and eating utensils are better when clean, regardless of whether they think all is matter, nothing is matter, or matter and mind are parallel processes.

We may take a famous philosopher's example. Imagine an anthropologist and a native who share no linguistic commonality walking through the forest. Suddenly the native points at what the anthropologist recognizes to be a rabbit and shouts, "Gavagai!" (Quine 1960). Quine's point was that the anthropologist's observation of the native's verbal behavior, in connection with his/her nonverbal behavior and the observable environment, will always be inadequate to specify whether "Gavagai!" means individual-physical-object-rabbit or particularization-of-the-form-of-rabbithood or momentary-phase-of-the-process-of-rabbiting. The native's ontology could be any one of these, and no native behavior or anthropological observation could discriminate between them. Quine called this the "indeterminacy of translation."

That famous phrase was overstated: the example shows only translation's *under*-determination. For there are all sorts of meanings ruled out, or made highly unlikely, by the behavioral situation. If the native is competent mentally, linguistically, and visually—hence an appropriate object of Donald Davidson's principle of charity—"Gavagai!" does not mean, "Dog!" or "Water!" or "Myself!" or "Don't look, nothing's happening there!" In fact, the possible meanings which cannot be ruled out or decided are a rather rarefied class, even if indefinitely large in extension. Leaving the ontology free to float, that is, *not* deciding whether the native meant "individual-physical-object-rabbit" or "phase-of-rabbiting" or "instance-of-rabbithood," in how many situations of interaction with the native is the anthropologist likely to go wrong? Very few, as Quine recognized. The anthropologist and native could identity and re-identify Gavagai, capture it, together make it a pet or a meal, without ever going wrong. The point is the native's or anthropologist's ontology may have no decisive role in fixing the contextual meaning of "Gavagai."[3]

In metaphysics localism decouples the understanding of anything from the description of the most inclusive order of being, whether that be Democritean atoms, Platonic Forms, a Hegelian Absolute, Husserlian lived experience, Whiteheadian actual occasions, Heideggerian *Sein*, Quinean physical objects, Derridean *différance*, or any conception of the Ultimate, the Comprehensive, or the Underlying. If globalism were true we would be in permanent trouble, for our knowledge of the ultimate *must* be less reliable than our knowledge of more robust scales. The extreme of physical reductionism, which would claim all existents are nothing but collections of or interactions among the simplest beings, *and* the extreme of idealism, whether Hegelian or Platonic, whether claiming all is a manifestation of Spirit or of eternal Forms, are equally violations of localism.

The rejection of globalism has two special consequences. Given the absence of reference to a Whole or Foundations we cannot assert the *a priori* or general priority of any one feature of reality, or any one method of investigation. Physics, phenomenology, cultural studies, pragmatism, biology, logic; quantum fields, experience, signs, social action, organisms, meanings—none is first *a priori*. We can of course make one of them first in our account of reality, but we must argue and give evidence for that. Its priority cannot be built into the conceptions with which we start our general metaphysics.

This is related to the second consequence. The loss of the Whole enables us to distinguish the language in which we discriminate beings—our starting point, a bit like a "meta-language"—from the language that we conclude gives the most likely true and intelligible account of them—like an "object-language."[4] We must be able to handle things in a preliminary way without predetermining our conclusions about them. We must select a starting point as *comparatively* neutral as possible with respect to major competing metaphysical theories, complete neutrality being impossible. The comparatively neutral language will be substantive; it will not reveal beings naked of our conceptual formation. However it will be relatively less substantive or partisan with respect to anticipated metaphysical disagreements than any other language. Also, like any theory, a metaphysical theory can be evidenced only if we can state the evidence in a language independent of the theory. We have to be able first to "name" things in a way that does not presuppose what we will decide is the best theory of them. Thus we need *two* languages: a more neutral language for setting out *what there is* to account for, and a less neutral language in which we *account for* it. We can use these two languages because we regard *any* language or theory as a hypothetical, limited reference point for maximizing probable truth and intelligibility, not a description of the Whole.

II. A Pluralist Language for Metaphysics

So, our metaphysics is going to be fallible, open to interdisciplinary and empirical information, a hypothesis to account for whatever we discriminate in a localist, pluralist preliminary language. We need the latter to begin. But what should it look like? Thomas Nagel famously denied that philosophy can seek a "view from nowhere," as if from outside the universe, because all judgments are made from somewhere or from some particular perspective (Nagel 1989). Presumably this also means there can be no view from everywhere, that is, from all possible perspectives at once. But a view from *anywhere* would be quite another thing. Eschewing the attempt to characterize the Whole it would provide a scheme by which anything can be analyzed locally.[5]

Such a language exists. It was developed by the American philosopher and scholar of Peirce, Justus Buchler, arguably the most systematic pluralist in recent metaphysics. In his "metaphysics of natural complexes" Buchler stipulated a principle of *ontological parity*, according to which nothing we can discriminate is more or less real or genuine than anything else (Buchler 1990). That is, he rejected entirely the various traditional philosophical distinctions between the "real," "true" (regarding things, not propositions), or "genuine," and the "apparent," "epiphenomenal," or "illusory." A fictional character, the possibility of my dying, the imaginary number *i*, and Heaven are all no less real than the computer keys under my fingers. Anything that can be discriminated, hence anything that is or was or will be in any sense, is a "natural complex." Complexes can be physical objects, facts, processes, events, universals, experiences, institutions, numbers, possibilities, artifacts, and all their relations and properties and functions. The theory of natural complexes is a natural complex. For Buchler the qualifier "natural" signifies that there can be no discontinuous realms of complexes, no worldly versus transcendent complexes, while the noun "complex" means that nothing is simple or incapable of further analysis. Like Peirce, Buchler denies that anything is either utterly determinate or absolutely indeterminate, or that the traits of any complex can be exhausted.

Pluralism and parity require Buchler to endorse *ordinalism*. The question "What is real?" is transformed into, to use Randall's phrase, "*How* is something real?" or for Buchler, "In what orders of relation does it function?" (Randall 1958, p. 131). This is what replaces our usual distinction between the real and apparent. A fictive truck and the truck bearing down on me are equally real, but the fictive truck functions in a literary order while the truck approaching me stands in an order of physical fact that includes my body. Every complex must be related to some other complexes—which is *not* to say related to *all* others, for things can be unrelated—hence is located in one or more contexts of relations or orders in which the complex functions and hence has an "integrity." Complexes and orders are related to others either strongly, to the other's integrity—hence an internal or constitutive relation—or weakly, to its breadth or scope in that order. A complex's identity or, in Buchler's terms its "contour," is the continuous relation between each of the complex's integrities and its total collection of integrities (Buchler 1990, p. 22). He extends parity to possibility and actuality. Possibilities are as real, as experience-able, as potentially causal, as actualities. They are traits of a complex in whatever order in which it functions. There are no fully determinate actualities, actualities without possibilities, nor "pure" possibilities without actualities.[6] Possible traits must be actualizable, hence commensurate with the identity of the complex; a baseball player has the possibility of striking out but not of scoring touchdowns.[7] Partly to accommodate possibilities and nonfactual orders, Buchler uses the language of "prevalence" and "alescence." A complex prevails in an order when it excludes other complexes from that order, or traits from its contour or identity; it alesces in so far as it admits traits into its contour and ceases to prevail. The rain prevails when it is raining, and alesces as the sun returns. A possibility prevails in an order, even though it is not "actual."

Last, Buchler rejects any substantive talk of the "Whole" or the "World." There is no "Order of orders" (Buchler 1990a,b). Such a Whole would fail to be ordinally located, it would not be related to anything outside itself. There can be no hierarchy of complexes and orders that is not itself ordinally located. Each thing is objectively contextual; we cannot determine non-contextual facts about contexts. We may use the phrase "innumerable orders" or the "provision of complexes" to speak of complexes indifferently.

I suggest that Buchler's is the closest thing we have to a metaphysics of any possible world. This is not to say it is presuppositionless, or neutral with respect to all other philosophical systems, but it is the nearest thing to it, the least suppositional, and the most neutral with respect to standard metaphysical problems. Conceive realities or worlds very different from our own, for example, a system of disembodied spirits, a quark-plasma, a set of Platonic forms, an Olympus of Greek gods, or a world in which beings otherwise like ourselves routinely communicate telepathically. Buchler's metaphysics would apply equally well to these worlds. Buchler deploys a scheme that is determined by only four parameters: a) pluralism, the claim that there are multiple "things," each exhibiting multiplicity or complexity, hence nothing is simple; b) ordinality or logical distributiveness, each integrity of a complex obtaining in a context of relationships, at least some of which are "strongly related" or in usual philosophical language "internal" to the complex's identity; c) continuity, meaning no orders are utterly discontinuous with each other, e.g., there is no transcendent or supra-natural versus mundane or natural orders; and d) parity, that there are no non- or supra-ordinal norms with respect to which complexes can be metaphysically ranked.[8] As long as a possible world does not violate these parameters it can be understood through the metaphysics of natural complexes. Buchler's is the most pluralistic metaphysics we have.

But if Buchler's metaphysics fits many possible worlds, then it is equally true to say that it does *not* pick out this world. His scheme underdetermines our reality. For example, as far as we can reliably judge, in our world lives, minds, selves, intentionality, and meanings require organisms, matter, bodies, neurons, and cultural objects, respectively. These relations of dependence are not symmetrical, for while living things presuppose the existence of atoms, atoms do not presuppose the existence of living things. Buchler's ordinal metaphysics thus allows all sorts of facts and processes that either cannot or at least do not occur in our reality as far as we can tell. Whether this is a vice or virtue depends on one's view of what the business of metaphysics is, that is, how far a metaphysics should go in fixing or entailing features of our world. My point is that Buchler's metaphysics is not by itself naturalistic in any strong sense; it merely denies supra-naturalism or transcendence.[9] It is a pluralism, not a naturalism.

The virtue of Buchler's scheme is that it provides us with an indispensably pluralistic background language for metaphysics. It allows us to speak of *anything* without needing to speak of *everything*. This will permit me to represent a kind of naturalism within or on the basis of his pluralism, one which will still reap distinctive conceptual benefits from the latter. In a sense, what follows is an attempt at a *new* metaphysics of natural complexes.

III. Naturalism

Suppose we now entertain a metaphysical hypothesis: *we, and whatever we robustly discriminate, can be included in nature.* This would be a kind of naturalism. We distinguish among all complexes or anything we might discriminate, those complexes we will call *natural* complexes, now in a strong sense of the qualifier (unlike Buchler's). We do not identify nature with the Whole or an Order of orders. Out of all complexes, I am focusing our attention on a restrictive class of orders, constituted by the most robustly accessible complexes, collectively called nature.[10] Nature is *an* (not *the*) order of orders.

Depending on how we characterize these orders, such an ordinal or pluralistic naturalism could fully accord with naturalism as commonly understood. While variously formulated, I will assume that to be "naturalistic" any contemporary view must accept the following three minimal constraints. First, a naturalism must hold that nature is one temporally enduring ensemble whose members are open to at least indirect mutual causal influence (subject to spacetime segregation, e.g., humans can neither affect dinosaurs or other galaxies). That is, no natural objects *and* their causal antecedents are in principle exempt from even mediated causal interaction with the rest of nature *and their* causal antecedents (past causal histories cross). No members of nature are causally isolated *in principle*. Hence nature is not divided into domains incapable of interaction, as in the Cartesian, Lockean, or Spinozan dualism of "mind" and "matter." Second, "nature" must include *at least* the physical, material, biological, mental, and cultural. This is a nonexhaustive shorthand list of some different kinds of entities, events, and properties. A naturalist must regard not only the objects of physics, the material and the biological sciences, but minds, intentionality, meanings, communications, societies, artworks, etc., all as natural or part of nature. How they can be included may vary, but if one is to be a naturalist, included they must be. Third, the conclusions of the natural sciences must have robust significance for the metaphysics of nature. This does not mean no other sources of knowledge exist, or that whatever the natural sciences say must be adopted at face value. Still, to fail to take the natural sciences seriously in one's metaphysics is not to be a naturalist.

But the present naturalism will reject two other claims common among naturalisms. First, my naturalism will *not* assume that everything that is or was or will be is natural. Our job is to describe those complexes that are natural. We make no *a priori* stipulation that nature exhausts all complexes. It might, but we do not presume it must. Whether we can discriminate any complexes that cannot be included in the orders of nature remains an ongoing question. The function of the rest of the book is to argue that kinds of complexes which might seem not to be natural, are. Second, nature is plural, having multiple kinds of entities, properties, structures, and processes. This means we will not assume nature is physical or material (more about which in Chapter 3). Nature must *include* the physical but we make no *a priori* presupposition about the ubiquity of physical entities, properties, or processes in nature. To claim nature is in principle all physical or material, or all natural events or properties are determined by the physical or material, is to be a physicalist or materialist, not a naturalist.

Let us briefly examine this. A pluralistic or ordinal naturalism accepts that all orders of natural complexes are ontologically on a par. It accepts both an entity-pluralism and a property-pluralism, hence what Wimsatt calls a "tropical rainforest ontology" rather than a Quinean "desert landscape" (Wimsatt 2007, p. 213).[11] This is associated with the notion of *emergence*, the claim that complex systems can exhibit irreducible properties (which will be examined in Chapter 3). In terms of scientific explanation, we shall see that nonreductive as well as reductive explanations are inescapable, because some properties of some systems are not explicable as linear products or aggregations of the properties of relatively isolable parts. Reduction and emergence are matters of degree, hence compatible. And since the justification of an ontology *is* its explanatory necessity, acceptance of multiple, irreducible sciences is *prima facie* reason to accept emergence and hence ontological pluralism. This will be argued in the following three chapters. For the moment we may say that an endorsement of emergence is nothing more than the combination of two ideas: that nature is pluralistic, composed of many different kinds of things and properties; and that some of those things and properties are ontologically dependent on others, e.g., the mental on the biological, the biological on the chemical. This means accepting a hierarchical view of natural beings and processes. Thus we are adopting a metaphysical naturalism that does not presume nature is one kind of thing, instead depending on a series of empirical studies and their philosophical analysis to see what nature is like.

Such pluralism also serves to unhook thought from the dominant bipolar disorder of modern metaphysics, the belief that there are at most two sorts of actuality, the physical and the mental. The former is intuitively identified with ponderable matter, but in philosophical practice epitomized by the objects of physics. The latter is intuitively identified with human consciousness, in philosophical practice epitomized by representational "belief-and-desire" states. The core issues of much of contemporary philosophy aggregate around the question of whether the latter can be reduced to the former (hence physico-material reductionism), or we are stuck with some kind of dualism of physical and mental existence (or even, at the other end of the spectrum, idealism or panpsychism), or we can accept nonreductive physicalist theories which hold that even if everything is in some sense physical, psychological explanation is true independent of physical explanation. The discussion generally assumes that there are no other relevant metaphysical kinds, that the objects of chemistry, the Earth sciences and biology are merely placeholders for the physical. Many concepts in current philosophy of science, theory of knowledge, and philosophy of mind and language, as well as metaphysics, presuppose this dualism. In contrast, the current naturalism will accept that the physical and the mental exist and are different, and the mental (like the cultural, biological, and chemical) is dependent (although, we shall see, indirectly) on physical entities, processes, and properties. But, I will argue, that does not justify physicalism, it merely justifies a naturalism which recognizes that dependence. The problems attendant on the bipolar dualism, the relation of the physical and mental, become more tractable when relocated from basic ontology to an empirical relation of dependence among two of several kinds of entities, processes, and properties.

IV. The Local Argument for Naturalism

Only now can I give the argument for my naturalism. Why ought pluralism lead to naturalism, any more than to any other "ism"? I will give a familiar and unremarkable argument for naturalism, one which will seem to beg metaphysical questions. But I will then argue that, given a *local* metaphysics, those questions are rightly begged.

First, locally speaking, no one doubts that in most of the events of personal and public human existence, cultural meanings and human minds subsist within, or on the basis of, biological, material, and physical nature. That is, no one doubts that human minds are dependent on human nervous systems, communicated meanings dependent on language and culture, life on chemical metabolism, etc. I do not mean minds depend *only* on brains, or that there is no downward causality or reciprocal dependence of brain on mind or environment. Obviously, while my biological cells depend on chemical macromolecules, my macromolecules also depend on living cells, which produce them. Nor do I even mean that there can be no disembodied minds or spirits *in principle* or *a priori*. I mean only that as far as we can tell, brain injury usually affects the mind of the person whose brain it is.

Second, we cannot practically doubt the validity of large areas of contemporary natural science, which reinforce and deepen the location of human life in nature. We can logically doubt it, of course, but we cannot fail to act as if it were valid, given our daily use of technologies whose functioning is explained and designed by natural science. We also cannot doubt that natural science is a prime example of communal, rational inquiry which subjects its conclusions to test, hence a commitment to such method must imply some confidence in the former. Repeatedly confirmed results of a community of self-critical inquirers is in principle a robust source of knowledge. What *can* reasonably be doubted is, first, the reliability or approximate truth of any particular scientific claim or theory, and second, the interpretation of accepted claims. Any scientific claim may be re-describable with a different set of ontological assumptions or in the language of another theory, or later be seen as true for a more limited domain of phenomena. We can certainly disagree with the ontological presumptions that attach to the formal or experimental claims of a theory, even admitting that this distinction is itself fuzzy. And we may believe natural scientific accounts of a phenomenon are insufficient to explain it, hence insist on supplementing the former, e.g., with divine intervention in human evolution, a dualist account of mind, parallel mental and physical explanations of behavior, a phenomenology of experience, a pragmatic account of human agency, etc. But nobody reasonably doubts that natural science gives us approximately true descriptions of how a baseball bat interacts with a baseball or electricity drives a motor, nor doubts the dependence of living organisms on their chemistry and human minds on human brains. This doesn't mean such knowledge is certain; it means we have reasons for believing it probably true.

Third, only in the last eighty years has science achieved a comprehensive and robust picture—the product of the dovetailing of multiple disciplines—of the temporal evolution of the universe.[12] The physical origin of the universe created an enormity of energy and the simplest kinds of gaseous matter, which only after billions of years formed stars, which then generated all the heavier elements, hence eventually solar systems with terrestrial planets and, at least in one case, a terrestrial environment in which life arose, itself evolving from the simplest forms to encephalized animals and later human beings. Unless natural science is *grotesquely* wrong, the minds we know are late achievements of the universe and cannot be imagined otherwise. This is the strongest argument against the claim that mind (idealism), experience (phenomenology), action (pragmatism), signs or culture or history (poststructuralism), or some "primary experience" or "symbiosis" prior to the subject-object distinction, is the fundamental context of reality.[13] Mind or experience or action or writing or culture—pick the one you prefer—has recently concluded that nature existed long before itself.

Philosophically, the foregoing may seem beside the point. The anti-naturalist may say, "Yes, of course, but the serious metaphysical question is about what underlies or causes or renders possible the *apparently* physical world studied by natural science, presupposed by social practice and to some extent confirmed in everyday experience. Those who reject naturalism do not deny that my mind depends on my brain, they believe the brain and its fellow material or physical objects must *depend upon something non-physical* that lies behind or supports or causes or constitutes or constructs the apparently physical world, whether that be something independent of humanity or a constructive process of human mind and/or culture."

It is part of the point of localism to deal with this objection. Whether one claims that reality in the most comprehensive or fundamental sense is physical or mental, some combination of the two, some third thing, or eschews all such questions, one must still account for the facts that my unaided imagination seems unable to alter the world, that human personalities exist in a context much of which they neither create nor control, that intentional meanings arise only through performances of neural systems of embodied acculturated organisms that are necessarily late and rare in cosmogenesis. Even if one locates the physical within, or claim it emerges from, something non-physical, one must still explain how individual minds and meanings emerge within or from their local physical, material, biological neighborhoods. Whatever the ultimate metaphysical context, however one may want to characterize the Whole or the Underlying, that problem remains. The absolute or Berkeleyan idealist, the German idealist, the Kantian or social constructivist, the dualist Cartesian or Lockean, the Spinozan psycho-physical parallelist, the Husserlian or (early) Heideggerian phenomenologist, the Derridean or Foucaultian poststructuralist, all must still explain the interactions of individual mentality or meanings or sign-use with local physical, material, and biological phenomena. The core local problems remain largely unchanged, like the lunching philosophers grabbing the spoon. Even if it were true that reality is fundamentally mental or semiotic or spiritual or ideal we would still have to explain how the apparently mental, semiotic, or spiritual interacts with what is apparently not mental, semiotic or spiritual. There is no cheap way to avoid at least a local naturalism here, short of global skepticism or solipsism. The point is that the local relation between orders is the issue that must be addressed, regardless of what one takes reality globally to be, which task we have declined. *Naturalism is at least locally true.*

That is the argument for my naturalism. Its validity will depend on its success at addressing the common objections to naturalism, and demonstrating just how much it can render intelligible with claims that are likely true. While the rest of the book is required for these tasks, we can at least suggest here how the common objections to naturalism can be addressed.

The most prominent, if not most comprehensive, objection is that naturalism is reductionist, and particularly that it produces an inadequate account of mind, self, and meaning. But, obviously, that holds only for a reductionist naturalism. If emergence and ontological pluralism are naturalistically respectable, then the objection disappears. The relevant question is whether an emergent, pluralistic naturalism can formulate a plausible account of them. I will argue that it can.

Likewise, the use of natural science in metaphysics will indicate to some a "privileging" of natural science and its methods over social, cultural, and humanistic inquiries. But as we will see, pluralism will mean that, *prima facie*, physical methods are robustly informative for the physical, as are material methods for the material, biological methods for the biotic, psychological methods for the mental, and cultural methods for the cultural. Any "privilege" or cognitive priority must be partial, relative to subject matter, not to mention fallible and tentative. Given a pluralistic view of nature, *all* those methods are "natural" which examine orders of nature. Each order is a domain whose best investigative treatment is a contingent matter. My focus on what are called "natural" as opposed to the "human" sciences is due to the fact that the former are more general, the latter being concerned with one biological species and its products.

A related objection is that naturalism is in principle unable to justify a normative ethics. We must postpone until Chapter 12 discussions of the naturalistic fallacy and related matters. But we can say something now. The objection is that naturalism can only say what happens in nature, what natural facts and processes are, and not justify normative judgments about them or anything else. This objection is sometimes put in the form that we cannot find values in nature. That, however, is false: as we shall see, there certainly are values, ends, and norms in biological nature, for organisms value certain ends, and part of what nature selects is that propensity to value (as we shall see in Chapter 7). At the very least, as long as biology's use of functional and teleonomic explanations are not reduced to physical modes of explanation, values obtain in nature.

However, this retort serves only to redirect, not resolve, the problem. The relevant difficulty is, I believe, twofold. First, naturalism raises the possibility of informing ethics with biology, e.g., sociobiology or evolutionary ethics, which seems to some to reduce the cultural to the biological. But that again presumes a reductionist naturalism. If a nonreductive account of mind and culture is possible within a naturalistic theory, so is a nonreductive account of human ethics. At this point, the critic may open a larger issue, that a naturalistic description of, say, values inherent in biological or human being cannot serve to justify why we inquirers *ought to value or disvalue* those described values. This is to claim that naturalistic description distinctively suffers from a "fact-value dichotomy." But whatever the validity of that dichotomy, or lack thereof (we shall see in Chapter 12), it is a difficulty regarding which most non-naturalistic accounts of reality have no *a priori* advantage. Absent a Platonic Form, or Divine dictation, of the Good, the problem of reasoning from, say, an ontologically distinctive conceptual or mental or spiritual or cultural realm to why we ought to value one thing over another is just as great for non-naturalistic as for naturalistic perspectives. If some non-natural source dictates moral values, then that is a fact from which, *if* there is a fact-value dichotomy, we are just as unable to infer what we ought to do as from a biological fact. My point is not to prejudge fundamental issues here, but to suggest that nonreductionist naturalism suffers from no unique, "in principle" disadvantage with respect to normative claims.

Lastly, if nothing else, naturalism is for some narrow in that it denies the supra-natural. Naturalism would seem to eliminate the divine. But a pluralist and local naturalism avoids this objection. It could in principle accept divinity understood naturalistically, as continuous with and causally interacting with other orders of nature. And local naturalism refuses to claim that everything is natural, but leaves that question open. Whether there are good reasons for positing divinity remains for it a serious question that would have to be approached from the standpoint of robustly accessible natural orders (as will be seen in Chapter 11). Now, such openness may seem anti-naturalistic. Shouldn't naturalism deny that anything can be supra-natural? Not if our approach to metaphysics is local and fallibilist. We can argue that what we robustly discriminate is in nature. But how to justify a claim that *nothing else exists*? We could make that claim only if we knew everything or the limits of everything. That is not something we should expect to be able to do.

My claim, then, is that given a localist practice of metaphysics, the way is clear for a pluralistic metaphysical language to claim that a naturalism which takes multiple sciences seriously is at least locally true. My working hypothesis is that the robustly accessible complexes *can* be incorporated into nature, thus understood. The task will then be to survey a host of contemporary studies, from physics to anthropology, to tease out the most robust and defensible notions of the kinds of beings, processes, and relations that characterize nature. Whether this approach advances understanding with respect to enough important problems to make it worthwhile can only be known by testing, empirically and conceptually, descriptions of robust orders and their relations, and their usefulness as staging grounds for extension to less robust orders, all within the framework of natural orders. We shall see.

2

A Selective History of Naturalism

This chapter serves not as an in-depth or even balanced scholarly examination, but as a stroll through key moments in the history of Western thought to highlight particular forms of naturalism and some of their most prominent opposition. My aim is not only to examine what I take to be promising forms of naturalism, but to indicate naturalism's historical debilities, to set the stage for showing how my preferred form of naturalism avoids these pitfalls. Since history can help make systematic philosophical efforts easier to understand, the hope is to put the current project in the context of related efforts. Of course, such a history can raise its own controversies and problems. Doubtless what follows will do that too.

I. Ancient Alternatives

It is useful to recall what we can, in the light of contemporary science, regard as the three most currently relevant metaphysical theories of ancient Greece, forged by the Atomists, Plato, and Aristotle.

Leucippus, Democritus, and later Epicurus held all things to be collections of a common set of tiny, simple parts, *atomoi* or indivisibles. Atomism held that all the many things in the universe are made of one kind of simple individual, and all other differences among things derive from the number and organization of those individuals, which differ only in shape, size, and possibly weight. All atomistic causality was presumably efficient and material (although Aristotle had not yet clarified the concept of causality). The only additional complication was that a venue for the movement of the atoms had to be posited, namely the Void.

There is little need to demonstrate that many of the great successes of modern physics have been achieved by constituting macroscopic phenomena from law-governed interactions of their simplest identifiable components. All macroscopic material solids, liquids, and gases, are in fact composed of tiny material atoms (or ions). There are differences, of course. The *atomoi* that Democritus imagined were impenetrable simples whose relations to others were purely external; that is, relations did not constitute the atom or change it. Today we do not have any confidence that there are simples in this sense; modern atoms are certainly not simples, nor are the protons and neutrons that compose their nuclei. Nor is space, or spacetime, a simple void; it is a dynamic metric field generated by the gravitational fields that occupy it.

In contrast, Plato held that the ultimate realities are not things but the forms, structures or laws which things instantiate or obey. The ultimate realities must be that which is most intelligible, hence the unchanging and universal objects of conception or intellectual intuition, like but higher than mathematical objects. With all this goes Plato's rationalistic belief that the principles of intelligibility cannot be inductively inferred from sense experience.

One can go quite a ways with the hypothesis that modern physics is Platonic. Certainly it is incorrigibly mathematical. Galileo famously remarked that true science must commit a "rape of the senses," meaning it must regard sensory information as a mere resource for mathematical, conceptual analysis and synthesis, the latter alone yielding intelligibility (Galileo 1957). Indeed, twentieth-century physics, particularly general relativity, the gauge theories of electroweak and strongly interacting particles, and quantum cosmology, arguably make mathematical patterns *the* ultimate realities. When we look at the growth of physical knowledge across what Kuhn called "scientific revolutions," Howard Stein claimed we see a continuous approximation of mathematical structures, the changing typology of fundamental particles being less important than the roles of and relations among them (Stein 1989). Recent "structural realists" have argued as much for microphysical ontology (as we will see in Chapter 5).

So, is today's physics Atomist or Platonist? The safe answer is: Yes. The spirits of Democritus and Plato are both served, if mutually compromised, in current physics.

But not so the third option in the ancient debate. For Aristotle all diverse phenomena were to be understood as properties or performances of a finite but very large range of qualitatively different kinds of relatively independent, non-simple individuals, called *primary beings*—that is, beings in the primary sense of the word—or for his Latin translators, *substances*. The basic notion is of something independent. Knowing a substance must include knowing its causes, which Aristotle famously pluralized, claiming that there are four distinct *aitia* or factors responsible for the existence of any substance: its "out-of-which" or material components, which he believed supplied possibility (material cause); the active or energetic source of its coming-into-being (efficient cause); the "what it is" or "what it means to be" the thing, which he believed supplied actuality (formal cause); and its "toward which" or end or purpose (final cause).

Substance has not fared as well as atoms or forms in recent philosophy of science or metaphysics. It is considered a faulty category, partly because it seems quantitatively un-analyzable and lacking in internal or constitutive relations to other things, and partly because of other Aristotelian notions connected to it, like final causes and the fixity of substance types, belied by Darwinian and cosmological evolution. The independence of substance has condemned the notion in the eyes of many philosophers of the last two centuries, for arguably nothing fits that description except the Whole (as Spinoza argued).

While true this is a bit unfair, ignoring as it does substance's substantial virtues. We must remember what counts as "independence" for Aristotle is essentially linguistic and commonsensical. In his metaphysics all the manifold types of being—which he listed in his ten logical categories, including place, time, qualities, etc.—are understood as grouping about and depending on one of the ten, which are the primary beings ("secondary being" refers to the *kinds* of primary beings). That is, all else is part of, predicated of, or a set of (if we include secondary beings) primary beings. The idea is that there must be a fundamental distinction between dependent properties and relatively independent entities. Aristotle chose as the latter physical individuals, having both matter and form, hence potentiality and actuality. Despite the problems of Aristotle's account, we will see that some version of qualified independence is an inevitable part of any plausible criteria of individuation, whether material or physical or biological. Aristotle's substance is also by definition more complex than either matter or form, since it includes both. His substance metaphysics is in principle pluralistic, since there are an indefinitely large number of qualitatively distinct kinds of substances. Substance is a distributive term, unlike "atom," "matter," or "form." Indeed, substance-like conceptions are not foreign to modern physical science: chemistry and biology both take as fundamental certain complex individuals and natural kinds, respectively, atoms and elements, and cells and species (cf. Chapter 4, note 2). Most of all, Aristotle avoided the globalist prejudice common to Atomism and Platonism, that until we come to the final term of a line of metaphysical thought, all the intermediate terms are in peril, because the simplest, highest, most foundational or most comprehensive so recasts the character of local orders as to make their independent description unreliable. Indeed, compared to the smallest and largest scales, the more familiar, observable scales of existence are not only more robust (in Wimsatt's sense), they are where the greatest complexity lies (e.g., in macroscopic ensembles and objects, living things, weather systems, ecosystems, etc.). Aristotelian substance metaphysics was designed to be, we may say in hindsight, a *metaphysics of the middle.*

II. Modern Physics and Its Metaphysics

We cannot here chart the development of modern physics. We may say, however, without too much dishonesty, that the intellectual sources of the new ideas of the modern scientific view of the world are four: Baconian empiricism, the "rationalist" Copernican achievement in mathematical astronomy, the Galilean-Newtonian dynamics, and the Cartesian-Lockean metaphysical dualism. Bacon raised the flag for empirical investigation, insisting that the experience of natural events, or we might say, Reason-Experience-Nature is the proper constraint on and test of belief, rather than the various "idols" that human society has come to revere, or Culture-Authority-Society (Cahoone 2005). While every earlier student of nature had observed, Bacon articulated a shift in epistemic ideals by which empirical innovation loosened the hold of traditional authority, the book learning of the tiny educated elite. And this has arguably been a lasting shift. Copernicus made the first great achievement of the mathematization of nature, in the realm where mathematics had been traditionally applied, astronomy, by replacing a clumsy Ptolemaic model of the solar system by a more efficient and elegant one. The great pattern of modern physics was thereby laid down: the analysis of a complex system into parts (or part-motions) susceptible of simpler mathematical modeling, which can generate the observations or predictions ("save the appearances") of that complex system, thereby exhibiting the causal structures underlying the observations.

Perhaps *the* great achievement, without which none of us would be having this conversation, belonged to Galileo. His breakthrough concept of inertia took the condition of zero acceleration, whether in a state of rest or uniform rectilinear motion, to be determined by zero net force, the two states being dynamically equivalent. This set the stage for the first post-Aristotelian *system* of physics, achieved by the genius of Newton. In a sense Newton followed the advice of Copernicus: first, seek an elegant mathematical model. But he applied that method to a mundane, Baconian subject matter, the dynamics of Earth-bound moving bodies, utterly breaking down the distinction between the celestial and terrestrial.

Modern physics emerged as the field of mechanics, the attempt to understand the motions of ponderable bodies as interactions with forces. In its view every change in motion is the result of nonzero force. Motion, as velocity and acceleration, and so momentum and force, were then understood in two related and very fruitful mathematical forms, as vectors and as analytic geometrical graphs of linearly related variables. The calculus, developed independently by Leibniz and Newton, permitted precise mathematical conclusions about continuous changes in motion and their relation to these forces. The application of these mathematical techniques to physical events rested on two related strategies, reduction and isolation. Reduction, or explaining the behavior and properties of something in terms of rules of behavior governing its components, is a vertical strategy. Isolation on the other hand is horizontal; it suggests that the best way to understand a system of interactions among many units is to generalize to the entire system the rules that seem to govern the interactions of just a few units in a spatial locale. This intellectual isolation of the system or bodies from other environmental phenomena then permits extension of the rules discovered in the isolated case to all similar systems, making up for neglected variables (like friction) later on.

The approach of early modern physics, and by implication modern science as a whole, is often called "mechanism" in a broader sense. That is a perfectly decent, if misunderstood, label. A machine, in the early modern sense, is a set of bodies connected by joints, wherein application of force to one part or region causes another to do work. Newtonian or classical mechanics understood phenomena as the deterministic result of (ideally) rigid bodies, in some initial state, that could be analyzed as collections of components (ultimately point particles) under forces acting with respect to a small number of properties of those bodies (mass, velocity, location), often in a mathematically simple fashion (e.g., for gravitation and electricity, on a straight line between their centers in proportion to the inverse square of distance between them). Together, the machine analogy and the success of Newton's account overlapped around an approach to natural events as the deterministic result of the simple mathematical relations among basic material properties of reductively understood ponderable material bodies due to a few universal force laws. The formula was that initial conditions of irreducible properties plus universal force laws yield deterministic results.

It is often claimed that nineteenth-century electromagnetic theory—the unification of electricity, magnetism, and light in Maxwell's field theory—and the theory of heat transfer—which Boltzmann founded in the statistics of atoms—were the beginning of the end for mechanism in physics, its final demise coming at the hands of relativity and quantum mechanics. But reports of mechanism's death have been exaggerated. Electricity and heat were both conceived as fluids in the nineteenth century, and light as waves in a fluid medium, making fluid mechanics the background for understanding these phenomena. Wave motion was understood on the basis of the simple harmonic motion of oscillators, like pendulums or vibrating strings, whose mathematics were themselves derived from circular motion. The tensors of Einstein's equation of general relativity are still modeled, like Maxwell's, on fluid flows, pressure and stress across a volume. It is not for no reason that the most revolutionary science of our age is called quantum *mechanics*. A field is still a spatial distribution of force vectors, and its points are modeled, in terms of their energy content, as tiny oscillators. Wherever you can explain a property of a system as the linear product of simply governed interactions among relatively stable and rigid components, you've got a mechanism. Not to mention that machines can be purposive—indeed, most are!—like a thermostat, whose behavior requires functional explanations. Nor do mechanisms have to be physical; there are arguably biological, psychological, social, and economic mechanisms. So while it is true that twentieth-century science eschewed mechan*ism* as a universally valid model or a picture of nature as a whole, and fundamental physics to some extent exchanged individuals for fields and entities for states, in many circumstances the mechanical intuition remains useful.

With physical mechanics came its philosophical interpretation. Descartes and Locke took different sides in the debate over how scientists could know what they knew—Descartes accepted the necessity of innate ideas, Locke inaugurated modern empiricism by rejecting them—but Locke fully endorsed Descartes' dualistic substance metaphysics: everything is a body or a mind. Bodies were for science, minds or souls were for ethics and religion. Indeed, the dominant metaphysics of the whole modern period of Western philosophy was Galilean-Cartesian-Lockean. To be sure, this hegemony was not absolute; in every generation some philosophers (in fact, the most prominent, at least in the rear-view mirror of historiography) reacted against it. But the dualist view remained dominant; it arguably was *the* modern view. Even thinkers who took a different path, like Spinoza, nevertheless drew near the same dualist destination, for while he endorsed pantheism by reinterpreting Aristotelian substance so that there could only be One Substance, God or Nature, Spinoza's mind and matter remained two utterly distinct attributes of the Divine substance, incapable of causal interaction.[1] Leibniz's atomic *monads* were themselves dual, on the one hand externally related to their environment in spatio-temporal terms, while internally manifesting a genetic process akin to mind, which represents or "perceives" that environment. Even here, mentality and materiality remain the opposing properties requiring explanation. The point is, by the late eighteenth and through much of the nineteenth centuries the "modern" educated view, emanating from central-western Europe, was of nature as matter-in-motion, whether atoms in the void or vortices in a plenum, eventually to be explained by calculus on Lagrangian and Hamiltonian equations in phase space.[2] The philosophical dualism that contextualized this model spread as well, for two very good reasons: the successes of science and technology (like the steam engine) kept giving people new reasons to believe nature to be a material mechanism; and it continued to accomplish its simple division of labor, rendering to Newton what was Newton's and to God what was God's.

III. The Revolt Against Mechanism

But there was at all times a background dis-ease with mechanism and dualism as a model for all science and philosophy of nature. One intractable problem with dualism, apparent to all, was its inability to understand the three inescapable interactions of mind and matter: how material events produce "ideas" in the mind (perception); how mind can cause bodily movement (action); and how ideas can truly represent material states (truth and knowledge). Second was the maturation of the other natural sciences. Chemistry contributed at times, but it was above all biology that remained the great inspiration for those discontented with physicalism or mechanism (even if parts of biology were and are mechanistic). Newtonian mechanics plus thermodynamics could not explain living organisms, making life the great exception to a mechanistic world. Lastly, empiricism, which began the modern era in parallel play with rationalism in the Cartesian-Lockean metaphysics, began to veer in off in its growing recognition that unaided experience could not provide an adequate logical support for the claims of science, including physics. The philosophically most powerful response to that inability was idealism and particularly the idealism generated in Germany. I will spend some time on this movement, which may seem strange in a chapter on naturalism. But it so happens that German idealism, apparently anti-naturalist, fueled the naturalism that followed it.

Modern idealism's first formulation came from an empiricist. Berkeley's many arguments against the existence of matter can be boiled down to one: the Cartesian-Lockean dualism of mind and matter is unsustainable, for no interactions between them, causal or epistemic or practical, are possible. The three interactions cannot make sense in dualism. (This was, of course, true.) If so, and if we stay within Cartesian-Lockean ontological options (which, actually, we shouldn't), either all is mind or all is matter. Since the mind of the inquirer cannot be doubted (Descartes showed), all reality must be mental, with God, as He so often did in this period, providing the service of uniting what metaphysics had torn asunder, in this case all our disembodied minds.

Kant's critical idealism offered a solution that was metaphysically more conservative and easier to combine with mechanistic science, but epistemologically far more radical. For Kant the mind actively constructs, or imposes conditions, on what can be empirical reality or experience, thus on nature *as* experienced, so our investigation of external reality must turn our attention to the subject's criteria for reality. Given that, the core of philosophy is the internal critique of Reason by Reason. Kant remained a friend of natural science by tailoring his transcendental machinery to construct principles congenial to Newton's physics, and by insisting that "things-in-themselves," like Locke's "unknown support" underlying each physical substance, must exist even if unknowable because outside the bounds of a possible experience. He thereby avoided Berkeleyan idealism. This view was crafted simultaneously to enable faith in free will, a future state, and God; they could be accepted without fear of contradiction because science, restricted to phenomenal appearances, cannot know things in themselves.

Most of the Germans after Kant, committed patrons of the French Revolution, wanted to reject materialism or mechanism for a metaphysics based in freedom and hospitable to religion. Fichte was the lynchpin (Fichte 1970). He expanded the transcendental argument of Kant to eliminate things-in-themselves, for once it is accepted that whatever we know of the thing-in-itself is dictated by the activity of the mind, the thing-in-itself becomes useless. Reality is the result of activity or will; the self is an act which posits itself *and* posits the non-self, nature. Nature, while other, is part of the self's progress, whereby the *I* ultimately recognizes itself *in* its objects. We can now make human freedom, rather than a problematic secondary feature of a dualistic or physicalistic metaphysics, the basis of reality, with nature dependent upon *it*.

This led to Hegel and Schelling. Both accepted the processural notion, already in Fichte, of reality as the self-actualization of God or Spirit—inspired by the mystical Pietist tradition of Meister Eckhart and Jacob Böhme—interpreted via a dynamic Trinitarian self-opposition and resolution (Magee 2001). The metaphysical Whole or Absolute sunders itself to become a self-articulated, self-knowing Whole, which it can only do via the progressive realization of itself through its parts, particularly, that part which was created for the purpose, human being. What Hegel uniquely added was the necessity of Spirit's self-development in human history, working through and overcoming the real historical abstractions and dichotomies, its partial or incomplete manifestations, on the road to the attainment of the perspective of the Whole. Human history is the working-through of the difference between Whole and part.

From a naturalistic perspective, Schelling holds special interest. Indeed, Schelling is perhaps not an idealist at all, but a kind of naturalist (Snow 1996). He was most fervently a Spinozist, but one committed to dynamism and freedom: there must be an original identity of nature and spirit, and freedom must have its basis *in* nature. His dynamism was organic. Life is higher than mechanical matter, and it requires internal opposition or difference to enable its development. No battle, no life. God must be alive, hence developing. Organic life is "self-organizing," purposive, its form and matter are inseparable, and is both "for itself" and "out of itself" (Schelling 1988, pp. 31–37). Following Aristotle he accepts that mind or soul is the principle of life, but claims this is not a "life-force." There is a "hierarchy" of freedom in nature: "nature reveals itself in manifold forms in progressive developments, in gradual approximation to freedom." For Schelling, "Nature becomes a circle which returns into itself, a self-enclosed system." This leads to his explanation of Creation. Like all things God must have a basis or reason (*Grund*) distinct from Itself, but in God's case this ground must be internal. Schelling distinguishes God's *essence* or existence (which are the same) from God's *ground*. In their original condition these two features are a matter of utter "indifference" called the *Ungrund*, a term borrowed from Böhme (Schelling 1936, pp. 86–87). The *Ungrund* is neither One nor Many, but a state in which numerical distinction is absent, a circulating "rotary" of forces "longing . . . to give birth to itself" (Schelling 1936, p. 34). The separation of those forces leads to expansion, which loosens the contained inner darkness of the Ground, which comes to be nature.

German idealism is regarded as anti-naturalistic, and excepting the problematic Schelling, that is basically true. But there are subtleties. After all, traditional naturalists also agreed that mind causes nature, namely the Divine Mind. Putting aside questions about God or the Whole, idealism is reliably anti-naturalist wherever it claims that the human mind or self (or its meaningful contents) are temporally, logically *or* causally prior to, and cannot be causally impacted by, the biological, material or physical (either on its own or because it is continuous with a supra-human Ground that is prior to nature). But even where they endorsed these claims, the Germans brought us an incipient pragmatism, the view that activity is the source of meaning and the nature of mind, and made process or evolution characteristic of all reality. Their idealist notions were proto-biologistic. Only two decades after Hegel's death such evolutionism view would gain further confirmation from the next great scientific shock.

IV. Naturalism After Darwin

It would be an exaggeration to say that Darwin changed everything, but not a great one. Darwin, Wallace, Huxley, and other evolutionists threw fuel on the concerns of nineteenth-century thinkers who, already inspired by Hegel and by the social, economic, and political revolutions of modernization, were concerned to account for evolution, primarily social evolution. Following Auguste Comte, Herbert Spencer attempted an evolutionary theory of life as well as society that would unify science. At the same time, Darwinism gave new impetus to the centrality of biology for any kind of naturalism. The biological perspective, which already had served as a recourse for anti-mechanistic thought, now had a unifying explanatory perspective of its own (even if its modern synthesis with the science of heredity would have to wait until the twentieth century). The result of all this would be the great age of post-Darwinian naturalism in philosophy, from 1870 to 1930, influenced in midlife by the discoveries of special and general relativity (1905 and 1915), and peaking in the decade of the 1920s. Like the seventeenth-century metaphysicians, these philosophers sought a metaphysics that could handle the best science of their day, but with two differences: they were evolutionary or process thinkers, and they had to deal with multiple sciences, not just physics. But just as this movement peaked, Western philosophy fragmented into opposing subcultures, "analytic" philosophy coming to dominate in England and America, and "continental" philosophy in central and western Europe. Naturalism thrived just before and during, but ultimately did not survive, that fragmentation, at least within professional philosophy.

One of the chief problems Darwin's theory raised for philosophers was, how could what appear to be qualitatively distinct biological properties evolve from the apparently random or "mechanical" (as they saw it) process of Darwinian evolution? The nineteenth-century philosopher who most prominently addressed this question was Henri Bergson. Bergson was a precursor of German phenomenology, affirming that the abstract space and time framework of physics could not do justice to a human experience accessible in raw form only to "intuition." He made two fundamental contributions to naturalism. Bergson criticized mechanics for analyzing time as if it were an infinitely divisible geometrical line composed of extension-less points. A temporal interval must be a unit of change, an "indivisible mobility," which Bergson famously labeled *durée*. Second was his notion of creative evolution, in which the essence of change is novelty. Evolution by its nature creates the novel (Bergson 1944). It should be noted that, while Bergson's *élan vital* or vital impetus seemed to link him with later vitalists in biology, who posited an irreducible life force, Bergson explicitly claimed that any "vital force" applies to life as a whole, not to individual organisms or species.

Bergson's notion of time and creative advance influenced many, among whom was an English mathematician and logician who eventually became the author of the sole systematic, mathematically- and physically-fluent metaphysics of the twentieth century, inspired by both relativity and early (pre-1926) quantum theory. Alfred North Whitehead's early metaphysics was based on a distinction of events and objects.[3] Reality is events, understood as durations, not instantaneous points. Objects and the characters or properties of objects are recognized "permanencies" among events. Extension for Whitehead is the relation of inclusion of parts in wholes; events can overlap and include one another. Our access to natural reality is perception, an event that establishes a "consentient set" or common spacetime framework for the whole of nature, with respect to which we can judge relationships. In Whitehead's mature "organicism," the fundamental process is the integration of each event or "actual occasion" out of its internal relations to others which it represents or *prehends*. This means the fundamental atomic entities include proto-life and proto-mentality; they are bits of experiencing. Completion of becoming is death, hence "objective immorality" as a spacetime object for other entities. In addition to actual occasions Whitehead posited "eternal objects" or possibilities which ingress into the occasions, much like Platonic forms. The manifold entities of our world are all "nexûs" or aggregations of indefinitely many actual occasions, coordinated to some degree. Highly coordinated groupings, like macroscopic living things, are governed by some dominant unifying character, hence are "societies." Nexûs and societies exhibit properties not characteristic of their components, properties that he calls "emergent."[4] Whitehead's contemporary followers are today some of the few philosophers continuing to practice a form of the early twentieth-century naturalist metaphysics.

It was the distinctive fate of the Americanist philosophical tradition, the most prominent members of whom were Peirce, James, Mead, Royce, Santayana, and Dewey—among which, by dint of his move to Harvard in 1924, at age 63, we might include Whitehead—almost entirely to coincide with this period of naturalist reinterpretation after Darwin. And indeed, with the exception of Royce, the Americans were nonreductive naturalists, as well as anti-Cartesians and anti-dualists. The naturalist fecundity of this tradition makes it impossible to explore in any detail here, and it has been chronicled elsewhere (Krikorian 1944, Ryder 1994, Marsoobian and Ryder 2004). Peirce, the inventor of pragmatism and a practicing scientist, was influenced by the German "philosophy of the act" but more so by Scottish empiricism (through Thomas Reid) and Darwin. Ostensibly a panpsychist, he nevertheless incorporated human mind in nature. James was arguably phenomenological in regarding experience as his ultimate category (in his "radical empiricism") but he remained a metaphysical pluralist. His extra-philosophical associations were social, rather than natural, scientific. Dewey was the most prominent nonreductive naturalist of them all. His 1925 *Experience and Nature* remains one of the great systematic naturalist treatises. Nevertheless, Dewey remained largely aloof from the natural sciences, and criticized Santayana, an important naturalist himself, for accepting materialism (Dewey, Hook, and Nagel, 1994). Dewey often looks like an Emergentist, but refused the term.[5] He became the centerpiece of the school of "Columbia naturalism"—he taught at Columbia University from 1904 to 1930—including the Aristotelian F. J. E. Woodbridge, himself an admirer of Santayana, as well as John Herman Randall and Justus Buchler, among others (see Krekorian 1944). Theirs was certainly a nonreductive naturalism, heavily influenced by Aristotle and Darwin, although not particularly informed by natural science.

But among the Americans it was arguably Mead—who taught with Dewey at the University of Michigan and then the University of Chicago—who most combined a consistent naturalism (unlike Peirce) with knowledge of the natural sciences (unlike James or Dewey). He called his perspective "social behaviorism." Humans are to be understood as social animals, and all phenomena associated with or revealed to an organism appear in the course of its acts. The species nature of every organism, committed to acts necessary for survival, selects an environment of relevance out of local physical reality. Mind and consciousness, while resulting from the central nervous system, are functions of the organism-acting-in-and-toward-an-environment. Mead endorsed both emergence and what he called "relativity." The former, which he ascribed to Lloyd Morgan, is the arising of novel properties in the performance of complex assemblages of simpler components. The latter is the dependence of a character of an object on an organism's relation to the object; e.g., the status of an object as "food" depends on the existence of an animal who can eat it. Mead ascribed the emergence of human mind to social communication, particularly the development of "significant gesture" (to which we shall return in Chapter 9). Mead also tried to bring special relativity to bear on his naturalism of the act. He accepted that spacetime must be understood relativistically (in the scientific sense), but unlike Whitehead held that any fixed order of relations among events must be established by the act of the organism, not a perceptual event. "The unit of existence," he wrote, "is the act, not the moment" (Mead 1938, p. 65).

It was in the Roaring Twenties, as a world still digesting general relativity was about to face the more challenging meal of quantum mechanics, that interest in multidisciplinary philosophical naturalism peaked. Most of all there developed a *school* of emergence, the British Emergentists, represented most famously by Samuel Alexander, Conwy Lloyd Morgan, and C. D. Broad, with the American fellow travelers Roy Wood Sellars and William Morton Wheeler. Just the list of the decade's prominent books is quite remarkable: Alexander's *Space, Time, and Deity* (1920), Sellars' *Evolutionary Naturalism* (1922), Lloyd Morgan's *Emergent Evolution* (1923), Broad's *Mind and its Place in Nature* (1925), Dewey's *Experience and Nature* (1925), Whitehead's *Science and the Modern World* (1925), Wheeler's *Emergent Evolution and the Development of Societies* (1928), Whitehead's *Process and Reality* (1929), and Mead's lectures of the 1920s, which employ emergence, and were published after his death in 1931 (e.g., *Mind, Self, and Society: From the Standpoint of a Social Behaviorist*, 1962).

The basic notion of emergence, that a whole or its properties may not be merely the linear sum or product of its parts, was influenced by John Stuart Mill and first named by George Henry Lewes (Lewes 1875, p. 412).[6] Emergentists claimed that there are wholes in chemistry, biology, and psychology whose properties could not be derived from or predicted by physics, even by, as Broad put it, a "mathematical archangel" or Laplacian demon (he actually had Rutherford in mind) (Broad 1960, p. 58–81). The central figure was arguably Lloyd Morgan.[7] Like other philosophers influenced by Darwin, Lloyd Morgan wanted to say evolution applies generally, not just to life, yet also maintain that novel properties of wholes are irreducible, without accepting Alfred Russell Wallace's spiritual infusions or Joseph Le Conte's "saltationism" (the notion that higher orders of nature are discontinuous with the lower). He took suddenness of change, a degree of discontinuity, or what would later be called "punctuation" in general evolution—including phase changes and the production of chemical properties from molecular parts—to imply the development of properties that mechanics, thermodynamics, and natural selection were not able to derive.[8] Alexander and Broad argued for downward causation of more on less complex levels. Broad and Sellars tended to be less speculative and "metaphysical" than Alexander and Morgan, decoupling evolutionary emergence from religious concerns. Broad explicitly attacked Driesch's *entelechy*, like Bergson's *élan vital*, and the theism of Alexander, putting emergentism in its position of splitting the difference between mechanism and vitalism. Sellars argued against Lloyd Morgan and Alexander that evolution is not teleological or universal, and applies only to parts of nature, not the Whole.

At this same moment, what would become the dominant schools of twentieth-century Western philosophy, analytic and continental philosophy, were being invented. Neither was particularly naturalistic. Analytic philosophy was born from two sources: Frege's new logic and the attempt to forge an account of mathematics, and secondarily, the attack on idealist metaphysics, an attack which, while concerned with grand philosophical notions, achieved its end particularly through a critical analysis of linguistic meaning. Both were championed by Whitehead's student and collaborator in logic, Bertrand Russell. The following generation of analysts, the positivists, changed that by following the new physics and aggressively pursuing physicalism. They accepted the explanatory and/or ontological primacy of physics, and sought to address the problems that arise when one tries to relate the public world of human experience and other sciences to that of physics (Carnap 2003, Reichenbach 1961, Nagel 1979). Nevertheless, positivism was primarily a philosophy of scientific methodology rather than scientific metaphysics. Eventually its foundational search for an "ideal language" philosophy undermined itself, yielding on the one hand to a practically-oriented "ordinary language" account of meaning (e.g., Wittgenstein) and on the other a pragmatically supported physicalism (e.g., Quine). At the same time continental philosophy after Husserl became generally anti-naturalistic, whether in the form of phenomenology, existentialism, structuralism, neo-Hegelism, hermeneutics, or poststructuralism. There were exceptions, of course: the polymath neo-Kantian Ernst Cassirer, Maurice Merleau-Ponty, who stretched phenomenology to a virtual alliance with naturalism as Schelling had done with German Idealism, and Hans Jonas, about whom more later. Certainly there were non-philosophers who contributed to naturalistic philosophy: Teihard de Chardin formulated a religious evolutionary philosophy of nature in his *Le Phénomène Humain* (1955); the Chicago School of American theologians adapted Dewey and the British Emergentists (Stone 2008); the biologist Jacob von Uexküll created an interpretive biology in *Theoretische Biologie* (1920); and Konrad Lorenz would help formulate naturalistic epistemology (1941).

But not long after 1930, emergence and the whole field of naturalistic metaphysical speculation on the basis of multiple sciences largely disappeared from the most prominent forums of philosophy. The reasons were undoubtedly many. It is likely that the combination of the most exciting and revolutionary thirty years in the history of physics (1900–1930), creating quantum mechanics and special and general relativity, made it seem that fundamental physics would eventually explain all natural reality. Quantum mechanics in particular convinced many that chemistry was fully reducible to physics, rendering emergence in the inorganic world superfluous. For Blitz the reason was the rise of logical positivism, specifically the work of Carnap, using the new logic to found science epistemologically, not ontologically, and on physics in particular (Blitz 1992). These reasons go together, of course, in that positivism was explicitly a physical reductionism. Henceforth philosophy of science largely became the methodology and epistemology of physics, to the exclusion of biology and chemistry. Philosophers uninterested in methodology and epistemology, or philosophy of physics, essentially dropped the scientific view of the world as colleague and viewed it instead as a competitor in the conflict of what eventually became "the two cultures" of C. P. Snow's famous 1959 essay (Snow and Collini, 1993). The division of Western philosophy into a positivist and then linguistic analytic philosophy, and a phenomenological then hermeneutic and structuralist continental philosophy, both converging on the view that language is the primary topic of thought, left little room for pluralistic naturalism. Systematic metaphysics, any attempt to formulate a general hypothesis about the world, came increasingly to be seen as wrong-headed by both analytic and continental philosophers. The comprehensive naturalisms of a Whitehead, a Dewey, or a Morgan seemed naïve attempts to find the unavailable "foundations" of truth and experience in a not-sufficiently-physicalist naturalism. Science was to be left to philosophers of scientific method, uninterested in metaphysics, and the rest of philosophy continued to act as if, to paraphrase Pope, "the proper study of persons is persons."

There were some dissenting voices, to be sure. German philosopher Nicholai Hartmann produced a metaphysics of hierarchic levels, Hans Jonas developed an "existential" biology, and Mario

Bunge created a metaphysics of systems (Hartmann 1952, Jonas 1966, Bunge 1979). But the prospects for naturalism were mostly pursued by interdisciplinary scientists, not philosophers. The notion that many natural systems cannot be adequately addressed by reductionism received attention from cybernetics, championed by Norbert Wiener (1948), and general systems theory, made famous by Von Bertalanffy (1968). Michael Polanyi was one of the few in the period to endorse emergence (1962). Developing a variant of systems theory, Herbert Simon argued that natural selection could produce leaps in complexity in a plausible time period only if there were stable intermediate forms capable of further hierarchical arrangement (Simon 1962). Such perspectives have been buoyed in the past three decades by the development of the physical and chemical study of "complexity," concerning chaos and critical point phenomena (e.g., Prigogine 1984) and self-organizing systems in biology (e.g., Kauffman 1993). Hierarchical systems theory evolved further to interpret such phenomena, e.g., Pattee (1973), Allen and Starr (1982), Salthe (1985, 1993), Ahl and Allen (1996), and Wimsatt (2007). Today we are witnessing a "re-emergence of emergence" in interdisciplinary studies of science (Clayton and Davies, 2006). Still, mainstream philosophy remains largely unaffected.

But suppose we were to find that a different kind of general metaphysics, and naturalism, are conjointly possible. Suppose it were generally recognized that emergence is metaphysically more conservative than reductionism, that whereas ontological reductionism aspires to the Promethean task of explaining all kinds of things by one kind, emergence accepts that the explanatory autonomy—which is not to say independence or mutual indifference—of different scientific disciplines mirrors the ontological pluralism of nature. Suppose we were able to conjoin a non-foundational method of metaphysics to that interdisciplinary notion of nature. We would then see that emergence is simply the recognition that nature is pluralistic and hierarchically organized, with sets of natural traits asymmetrically dependent upon others, all equally real, with causal arrows going potentially in any direction, to be determined by successful explanatory practices and not metaphysical fiat. Toward that end, in the following chapter we will explore a pluralistic conception of scientific explanation (Chapter 3), then place it within a pluralistic background language for describing nature (Chapter 4).

3

Reduction, Emergence, and Physicalism

The nature and limits of reduction, and hence whatever one takes to be the absence or opposite of reduction, supervenience or emergence, is normally viewed as a problem in epistemology and the philosophy of science. But it is not hard to see that it is a fundamental issue of metaphysics. The attempt to understand natural systems in terms of their smallest constituent parts is as old as Democritus. Modern notions of reduction are rooted in Galileo's twin methods of analysis and synthesis, and made significant by the great (albeit partial) success of mechanistic explanation in the modern sciences as noted in the last chapter. The modern philosophical conception of reduction is grounded in the work of the logical positivists, several of whom hoped that, linguistically or ontologically, the "unity of science" could be achieved by grounding the terms, observational language, and explanations of all other sciences in those of physics. This was fueled by the apparent absorption of Newtonian physics by Einstein's physics, and quantum mechanics' explanation of the electronic configurations of the chemical elements. The program was furthered by Ernest Nagel's "bridge" laws of translation between disciplines (Nagel 1961). All the sciences except physics came to be denoted "special sciences."[1] A special hope was to avoid dualism in the philosophy of mind. But the strong reductionist program of actually replacing statements about wholes or higher level phenomena with statements about physical parts and processes proved unsustainable.

In the opposite corner, as noted in the preceding chapter, was emergence. The 1920's Emergentists attempted to limit mechanism without turning to "vitalism" in biology—which posited a special life force—by asserting that supra-physical orders, like chemistry and biology, had novel properties irreducible to microphysical reality. The notion of emergence soon came to be seen in Anglo-American philosophy as anachronistic. Later, as the strong reductionist program began to sputter, Davidson's notion of supervenience articulated the dependence of mental events on physical (i.e., neural) events as the impossibility of a higher-level (mental) difference without a lower-level (physical) difference. This seemed to enable a "nonreductive physicalism" without the baggage of emergence (Davidson 1980). It would not be wrong to say that both emergence and supervenience stand among a series of attempts by philosophers and non-physicists since the eighteenth century to assert that the methods and concepts of physics are insufficient for other disciplines and their objects, while nevertheless retaining some priority for physics (Bechtel and Richardson 1992).

The literature on reduction and emergence can hardly be explored at less than book length (e.g., see Beckerman, Flohr, and Kim 1992b; Boyd, Gaspar and Trout 1992; Sarkar 1998). The number of distinct types of reduction, supervenience, and emergence threatens to approach the number of contributors to the discussion. Part of the reason is that a host of issues need resolution. First of all there are two distinct scenarios for reduction, inter-theoretic or *successional* reductions of one theory to another, e.g., Newtonian to Einsteinian dynamics, and componential or *inter-level* reductions of one scale of phenomena to another, e.g., Mendelian to molecular genetics.[2] There is a distinction between *ontological reductionism* of systems to components or lower-level phenomena—what ultimately exists?—versus *explanatory or theoretical reduction* of our explanations or theories of higher level phenomena to lower-level science—e.g., biology to chemistry. Hard-line reductionists assert the *identity* of the explanandum (the explained phenomenon) and explanans (what does the explaining), and some of those believe identity justifies *elimination* of the former and replacement by the latter. Identity of higher- to lower-level entities or properties can be weaker *token-token* identity (of particulars) or stronger *type-type* identity (of classes). Reductionists used to assert transitive explanatory reduction—psychology can be reduced to biology, biology to chemistry, so psychology reduces to chemistry. More common today are *nonreductive physicalists*, asserting that reality is physical but multiple forms of explanation are necessary, even acknowledging novel phenomena at higher levels (the calling card of emergence). Supervenience and emergence have "strong" and "weak" versions. There are many different linguistic analyses of reduction, differing on whether reduction "derives" or "deduces" or "constructs" or "explains" or "predicts" composite from simpler phenomena. And there are very different kinds of dependence among natural phenomena. The dependence of an engine on its parts, of the color of a stone on its chemical composition, of a rat's behavior on its genotype, and of my thoughts on my brain are very different. Different kinds of examples lend credence to different analyses. This is a prescription for conceptual complexity: a variety of approaches chasing a variety of phenomena, but all claiming to answer the same question.

Today there are few hard-line transitive reductionists in the philosophy of science. Statements like Rutherford's, "All science is either physics or stamp collecting," are now viewed as embarrassments (Blackett 1962, p. 108).[3] The question at the heart of the current debate is, how far can the notions of reduction, explanation, and identity be refined so as to avoid the older, stronger versions of reductionism while yet remaining committed to physicalism?

Physicalism is based in some compelling intuitions, its appeal buoyed by the success of the natural sciences, especially physics. Its intuitive base is a recognition that physics and its objects occupy a special place in our account of the world. Its simplest form might be that everything is physical, or is determined by the physical, or there is "nothing over and above the physical" (Hellman and Thompson 1975). But its formulation is by no means easy, and its study continues to produce multiple versions (e.g., Poland 1994, Melnyk 2003, Dowel 2006, Witmer 2006). Taking this variety into account, we might say that physicalism holds that all entities, events, or properties are (or are identical to the) physical, *or* can be shown to be constituted, determined, explained by, reducible to, realized by and/or supervenient on physical entities, events, or properties. "Supervenient on" and/or "determined by" are the relations most commonly asserted, but all the verb phrases listed are represented in various forms of physicalism. Reductionists like Jaegwon Kim, who deny mental causation, hold that "all things that exist in this world are bits of matter and structures aggregated out of bits of matter, all behaving in accordance with the laws of physics, and that any phenomenon of the world can be physically explained if it can be explained at all" (Kim 2005, pp. 149–50). Nonreductive physicalists, like John Post, assert that mental states or properties that are not identical to physical states or properties can cause events. But Post still holds that any physicalist must say something like "everything is physical," or "all truth is determined by physical truth," or "no difference without a physical difference," the last meaning all other realities supervene on the physical (Post 2007, ch.4). So how far can physicalism be stretched?

Rather than explore this literature here, we can only suggest a different approach that may make a useful contribution. I believe it is the success of the natural sciences that motivates the debate. Therefore, rather than analyze the philosophical conflicts between reductionism, dualism, physicalism, and emergentism we need to focus on the role of reductive and nonreductive explanation in science, indeed, in multiple sciences, not just physics. That is what we will get (in Section II) from the work of William Wimsatt. My first section will present notions of reduction, emergence, and supervenience that need to be avoided. I do this because among many philosophers they still carry weight. Then my final section draws my own conclusion, that given our renunciation of globalism and embrace of metaphysical pluralism (Chapter 1), and the compatibility of reductive explanations and emergence (Section II below), the way is open for a naturalism that incorporates the physical but is *not* physicalist, in either a reductive or nonreductive sense.

I. Reductionism versus Emergentism

As a preliminary definition we can say that inter-level reduction, hence reductive explanation, is an attempt to explain a system's properties or performances through the properties or performances of its parts, and, in general to understand phenomena at scale N through phenomena at scale N-1. Emergence is then simply a name for the fact that some properties and performances of systems are not reductively explainable. Both, I will argue, are factual: there really is emergence, and there really are reductively explained phenomena. However, this section is concerned not with reduction and emergence but with reduction*ism* and emergent*ism*, particularly the former.

As noted there are today few hard-line, transitive, explanatory as well as ontological reductionists in science or the philosophy of science. Still it is instructive to remember what the consequences of such a view would be. What would it mean? First, that the only stratum of nature rightfully explaining *all* phenomena—even if we cannot achieve it yet—would be that of the smallest components. This view would *not* mean everything should be determined by physics, nor (leaving out spacetime structure and gravitation) by quantum mechanics, but rather by quantum field theory. And that is just for now. For it would mean the yet-unachieved adequate theory of quantum gravity is the determinative-explanatory level, with respect to which the rest of physics would be no more decisive than chemistry or biology. Not only would the folk psychology of human agency cease to be explanatory, so would biology, chemistry and nonrelativistic quantum mechanics. Natural selection, which explains change in organisms or their genes by interaction between phenotype and environment, would cease to be explanatory.[4] Such a reductionism would be an endorsement of Rutherford's quip above, albeit more extreme, applied to only *one part* of physics. Finally, note the metaphysical implication of such a view. It would mean that nothing of metaphysical interest has occurred in the evolution of the universe since the initial stage of the Big Bang, since the Planck time or at the very most the first second. For everything else, including stars, all heavy elements, and macroscopic material objects, are later products of what existed at that time. They would be no more *ontologically* interesting than the difference between a wooden crate and a pile of wood and nails.[5]

Today almost no one holds such a view, but many hold that ontological reduction is true while eschewing theory or explanatory reduction. That combination is supposed to allow us to claim that everything *just is* its most elementary components and their interactions—so our metaphysics can be physicalist—while recognizing that nobody is likely ever to reduce all higher-level theories to "lower" or the "lowest" level theory. Now, methodologically such a view implicitly denies that the justification of an ontology is its explanatory power. But leaving that aside for the moment, ontological reductionism of a simple form is *false*. It is not true that a bacterium or a glass of water is "nothing but" quarks and leptons (e.g., electrons). There are also relations and events, or structures and processes. The point is obvious, but among philosophers of science ontology is often conceived in purely entitative terms, as Bickhard points out.[6]

Now, the reductionist can rightly object that I have oversimplified her view. "Of course," she may say, "I did not mean to exclude a system's structure, process, environmental interaction, and function from consideration! But those are relevant only where their effects can be found registered *in or on its parts*. How the parts are organized is itself a property of those parts. All processes, relations, and properties of the parts fund the reductionist explanation." Now, I agree that the parts plus their relations plus all processes relevant to them *are* the whole (Teller 1992). But now the question becomes, how to explain the parts' possession of those "relational properties"? If those properties are entirely explicable "from below," from part-to-part interactions, fine, that is a fully reductive explanation. But where the relational properties of the parts are themselves determined and explained *by the location of a part in the whole system*, then we have now turned round to explain the parts by the whole, which too is fine, but it is no longer a reductive explanation! Practically, to "download" complex relational properties onto the parts is to miss the point of reduction, which is to transform a complex problem into a simpler one. Inter-level reduction seeks to explain the properties of complex wholes as the result of relatively simple interactions among relatively isolated parts, relying only on a background theory or interaction rules which would govern those components *if they existed outside* this particular whole. If the whole is the product of complex, context- or environment-sensitive relations among these components, then the whole must be referenced to explain those relational properties. The whole is being explained by parts whose properties are themselves being explained by the whole.

Some insist that, however explanatorily irreducible are the other sciences, physics is at least the most *fundamental* science. This includes some of the most sophisticated formulators of physicalism (e.g., Poland 1994, Melnyk 2003). In his famous paper "More is Different" Philip Anderson disagreed:

> the more the elementary particle physicists tell us about the nature of the fundamental laws, the less relevance they seem to have to the very real problems of the rest of science . . . at each level of complexity entirely new properties appear, and the understanding of the new behaviors requires research which I think is as fundamental in its nature as any other. . . . (Anderson 1972)

In response, physicist Steven Weinberg, considering whether "new kinds of law . . . govern complex systems," admitted "Yes, of course . . ." (Weinberg 1994, p. 62). He then rhetorically asked, "But just as fundamental?" For Weinberg they are not.

Are the rules governing the electron more "fundamental" than the rules governing, say, pond eutrophication? They are only in two relative senses. Ecological and biological processes are *asymmetrically dependent* on the existence of elementary particles: no electrons, no eutrophication, whereas the reverse does not hold.[7] Also, physics is more extensive in scope than ecology and biology; the laws governing the free electron have something to say to all occurrences of electrons, in and outside living things. But are these rules "fundamental" in the sense of being either comprehensive or more causally or explanatorily decisive? The answer to both questions is *no*; they do not comprehend or include eutrophication, and they are not determinative for it. In a pond undergoing eutrophication the differences and changes in the behavior of electrons will have something to say, but it will be something negligible, because they are swamped by the role dictated to the electrons by the atoms and molecules they inhabit and, at a higher level, by the pond and ecosystem in which it functions. The enormous number of different electron configurations and energy states that underlie the region of a pond's phase space which signify eutrophication will likely show no systematic difference from the electronic configuration of other regions of its phase space.

Let us be more specific. What is informative and what is misleading in the claim that "the pond is quarks and electrons governed by their interaction rules"? It is true that: the pond has quarks and electrons in it, it is asymmetrically dependent on them, indeed, the parts of its parts of its parts . . . are quarks and electrons. But it is *not* true that: a) there is nothing else in the pond (even the pond *water*) beside quarks and electrons; b) the interaction rules explaining quark and electron behavior explain the pond's behavior; c) the state of the pond is determined by what happens to its quarks and electrons; or, d) what happens to the pond happens to its quarks and electrons. For there are spacetime structures, relations, processes, and events characterizing the system that are not "in," not "properties of," nor "explained by" the system's quarks and electrons plus force laws. The states of the pond are tuned to remain stable despite most of the atomic and subatomic chaos underlying them; most microscopic states and changes have *no* consequence for the enduring macroscopic systems they compose. Concomitantly, things can happen to the pond that cannot in principle happen to its quarks and electrons—the pond can freeze, but its quarks and electrons cannot, because freezing cannot happen to something smaller than a collection of molecules. There are regularities governing the pond system that are inapplicable to its elementary particles *by definition*.

The error of reductionism is not ignorance, just as its virtue is not a unique compatibility with science. Its error is a characteristic philosophical one: stubborn foundationalism, generalizing from paradigmatic cases to characterize "The Whole," or "The Explanation" or "The Basis" of everything. The relevance of components and lower-level phenomena does not justify a philosophical claim that they constitute the primary reality, first causes, or most relevant part of any explanation. Reduction falters when it becomes an epistemological or ontological doctrine. Reductive explanation *yes*; reductionism *no*.

If the apparently antithetical "emergentism" has fewer problems than reductionism, it is only because by definition its claims are less expansive. No emergentist ever claimed that everything is emergent; nobody doubts the validity of some reductive explanations. Today we have a variety of brands of emergence (see Clayton 2004). But it is useful to understand its limits. Emergence is true of systems' properties where reductive explanation is inadequate, hence needs supplementation, where whole properties play a causal role. What would a "pure" case of emergence mean? It would mean the causal irrelevance of the parts to a whole, or absence of dependence of any whole-properties on most or all part-properties.

That would be relatively true of the shapes in Figure 3.1. A and B share the property of being a triangle; C does not. But A and C have identical components, and A and B share no components. Their shape obtains regardless of a large—not endless—equivalence class of typographical components. But the relative dependence of whole properties on part properties is rarely so negligible in wild (versus humanly constructed) nature. The kind of components usually matters.

While emergence is a fact, a philosophical endorsement of emergent*ism* is misleading. First, emergence by itself explains nothing; it merely states a fact about the world and our account of it. Second, emergentists in the human sciences must recognize that emergence is capable of supporting social organicism, the claim that there are social entities or properties which are not explainable by reference to human individuals. This may well be true, but it is not something many emergentists seem eager to talk about. Third, as supporters of emergence often fail to see, emergence *needs* reduction, for only a partially successful reduction can give evidence for emergence. Only if we perform a decomposition of a system, which will then allow us to explain something about the whole system through the analysis of parts, can we then say that other properties or performances of the system cannot be explained by part properties and performances alone. A claim of emergence requires a prior reductive analysis. Last, emergentists have also been wrong in determining the proper name of their opponents. One who accepts emergence can also accept reductions or reductive explanations and the existence of mechanisms in nature. Only the claim that *everything* is a mechanism or is mechanistic, like the claim that all things are or should be explained through reduction, violates emergence.

```
      *                 +             *  *  *
    *   *             +   +           *  *  *
  *       *         +       +         *  *  *
*   *   *   *     +   +   +   +

      A                 B                C
```

Figure 3.1

One last preliminary point should be made about a concept commonly used by physicalists. Supervenience not only is non-equivalent to emergence, it has its own problems.[8] As noted, the modern version, from Davidson, asserts "no higher level difference without a lower level difference." This is a very particular and narrow form of dependence-relation. In nature, we are likely to find there are manifold kinds of dependence; e.g., the mental is dependent on the biological, but certain kinds of mental or intentional phenomena depend as well on the social, and some (we shall see) depend on physical and material signs. Are these all cases of supervenience? It is true in that in many cases a phenomenon of scale or level N, or a difference between two N-level states, cannot obtain without some equivalence class of states or processes at the lower N-1 level. But the more complex the phenomenon, the more likely there is dependence on multiple lower- (and higher-) level processes making it impossible to say the phenomenon depends on *these* lower-level states *alone*. If what supervenience theorists call the "subvenience base" includes complex internal interactions (e.g., not just a "brain state" but a brain-plus-central nervous system-plus endocrine system state), complex environmental interactions (e.g., social and ecological state), or even worse, historical conditions like the organism's past learning, then it becomes difficult even to specify the extent of the subvenience base. If the subvenient bases cannot be specified then how can we decide whether two of them are identical?[9] So, while in what follows we will have many occasions to say that one kind of thing is dependent on another, I will generally avoid the term "supervenience." (We will return to this issue in Chapter 8.)

II. Reduction *and* Emergence

Contrary to most philosophical opinion, Wimsatt argues that emergence and reduction are not in conflict. For him, "A reductive explanation of a behavior or property of a system is one that shows it to be mechanistically explicable in terms of the properties of and interactions among the parts of the system" (Wimsatt 2007, p. 275). The crucial point to recognize is that in actual scientific practice, reduction explains: a) *only some* properties or performances of a whole system; b) on the basis of a *perspectival* (hence selective) *decomposition* of the system, i.e., a particular way of cutting it into parts; c) by using an *idealized model* of the parts and/or their interactions, resting on or employing significant approximations. We may succeed in explaining one property of a system out of a several properties we would like to explain, once we decompose the system in a particular way and presuppose an idealized model of the interactions among parts (e.g., thinking of them as point-masses or spheres or oscillators or pumps).

The recognition of multiple perspectival decompositions is particularly important. Wimsatt points out that even in analyzing a piece of granite we can produce different decompositional maps of its parts or regions based on chemical composition, thermal conductivity, electrical conductivity, density, tensile strength, etc. In the nonbiological realm these different maps are likely to divide up the system into parts with what Campbell called "coincident boundaries" (Campbell 1960). The variations in density, tensile strength, etc. are localized in a fairly consistent way; the different decompositions typically match up. But this is *not* so when we turn to *drosophilia*, the biologist's favorite fly (Wimsatt 2007 p. 71). In this case the same physico-chemical decompositions, although more complicated, roughly maintain their boundary coincidence. But regarding decompositions like anatomical organs, cell types, developmental gradients, types of biochemical reactions, and cybernetic flow, *boundary coincidence across decompositions disappears*. In such "descriptive complexity" each decomposition produces its own unique map of the "parts" of the organism. The parts do not line up; there is no one invariant list of parts to work with. Furthermore, we also find "interactional complexity" or interactions *between* these perspectival decompositions. The more interactive the system is, the smaller the percentage of the system's total properties that can be captured by any particular decomposition. We can juxtapose multiple decompositions, but relations among them are dictated *by the whole* and its dealings with its environment, including encompassing systems. The reductions inform, but cannot avoid causal reference to, the whole.

According to Wimsatt, the endpoint of a complete reduction, which would justify the claim that a system property or performance is explicable as "nothing but" its part properties, is achieved in those cases where the system properties or performances are *aggregations* of part properties or performances.[10] There are four conditions of aggregativity: *intersubstitutability* or invariance of the system property under rearrangements of the parts, so serial or aperiodic ordering does not play a role; *qualitative similarity* under scaling, where addition or subtraction of parts leaves the property only quantitatively changed, bigger or smaller but with the same properties; *re-aggregativity*, or invariance of the system property under decomposition and re-composition, so it will be the same if we take it apart and rebuild it; and *linearity*, where change in output is proportional to change in input, with no feedback, either cooperative amplification or inhibitory damping (Wimsatt 2007, p. 281). Only when all these hold can we say the system property is nothing but the aggregation of decomposed parts and their interaction rules, a linear sum or product of part-properties that have minimal interaction with each other and can themselves be treated as isolable.

This is rare. Mass is one of the few properties of physical systems which is just the aggregation of the same property of the components (e.g., my mass equals the sum of the mass of all my chemical substances which equals the sum of the mass of all my atoms, etc). Volume is not, for in some chemical reactions volume changes. Wimsatt makes the interesting suggestion that aggregativity tracks the conservation laws of physics, that the properties which are the subject of conservation laws—mass, energy, charge, spin—are those whose values are indeed invariant in all interactions. This shows how fundamental, and yet how narrow, the band of aggregative properties is. Reduction explains *something* about almost everything, but everything about *almost nothing*. Since the usefulness of reductive explanation goes much further than the few cases of aggregativity, since its adequacy is a matter of degree (e.g., explaining one property of a system but not another), in most scientific analyses we see a necessary combination of reductive and nonreductive explanation.

Emergence can simply be defined as *non-aggregativity*. It occurs when a true explanation of an event at level N cannot be reconstructed as a causal sequence of entities, processes, and/or forces of level N-1 without employing reference to processes, structures, and entities at the N or N+1 levels. In "phenomenal" explanations, explanandum and explanans are at the same level (N); for example, we explain the dent in the fender by the impact of the other car. Most explanations are phenomenal.[11] "Functional" explanations explain a system's properties by reference to higher-level or encompassing systems at N+1 (or higher, +2, +3, etc). The desideratum is to what extent the structure/processes of the whole system are caused by more or less isolable, environmentally-uninfluenced properties of parts and their simple interactions, usually meaning no more than two-body interactions. To the extent the former holds, we have a more complete reduction. To the extent it does not, we are saying the whole properties are the result of part properties and interactions *plus* properties of the whole or an encompassing system, requiring reference to the latter to fix or explain (phenomenally or functionally) the causal contributions of the parts.

Nothing is inexplicable or mystical here. It just means reduction must be supplemented by other explanations. Reduction remains a crucial component of our explanatory practice. Wimsatt is happy to call himself a "kind of reductionist," given that,

> The aim of what are called reductionist explanations in science is not to atomize phenomena . . . but to be able to articulate and understand entities, events, and processes at diverse levels . . . We can get a robust . . . appreciation of processes at higher levels of organization in their own terms that is not compromised by having lower level accounts. (Wimsatt 2007, p. 4)

When and where reduction works, it is tremendously simplifying, and generates the most context-independent entification. So we understandably keep trying to decompose systems and idealize their components' interactions in just such a way as is likely to yield workable reductions, "decomposing, cutting, pasting, and adjusting until these conditions are satisfied to the greatest degree possible . . . [tending] to regard decompositions meeting the aggregativity conditions as 'natural,' because they provide simple and less context-dependent regularities, theories, and mathematical models . . ." (Wimsatt 2007, pp. 286–7). There is nothing wrong with this, as long as we recognize its merely partial success most of the time, and do not presume it is the only or the ideal form of explanation, or worse, our definitive ontological criterion.

Some of Wimsatt's most persuasive analyses of emergence concern nonbiological systems. He lists a series of systems and properties as to their degree of aggregativity (Wimsatt 2007, p. 278). A system property may be invariant under some of the four criteria of aggregativity but not others. As noted, volume of chemicals in a mixture may be invariant under size change or recomposition, but not under some rearrangements (reactions). Critical mass of fissionable material may be invariant under rearrangement and recomposition, but not under linearity, for its threshold is determined not only by amount but organization, yielding amplifying effects on the decay process (Wimsatt 1986, pp. 267–69). The most counterintuitive example is the stability of a rock pile. If anything is "nothing but" its components, that would seem to be it! But is the pile's stability linear, is it open to recomposition, qualitatively invariant over scale change, or invariant over rearrangement of parts? Only the last answer is possibly affirmative, if the rearrangement replaces individual parts with rocks that are *very* similar. For the pile is prone to collapse as a result of minor disturbance, cannot be easily taken apart and recomposed, and adding or subtracting a rock is liable to make it fall.

Wimsatt makes especial use of electrical cases, for example, an oscillator that produces a periodic electrical signal.

> There's nothing anti-reductionist, mysterious, or inexplicable about being an oscillator. . . . You can make one by hooking up an inductance [an electric circuit making a magnetic field flux], a capacitor [which stores electricity between plates], and a resistor [which causes a voltage drop between poles] in the right way with a voltage source . . . the system has the property of being an oscillator although none of its parts . . . exhibit properties . . . like this. . . . (Wimsatt 2007, p. 276)

"Being an oscillator" is an emergent property, and this is demonstrated by the reductive analysis and its limits. The reason is that the circuit's components, their properties in isolation, and the interaction rules that govern them independently of their inclusion in the oscillator's structure, are insufficient to explain or produce the oscillator. Something must be added: a highly peculiar organizational structure. The relations among the components in that structure endow them with the ability to make their contribution to "being an oscillator." None of the components, under the interaction rules governing their behavior, would generate that structure; we must impose it.

So the success of reductive explanation holds relative to: a) some property of components and wholes; b) for some decompositions (or equivalence classes thereof); c) with respect to Wimsatt's four aggregativity criteria; d) for some idealized mechanism of those components to some approximation. As Wimsatt argues, the appearance of "common and unqualified aggregativity," in which we ignore the fact that there are alternative decompositions, that an idealized mechanism was used to model the components and their interactions, and we need approximations up and down, "is a chimera." Our stubborn tendency to seek reductions, when unqualified, leads to typical kinds of errors, "a product of un-inspected assumed constancies, idealization, and overlooked possible dimensions of variation. . . . Such kinds of errors are so easy to commit that they are almost the rule rather than the exception, contributing to design failures in engineering, modeling errors, etc. . . ."(Wimsatt 2007, p. 286). The fact of reduction's success and limits teaches important metaphysical lessons, as we shall see.

Philosophers who take a computational approach to mind often claim that a characteristic of mental processes is that they can be realized on the basis of a variety of lower-level or "hardware" states, hence are "multiply realizable." Wimsatt point out that multiple-realizability of higher-level properties, hence the dynamic autonomy of those higher levels, "is a general fact of nature" (Wimsatt 2007, p. 217). Across not only mental, but biological and material systems, macrostate stability often rests on microstate flux. Again, there is nothing mystical here. It simply is the case that "the [relative] stability of macro-states . . . further entails that the vast majority of neighboring (dynamically accessible) micro-states map into the same or (more rarely) neighboring macro-states." For the tree in the forest, the minerals and water it absorbs, and the atmosphere surrounding it, stable macrostates have to be "tuned" so that the chaos at lower levels within a numerically identical system does not lead to deviation-amplifying effects that would destroy macroscopic stability. While the macro-properties must be sensitive to certain kinds of micro-changes, "it is crucial that most differences [at the micro-level] do not have significant [macro-level] effects most of the time" (Wimsatt 2007, p. 218). This has a simple but powerful consequence: there *cannot be* purely micro-level explanations for most stable macro-level properties. In such cases we cannot regard the micro-properties as a sufficient cause of the macro-properties. He writes,

> In giving extensive micro-level detail in an explanation, there is an implication that the detail matters—that the event or phenomenon in question would not have happened but for the cited details, that if just one detail were different, the outcome would have been significantly different. But if a process shows multiple realizability and dynamical autonomy this is just what is denied for the relation of most microscopic events to their macroscopic descriptions. (Wimsatt 2007, p. 220)[12]

As a philosopher of biology, Wimsatt has a special interest in clarifying functional explanations and their relation to the thorny question of teleology. Many theorists accept that functional explanations are necessary to biology. Kitcher explains that some part or trait S has a function F where S was selected "to do something" (Kitcher 2003). For Wimsatt, functional explanations are inherently teleological, and teleology has an ineliminable reference to *purpose*. That is, "purpose and teleology are correlative concepts" (Wimsatt 1972, p. 12). None of this implies any necessary reference to consciousness, subjective agency, mental intentionality, or divine design.[13]

Wimsatt explains that "function" is what a trait or being is "selected for." We can only infer the function from a complex set of factors, leading Wimsatt to formalize function as F[B(i), S, E, P, T] = C, meaning that "According to theory T, a function F of behavior B of item I in system S in environment E relative to purpose P is to do C" (Wimsatt 1972, p. 32). The P-variable, purpose, is essential for it picks out criteria for an "intensionally defined class of state-descriptions of the system and its environment," that is, a set of conditions of which B(i) "promotes the attainment." Furthermore, there is already feedback by which B(i), S, and E are themselves partly the product of P as it fits into a system understood by T (Wimsatt 1972, pp. 39–40). In short, "It must be appropriate to speak of a purpose if function statements in the teleological sense are to be legitimate . . ." (Wimsatt 1972, p. 62). Function explains the functional property or entity, and does so by ascribing *causation.* The functional consequence is "causally responsible for the selection and presence of the functional entity . . ." It answers the "why" question, "giving a causal answer to a causal question" (Wimsatt 1972, p. 70). If this view is correct, then, Wimsatt concludes,

> the usual disputes over reduction in biology have no bearing upon the status of teleology [i.e., teleonomy] in biology. Whether macroscopic biological theories in terms of tissues, organisms, and species are eliminable in favour of molecular descriptions . . . is irrelevant. Selection processes occur and are felt at molecular as well as macroscopic levels, and this guarantees that molecular theories of the evolutionary process are also teleological. Only by denying that evolution has occurred as a result of selection processes or that it is of any scientific interest could teleology [i.e., teleonomy] be eliminated from biology. (Wimsatt 1972, pp. 66–67)

Wimsatt does not hesitate to draw conclusions about what the world must be like to generate the mixed success of reductive explanation. Remember that reality for Wimsatt is what is known *robustly*, accessible by multiple means, or what is invariant across multiple methods of access whose probability of failure are independent. He employs three special ontological categories (Wimsatt 2007, pp. 196–7). *Entities* are stable loci of causality. *Perspectives* are organized, selected structures of phenomena which abstract kinds of values from a set of systems in order to define problems. It is perspectives that fix decompositions of a system. The largest perspectives are "sections" or disciplines. Last are the compositional *levels* of natural phenomena. A level is a "hierarchical division of stuff (paradigmatically but not necessarily material stuff) organized by part-whole relations, in which wholes at one level function as parts at the next (and all higher) levels" (Wimsatt 2007, p. 201). Levels are usually, but not always, consistent with differences of scale. They are in effect a type of perspective, those ordered by "hierarchical part-whole composition relations."[14] (In Chapter 4 I will give my own terminology for what Wimsatt is discussing here.)

Wimsatt is able to define natural levels rather objectively as "peaks of regularity" or "local maxima of regularity and predictability in the phase space of different modes or organization of matter" (Wimsatt 2007, p. 249). As such, they are real objects in the world; we perspectivally select them for description, explanation, and prediction because *that is where the explanations are*. The range of entities with which an entity interacts is a non-arbitrary and informative fact about that entity; levels are collections or orders of interacting entities. Size or scale is a common, not always sufficient, indicator of level, for "size is . . . a robust indicator for many other kinds of causal interactions." Entities are generally at levels; levels are "where the entities are," ranges of scale where one finds the greatest "density of types" of entities. There are processes and phenomena between levels, but levels naturally act as "attractors" for entities. Levels emerged during the evolution of the universe, perhaps selected by their achievement of stable equilibria, defining "niches" for composing entities and co-evolving with them (Wimsatt 2007, p. 213). Wimsatt summarizes,

> *levels of organization are a deep, non-arbitrary, and extremely important feature of the ontological architecture of our natural world, and almost certainly of any world that could produce, and be inhabited or understood by, intelligent beings. . . . They are constituted by families of entities usually of comparable size and dynamic properties, which characteristically interact primarily with one another, and which, taken together, gives an apparent rough closure over a range of phenomena and regularities.* (Wimsatt 2007, pp. 203–4, his emphasis)

For purposes of clarification we can imagine worlds with different distributions of levels. Wimsatt distinguishes five "waveform" diagrams of possible worlds, each a horizontal oscillating wave drawing indicating a distribution of levels (*a* through *e* in Figure 3.2). The degree of entification of a level is given by the height of its

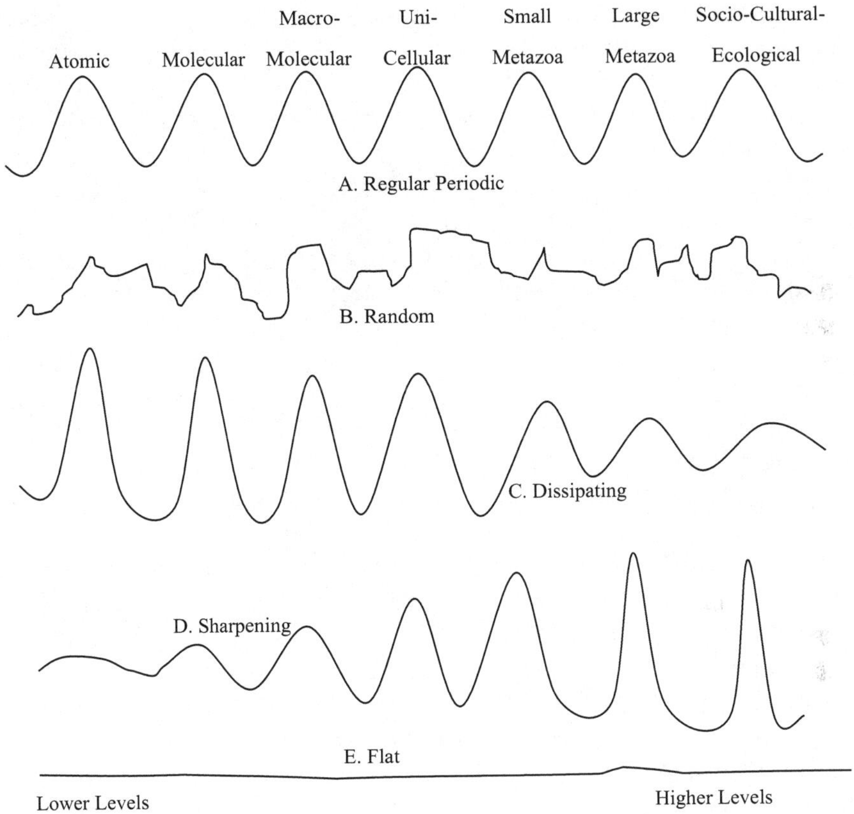

Figure 3.2. "Waveform" Diagrams for Levels of Organization. Modified from © William Wimsatt, "The Ontology of Complex Systems: Levels, Perspectives, and Causal Thickets," *Biology and Society: Reflections on Methodology.* Ed. by Mohjan Matthen and R. X. Ware, *Canadian Journal of Philosophy*, Supplementary Vol. 20, 1994:207–74, p. 230.

wave peak, the range of scale of those entities by its width. One possible world would be a perfectly regular periodic wave (*a*) with (left to right) peaks at atomic, molecular, macromolecular, unicellular, small metazoan, large metazoan, and socio-cultural-ecological levels (unique human properties lie in the last of these). In such a world, all levels have equal degree of entification (peak height) and equal scale ranges (peaks are evenly distributed from left to right).

Second is a world of random fluctuations with some regularity peaks but no well-defined levels at all (*b*). Third is a "dissipative" wave of periodic levels that move from narrow and high at the atomic end to flatter, wider peaks at the socio-cultural end, signifying less clear entification (*c*). Fourth is the opposite, a "sharpening" wave, flat and wide at the atomic level but narrowing at higher levels (*d*). Last is a "flat wave" with no levels, no entification (*e*). Wimsatt suggests our world seems closest to *c*, a dissipative wave, with regular peaks but differing in height and width, or sharpness of entification, so that at more complex levels entification flattens and causal complexity becomes endemic. That is, rather than regular "periodic" entification at each level, the causal structure of higher levels is itself more complex, less localizable.[15]

The point is not to catalogue possible worlds but to identify important features of our own, most importantly, that we live neither in *a, b,* or *e*, neither a world of equally distributed levels (*a*) nor a world of no levels (*b,e*). At the higher levels complexity outstrips entification, and componential analysis becomes untrustworthy. Wimsatt calls these "causal thickets," peaks of complexity where causality can no longer be segregated by perspectives. It is then not clear either how to decompose and/or which way the causal and compositional arrows should go among decompositions. Wimsatt identifies several types of causal thickets: the "bio-psychological thicket" of an individual, minded, animal organism; the human "socio-cultural world;" and the human and nonhuman "socio-ecological world" (Wimsatt 2007, pp. 232–3). He writes, "The neurophysiological, psychological, and social realms are mostly thickets, which are only occasionally well-ordered enough for local problems to be treated as perspectival or level-relative problems" (Wimsatt 2007, p. 239).

III. Naturalism, not Physicalism

A full analysis of the use of reductive, phenomenal, and functional explanation that would justify Wimsatt's account cannot be provided here. Nevertheless, it can at least be argued that his analysis has a number of virtues we would want to find in any such account. It is sensitive to the practice of multiple sciences. It is realist yet fallibilist in its understanding of human reason and the prospects for inquiry. It maintains the dependency relations we find in nature but rejects the tendency to regard reduction and emergence as opposed philosophies. It correctly links emergence to a hierarchical conception of nature. Last, and of no small merit, what I have called the bipolar disorder of modern thought, the belief that reality is divisible into only the physical and mental, is absent from Wimsatt's "tropical rainforest" ontology.

We can combine this account with the metaphysics begun in Chapter 1 to make a more expansive point. Emergence is, as for Wimsatt, an attractive term for those traits of systems that cannot be reductively explained. But we can now see that to accept "emergence" is just to accept three ideas. First, complexity of organization and/or processing of components in some cases seems to yield irreducible (non-aggregative) entities and/or properties, so our adequate causal explanations cannot be solely reductive. Second, in nature certain kinds of things asymmetrically depend on others, e.g., the psychological depends on the biological, both depend on the chemical, which depends on the physical. Third is a limited ontological pluralism: our successful explanatory practices will tell us how many different kinds of entities and properties we need. Given multiple irreducible sciences, nature contains at least that many kinds of entities, properties, structures, and processes. Combining these ideas yields the notion that nature, like its study, is hierarchically stratified.

This yields a major conclusion: *there is no reason to endorse physicalism, either reductive or nonreductive.*[16] The reasoning is straightforward. First, the primary justification of an ontology is its explanatory necessity. If so, then holding on to physicalism or ontological reductionism while dropping explanatory reductionism and accepting the irreducible validity of multiple sciences makes little sense. Multiple irreducible explanatory schemes, or sciences, give us *prima facie* reason to accept a pluralist ontology. Second, the claim that all is physical or determined by the physical—hence, once all physical states are determined, everything is determined—has a variety of problems. I will mention one: the definition of the physical

Sometimes physicalists define the physical as spatial extension (which was Descartes' definition). But if so, the physical has no more to do with physics than biology, behaviorist psychology, or art history, all of whose objects are extended. Physicalism then becomes the not very interesting claim of the supervenience of the nonspatial on, or its realization in, a very pluralistic class of all spatial objects, from quarks to the *Guernica*. More commonly, the physical is defined as the objects of physics (general or fundamental, present or future). But if the physical is defined in terms of the objects of physics, the *unique* objects of all sciences other than physics are not physical. If one then accepts the "causal closure of the physical"—that non-physical cannot cause physical events (Kim 1993, p. 356)—then the unique objects of other sciences cannot cause changes in the features of reality described by physics. Hence the chemical properties of the Earth's iron mantle could not generate the Earth's magnetosphere, cyanobacteria could not have produced our oxygenated atmosphere, and historical human activity could not have caused the microphysical structure of bronze.[17] The only way to avoid this absurdity is to accept a full reductionism of all science to future physics, claiming all such objects will someday be adequately explained by physics. In short, the definition of the physical, combined with causal closure, drags nonreductive back into a very reductive physicalism.

But physicalism is not our only non-idealist, non-dualist option. There is also naturalism. How can our naturalism remain coherent with the natural sciences without physicalism? Simply by maintaining the core notion of physicalism as an hypothesis about nature *within* our naturalism. Physicalism is based in the intuition that the objects of physics have some kind of priority. That seems to be right. But what kind? As noted, it is a nomological fact that all the kinds of things we call cultural, psychological, biological, and chemical depend for their existence either directly or indirectly on the physical (which, again, is not to say these are the only relations of dependence, nor that causal arrows cannot go down and side-to-side as well as up). This is just the claim of asymmetric dependency, transitively applied. But once we adopt a local conception of metaphysics and renounce globalism, that claim becomes an empirical hypothesis about the relations among natural orders, *not an a priori claim* that privileges one order metaphysically. After all, why ought the dependence of B on A mean A is more real, or B-entities *are* A-entities, or the B's should be understood through an A-metaphysics? Many complex natural systems also directly or indirectly depend on the material, which (we shall see) is different from the physical. Cultural meanings and intentional events (e.g., feelings) are not themselves physical. Neither is natural selection, which is essential to biological explanation. There is no need to give a collective metaphysical characterization of *all* complexes or natural complexes—calling them "physical," "material," or anything—or privileging one science above all others, no need to pick one feature of nature as definitive of the real or determinative for the rest.[18] All we need is to show that robustly known complexes are natural, and of what we find to be the various orders of nature, the physical is the widest and simplest, on which other natural complexes depend directly or indirectly. Nonreductive physicalism may be less wrong than reductive physicalism, but it continues to make one domain of nature metaphysically prior. The debate between more and less reductive versions of physicalism is avoidable and ought to be avoided.

Nothing about this rejection of physicalism is romantic, idealist, or vitalist. Nor is it anti-scientific. It simply bases itself in multiple sciences rather than one. Such pluralism is sometimes referred to as claiming the "autonomy" of the individual sciences. That is an apt description only as long as autonomy is not confused with either lack of relation or independence. Progress in the sciences comes both from their own internal development and from repeated cross-fertilization. Explanatory strategies and analogous concepts often migrate across disciplinary lines. And complex wholes with distinctive properties that depend, and in some cases evolve from, simpler components must be informed by the science of those components. The task of understanding reality is one of understanding the relations among, and processes of interaction between, the objects of the natural, and social, sciences. The program of such inquiry is not the *unity of science* (Oppenheim and Putnam 1958), nor the *disunity of science* (Fodor 1974 and Dupré 1993), but rather the *integration of sciences*, understood as the endless project of interrelating our natural and social scientific conclusions. Naturalism provides a framework for that project.

4

Concepts for a Pluralistic Nature

We are now ready to formulate a vocabulary of concepts that analyze nature understood as a plural set of interacting orders studied by distinctive but related sciences, a nature whose orders and features exhibit (following Buchler) ontological parity arranged in a hierarchy of emergence and dependence (described by Wimsatt and others). Again, there is no claim that all beings are comprehended by this scheme, or that its validity is deductive or a *priori*. It is a set of ideas which will be used to understand and relate orders of nature, hence sciences, to be analyzed in Part II of this study. Its merit will lie in its usefulness and coherence in that endeavor. We aim to show that the ontological priority and physicalism common to other naturalisms can be avoided while still anticipating the needs of science and philosophy.

We may start with Buchler's term "complex," described in Chapter 1.[1] Everything discriminated in any sense is a complex. Every complex obtains, or is located, or functions in some order, and more likely multiple orders of relations with other complexes. We will try to identify, relate, and understand *natural* complexes, seeking to account for as much of reality as is possible by inclusion in or relation to the robustly accessible orders of nature. But in applying Buchler's notion of complex to nature as understood by contemporary science two qualifications are necessary.

Linguistically, the scientific and Buchlerian notions of complexity are different. For Buchler, "complex" does not admit of degree. He writes "No complex is inherently 'more of a complex' or 'more complex' than any other. . . . The whole is not simpler than a part, nor a part simpler than the whole" (Buchler 1990, p. 24). This is because for him complexity refers to the potential as well as actual ordinal locations of a complex. A complex is "analyzable and interpretable without end; or . . . manipulable in an indefinite number of orders," and novel analysis or manipulation provides more orders and complexity (Buchler 1990, p. 6). We cannot say one complex is less complex than another, since its analysis can never be foreclosed. In science, however, complexity means how much structure or information a system exhibits (more about this below). Rather than using another term for scientific or organizational complexity, in what follows "complexity" will have its scientific meaning, under which things can be more or less complex, even if, following Buchler, none can have zero complexity, there being no simples.

The last point may seem to be violated by elementary particles, or quarks and leptons (e.g., electrons). But this objection is not compelling. First, we may someday find complexity within electrons and quarks; it would be surprising if today's elementary particles do not exhibit complexity in a future adequate theory of quantum gravity. But even if quarks and electrons turn out to be the *most* elementary of components, the least complex, to call them "simples" is another matter. Quarks are confined, meaning they only appear in clusters, hence are constituted by relations to things outside themselves, namely other quarks. That still leaves electrons. But electrons, like all quantum particles (including quarks), are subject to non-locality or entanglement with others, meaning their states are internally related to states of other entities. Last and most basically, quantum field theory conceives all particles as field excitations in an underlying ontology of fields, and whatever fields are, they are not simples. Cao and Smolin have separately warned against the search for "structureless" components or simples in microphysics (Cao 1997, Smolin 1997). So quarks and leptons are not simples, even if they are the *most* simple of physical entities.

I. Systems as Natural Complexes

How to characterize the complexes of nature most widely? The natural sciences have a term which applies to a broad class of natural complexes: *systems*. Indeed, the term is almost as ordinal as Buchler's term "complex," since it can be recursively applied to parts and wholes: a system is constituted in some sense by its component systems and by its inclusion in or relations to more encompassing systems. Without prejudging alternate analyses, we can tentatively say the objects of natural science are systems. ("System" is almost a synonym for "entity," as we will see below.) Like complexes, systems have traits. Anything predicated of a system is a trait—including properties, possibilities, pasts or alescent actualities, performances, structures, processes, components, anything the system is, has, or does, in present, past, or future.

This does *not* mean that all natural complexes are systems. Properties and performances of systems, like temperature and velocity, or rotation and melting, are not. Natural kinds are not: a hydrogen atom is a system, but not hydrogen *per se*.[2] Last and most crucial, structures and processes such as a crystal lattice and the double-helix, or oxidation and photosynthesis, are natural complexes but not systems. All of the above can however be conceived as traits of systems or kinds of systems. So, for the moment at least, we are orienting our consideration of any and all natural complexes in a neo-Aristotelian way, dividing natural complexes into particular systems, kinds or second-order classifications of systems, and the traits of systems (like Aristotle's distinction of primary substances, secondary substances, and properties of substances). Everything natural is a natural complex, but our analysis is focusing our attention on natural systems, their types and traits, as members of orders of nature.

It may seem we have adopted an entitative or substance ontology. However, we can avoid that result if we conceive many systems of nature—in fact, those we know most robustly—as simultaneously and co-primordially a set of lower-scale entities or *components*, a *structure* of relations among components, and a *process* (or processes) that constitutes and maintains the structure. (*States* are snapshots of processes, system processes as change in or interval of time, Δt, approaches zero.) We may say that components (themselves systems), structure, and process exhibit ontological parity (see Figure 4.1). A system equally *is* its parts, *is* a structure, and *is* a process (*if* it has all three, which not all systems do, we shall see). I do not mean each is identical to each other and to the system, but that, *ceteris paribus*—all things being equal—all three have an equal ontological claim on constituting the system. Knowledge of the system requires knowledge of all three.

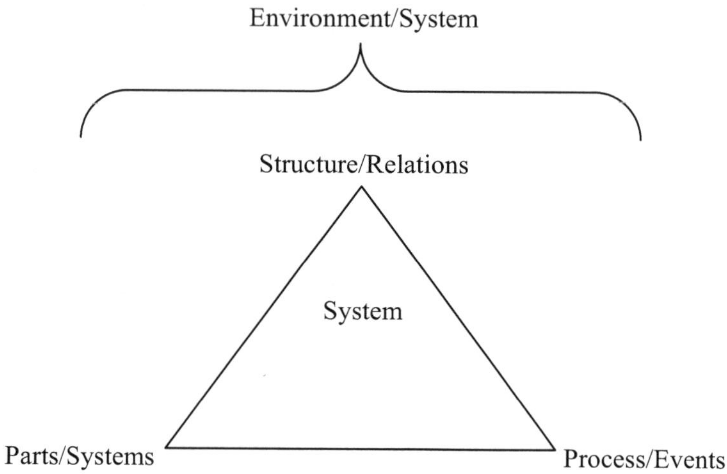

Figure 4.1. Ontological Parity of System, Process, and Structure

The parity of parts, processes, and structures calls attention to something often unrecognized in philosophical discussions of reduction. What virtually all participants mean by "reduction" is compositional reduction, reduction of wholes to parts. But the parity of entities, process, and structure suggests that to consider an individual or system or order solely a "process" or a "structure" is also reductionist, at least in spirit. Some metaphysicians and cosmologists, in reaction against the historical supremacy of entitative metaphysics and componential reductionism, have attempted to conceive structure or process as the ultimate reality, replacing entities. From the perspective of ontological parity this is no improvement. *If there is no a priori reason to privilege entities there is also no a priori reason to privilege either processes and events or structures and relations.* In the robust orders of existence to which we have greatest and multiple means of access, just as we find no entities that are not structured and undergoing some kind of process, we find no structures without something structured, and no processes without something undergoing the process.

While the meanings of the terms overlap, "entity" will have a slightly narrower reference than "system," for two reasons. One is the existence of strata or levels. Wimsatt cautions that entities can only be discriminated at levels. Hence entities remain scalar or scale-relative. More precisely, entities are systems at *one* stratum. "Entity" refers to a natural complex as a whole in its environment at its scale, while "system" refers to the same complex as something that, as containing components, may cross strata. Ontologically there are two ways to refer to any system that crosses strata: the entire stratum-crossing system in question, or the entity at its distinct level. The second reason is that, as we will see below, being an entity is a matter of degree. All entities are systems, not all systems are entities, and one system can be more "entified" than another (which does not mean more "real"). I am a human, and much of the time explanations of my behavior will be phenomenal at the social or cultural level of inter-subjective meanings and agency. But as a system that crosses strata, I am biological, chemical, and physical, and in various situations (e.g., medical) we may have to inquire into all at once. Still, Lawrence Cahoone does not appear at the macromolecular level; only my macromolecules do.

Systems and entities, in distinction from structures and processes, are particularly useful because of key features we shall discuss more fully in Chapter 5. A system is not merely a node of transmission for causation, but a *possessor* of causal potency under relevant conditions (Cao 1997, pp. 10–14, 24–26). Spatially, a system cannot be nowhere or everywhere; it must be somewhere. Temporally, entities must be characterized by temporal endurance, hence subsistence or stability *as something*, meaning possessing traits or content invariant over some range of changing relations to other beings or conditions. The stability is relative; "stable" means changing at a rate relatively slowly for the stratum in question. As Wimsatt argues, causal potency is what draws our interest, and strata or levels are where the entities, the causal nodes, lie; spacetime location tells us *with what* this causally potent entity can interact.

Entification is both a process of entity-formation and a continuum of degrees of state. Given the work of recent physical science, it will be convenient to distinguish three types of systems, without assuming that they are the only types. The three are *fields, ensembles,* and *individuals.* Fields, like the fields of the strong, weak, electromagnetic and gravitational forces, are distributions of some quantity across spacetime. Ensembles include volumes of gas and liquid, societies, ecosystems and the biosphere itself. Individuals include at least atoms, molecules, solid-state entities, organisms, and stars. As such, individuals are the peak of entification, the paradigmatic case of entities, because they are *simultaneously* components, structure, and process, whereas fields do not have components and ensembles have negligible structure. It is this combination that makes individuals capable of modular hierarchical construction, hence cosmic, chemical, and biological evolution (Simon 1962). Certainly there are transitional states and transitions between fields, individuals, and ensembles. Subatomic fermionic particles, exhibiting quantum entanglement, are arguably quasi-individuals at a transitional level between being properties of fields and true individuals. Freezing turns an ensemble of water molecules into an individual piece of ice, melting turns it back.

II. Law, Causality, and Complexity

Before proceeding, some preliminary topics must be clarified. As we saw earlier, Buchler used his account of possibility for analyzing universals. In his pluralistic metaphysical language, possibilities are traits of a complex in whatever order in which it functions, as real as actualities. There is nothing strange or extravagant about this for science; many systems are defined partly by their possibilities (e.g., a chemical substance by its boiling point, an immature organism by its potential adult form). A *similarity* between complexes is an actuality of sameness in some respect, e.g., the red apple and the red wall are red. A general or *universal* is "the possibility of different complexes having traits that are similar in a given respect," so redness is that universal in terms of which a red apple and red wall are similar (Buchler 1990, p. 180). A recurring pattern of shared traits is a *kind*, e.g., the set of apples of any color. If a complex's possibilities in a range of orders exhibit invariance, either continuous or periodic, that invariance is a *law*. A law is thus a possibility that is recurrent or continuous, expressed by words such as "For all X's, under conditions Y, Z will hold." Laws dictate what relations among universals will obtain. Laws have to be possibilities, not actualities, for they alone can *become* exemplified or actualized, which cannot happen to an actuality. They hold even when the conditions of their prevalence are absent, that is, when they do not apply (Buchler 1990, p. 178). In what follows, "form" might be the best term for invariance of structure and "law" for invariance of process, but in some contexts the difference is irrelevant and we can speak of them interchangeably. Laws and forms are universals of universals, possibilities of possibilities. The laws of physics specify the mathematical relations that must hold among possible physical states and properties of systems.[3]

Note that I am only using the term "law" in a minimal and broad sense. It is a major question in the philosophy of science what is a law and whether there are any laws outside of physics. In what follows I only mean by "law" relatively well-confirmed hypothetical statements dictating the possible relations among the possible values of variables. The ecological formula relating population size (N), reproduction rate (r), and an environment's carrying capacity (K) over time (t), namely $dN/dt = r\,N\,(1 - N/K)$, is a law; the Periodic Table is a set of laws (or in my terms, a mix of laws governing process and forms governing structure). The laws of physics may have a special status, but they govern only the physical features of systems, like energy, mass, momentum, charge, and spin. To claim they hold for, say, biological entities, is misleading. While it is an important contingent fact that they cannot be violated by the physical features of biological systems, they do not even refer to the uniquely biological features of such systems, and so cannot be violated *or obeyed* by the latter.

We must likewise say something about the concept of causality required for a pluralistic naturalism. We cannot hope to analyze it adequately, or adjudicate among different philosophical models. But the current notion of causality in analytic philosophy, especially philosophy of science and mind, is narrow; it rests on presuppositions, several of which I reject, like physicalism. I must state a broader and more minimal notion of causality, one that presumes less than the standard account.

We have come a long way from Aristotle's doctrine of the four causes, the four different kinds of factors he claimed to be simultaneously "responsible" for the existence of an entity: material, efficient, formal, and final (see Chapter 2). Modern philosophy and science performed a service in rejecting Aristotle's substantial forms and purposeful final causes from physics in the seventeenth century. This was followed by the notion, especially among empiricists, that matter is in itself inert. Hence efficient causes seemed the only causes left standing. Efficient causation was conceived in abstraction from matter or substance, form or essence or natural kind, and function, hence as the mere relation between two subsequent events or states sharing only a temporal connection. As such it was recognized by Hume to be devoid of necessity, setting up an ongoing philosophical problem of interpreting causation.

The recent philosophical discussion of causality has been dominantly concerned with questions like: *modality*, whether causality implies necessity or, as Hume argued, is a non-modal regularity; *causal selection*, or how "the" cause of something can be isolated from other conditions equally necessary for it; and *determinism*, or whether all events are caused. Various analyses have been given, particularly the "regularity" or Humean interpretation, and necessitarian or modal accounts, understood in terms of counterfactuals, the derivation of necessity from laws, or the ascription of causal powers or dispositions. David Lewis' classic argument for a counterfactual interpretation began by saying, "We think of a cause as something that makes a difference, and the difference it makes must be a difference from what would have happened without it" (Lewis 1973, p. 557). So causes are necessary conditions where all other conditions are as identical as possible.[4] While he assumed for the sake of argument that causality concerns events, he cautioned "Not that events are the only things that can cause or be caused . . ." (Lewis 1973, p. 558). Nevertheless, once we combine Lewis's causality with physicalism, hence believe all events are physical events, we have the doctrine that all causes are efficient physical events. Causality is thus understood as the necessary dependence of something on an antecedent event. Of course, matter still *matters*, as another kind of ontological dependence, e.g., of a whole on its parts (mereological dependence), and forms, in the sense of natural kinds, structures, and laws, also can matter but not causally. While it is widely accepted today in philosophy of biology that functions can be explanatory, many think that matter, form, and function cannot be *causally* explanatory.

These notions have profound effects. For example, when the idea of causality as physical and efficient is assumed, by definition there can be no mental causality. As we will later see, mental causality is controversial. It might not exist. But if it does not, should it fail to exist *by definition*? Also, if all causation is efficient it is not clear how fitness or function in natural selection can make a causal difference. We commonly say the fact that the eye accomplishes a function for the organism possessing it has made a difference to the evolution of that organism. Last, many events do not have efficient causes, such as radioactive decay. We know a radioactive substance will emit a particle every unit of time, but no efficient cause triggers it. If there is a cause, it is the nature of the substance and its constitution, not an event. The restriction of causes to efficient events seems too narrow for science.

Some recent theorists have tried to rehabilitate Aristotle's four causes. Salthe points out that, given a "focal" system (level N), its components or lower scale systems (level N-1) typically provide "initiating conditions," while encompassing or higher scale systems (level N+1) typically provide "boundary" or constraining conditions (Salthe 1985, pp. 86–93). He understands the material cause as physical or material constituents, including whole systems (which can include the nonmaterial causes of those systems). Efficient causes are processes or events which, in Salthe's terms, *trigger* an event or process.[5] For the formal cause we can understand a natural kind, structure, form, or law, hence a rule-governed structure or set of relationships. A final cause is the function or role played by the focal system or property in an encompassing system or environment which selects that system or property, or in Salthe's terms, manifests a disposition to constrain the focal system's output. This need not imply purpose or goal. Hence material, formal, and efficient causes can come from a higher scale ("downward causation") *or* lower scale, while final causation typically comes from "above" or higher.

In what follows I will not assume that only events are causes. I will accept that "C causes E" means, at a minimum, that *C is necessary to the occurrence of E*, given sameness of all other relevant conditions and/or constraints. C could be an event, constituent, form or law, or a function. Such usage is very broad, but I believe no harm is done as long as types of causes are made explicit in context and kept distinct; I will make clear when I treat as a cause something non-efficient. Probably most macroscopic events and systems to which we have robust access exhibit at least efficient, material, and formal causes, and sometimes final or functional causes as well. But

I will *not* assume that any complex *must* have or exhibit all four causes. In what follows we will be dealing with many disciplines, from physics to biology to anthropology, employing many kinds of explanation. Naturalism cannot dictate *a priori* the proper form of explanation to working scientists. All causes serve as the basis for explanations, but there can also be explanations that are not causal. For current purposes these issues can be left vague.

Last, the notion of complexity will have an important role in this study. Physicist Paul Davies claimed that the recognition of complexity constituted a third revolution in twentieth-century physics, in addition to relativity and quantum theory (Davis 1992). The term refers to a family of recent research programs in a variety of disciplines, including the study of physical and chemical systems exhibiting unexpectedly complicated forms of order (e.g., chaos, catastrophe theory, fractals, dissipative structures), cybernetic, information-theoretic, and computational approaches to natural systems, the use of the concept of self-organization as an adjunct to natural selection in biological evolution, and systems theory, including hierarchical systems theory.

But what is complexity? In the last twenty years the term has acquired multiple, sometimes incompatible meanings (see Edmunds 1999, Appendix 1, and Mitchell 2009, ch.7 and 19). This has led some to question the usefulness of the concept (McShea 1991). In the commonly met Kolmogorov or algorithmic conception the complexity of a system is the length of the shortest, or "incompressible," string of symbols that determines all its states. Some connect complexity with "information capacity," the number of possible states of a system that could be selected as a message.[6] Both imply that randomness is a high complexity state. Others use the term for highly structured, negentropic or self-organizing systems, systems that are the opposite of random. Some consider complexity a trait not of the system studied but of the language in which we study it.[7] Clearly some complexity can be "gratuitous" (Gray et al. 2010).

We cannot attempt to resolve these matters here. We may grant that there are many kinds of complexity, hence multiple legitimate uses of the term. It is enough to distinguish two families of definitions, or kinds. First, whatever is quantitatively more is, *ceteris paribus*, more complex. A system with more parts or states or properties or behaviors, or more kinds thereof, or undergoing more processes, is more complex than one with or undergoing less. We may, modifying a term of Salthe's, call this "extensional" complexity. The concept is simple, although applying and quantifying it can be difficult, for three reasons. One is that we must decide the identity and boundaries of the system, which will determine what is a part and what is an interaction with environment. Second, we must decide how to decompose the system, what to count as its unit parts. Both considerations will change with our interest, like causal relevance. Last is the problem of weighting different kinds of complexity, i.e., where system p has more complexity of type x and less of y, and system q has more of complexity type y and less of x. Nevertheless, the rule of thumb that "more" implies "more complex" is still *ceteris paribus* true.

But there is a second set of considerations, embodied in a number of other definitions, that one system may be more highly organized than another, for example, have more internal constraints on energetic pathways restricting components to exhibit a narrow set of their possible configurations. Salthe calls this "intensional" complexity. There are various kinds of system organization: a) degree of structure and internal processes maintaining it as a whole and determining its behavior; b) multiplicity of alternative decompositions of the system, or functionally significant sets of relations among the parts, without boundary coincidence (Wimsatt's "descriptive" complexity); c) interaction among such relational sets or decompositions (Wimsatt's "interactive" complexity); and, d) hierarchical ordering of multiple levels or strata of components, structure, and processes (Simon's "architecture" of complexity). The motivation which unites these concerns is a particular hypothesis: equilibrium states, as the "default" condition toward which closed systems go under the Second Law of Thermodynamics, are least improbable, least difficult to achieve, and hence *not* complex.

These two versions of complexity often conflict. A cubic kilometer of atmosphere may be more quantitatively complex than a living cell, while far less complex organizationally. But if the versions of complexity conflict, they also overlap. Regarding paradigmatic cases of typical entities of different orders of nature we will find entities that are more complex than others *both* in the sense of being more quantitatively or extensionally complex *and* in the sense of being more organizationally or intensionally complex. A living animal is more complex in both senses than is its weight in lead. Easier still, animal and lead are more complex in both senses than their *own* parts. So there are cases where we can say something is more complex without deciding between the different types of complexity.[8] It is those that will be my focus.

III. Orders of Nature

We can now expand on the definition of local naturalism from Chapter 1. "Nature" is one temporally enduring ensemble of complexes, whose members are open to at least indirect mutual causal influence (constrained by spacetime segregation, as described), manifesting plural kinds of entities, structures, processes, and properties, robustly posited by our successful explanatory practices. We have seen that these objects, as described by multiple sciences, exhibit peaks or "levels" of entities and processes, and that the relations among some of these levels exhibit asymmetric dependence. Not our only, but our main task then becomes the location of natural complexes in or at levels and the understanding of such levels and their relations. That is where metaphysical problems accumulate.

What does it mean to distinguish levels or strata of nature? Salthe contrasts two types of hierarchies (Salthe 1993). Compositional or scalar hierarchies relate nested systems of differing whole-part relations, while specification or subsumptive hierarchies relate types of systems of differing intensional or informational complexity that are simultaneously taxonomic and evolutionary. The former are "triadic," locating any focal system within more encompassing systems as the environment for composing systems (Figure 4.1 above). These relations are not transitive; it is misleading to say that molecules are parts of an organism—they are parts of parts of parts of parts . . . of the organism. Typically, higher and lower level entities provide boundary and initiating conditions. In a specification hierarchy the higher levels are characterized by the greatest cumulative informational constraints, integrating the behavior of particular systems operating at lower levels, e.g., the series [physical [material [biological [mental]]]. The series is transitive: the later "higher," embedded strata are dependent on the former, while the former are more widespread and arise earlier in evolution.

These two types of hierarchization must be kept conceptually distinct, but can be combined or interwoven, e.g., local compositional levels can be arranged inside a global specification hierarchy. We saw that Wimsatt describes levels or strata as peaks of regularity and entification corresponding to whole-part composition. Strata must have some properties that are irreducible (e.g., to the N-1 level), hence require phenomenal (N-level) or functional (N+1 level) explanation. They will also be cognitively selected by a particular suite of methods of investigation. The complexes of a stratum show a greater tendency to interact with each other; indeed, levels must often be at least partly screened off from each other so that perturbations at lower levels do not disturb higher-level processes. Typically, Salthe points out, compositionally lower levels have faster reaction and shorter relaxation times. We saw that Wimsatt distinguishes the atomic, molecular, macromolecular, unicellular, smaller metazoan, larger metazoan, and socio-cultural-ecological strata. Adding subatomic strata to Wimsatt's diagram (Figure 3.2) would convert nature's wave function to wave undulations along a normal curve (an inverted U), from subatomic levels where entification is low because of quantum features, to high entification from molecular levels to macroscopic material objects and organisms, to dissipating entification at strata of highest complexity (not scale), the causal thickets of ecological, biosocial, and biocultural phenomena.

Salthe points out that there can be—and need be—no canonical listing of levels. There are far more instances of emergence, far more componential constructions of wholes with novel properties, than we can list, so emergence is a necessary but insufficient criterion of strata. The criteria that can legitimately be used to distinguish strata are many. They include at least: a) scale of mass, energy, volume, or speed of processes; b) regularity peaks; c) degree of entification; d) compositional level, or what is a part of what whole; e) degree of complexity; f) emergent properties or performances; and g) the need for distinctive scientific methods and concepts. Fortunately several of these criteria tend to cluster together. Regularity peaks are often entification peaks at scale. Some emergent properties and performances call for distinctive scientific methods and concepts. Thus the different sciences mark particularly robust points of emergence. And a series of these sciences do correspond to rising complexity. Thus it is possible to highlight as "orders" of nature a specification hierarchy of phenomena studied by distinctive sciences, recognizing there are many local levels of emergent strata within them. The result is a small set of wide strata with properties distinctive enough to be the objects of differing sciences arranged in a hierarchy of dependence and complexity. These I call the orders of nature: the *physical, material, biological, mental,* and *cultural*. Each is a set of systems and system process and properties, whose systems include the "lower" orders on which they depend. In some cases the relation between orders is compositional, in some cases not.

For the sake of comprehension these can be depicted in a diagram (Figure 4.2).[9] The cosmological (i.e., *physical*) evolution of the universe from Big Bang to stars to black holes and clusters of galaxies is portrayed from **A** to **F**, while our local branch of material, biological, mental, and cultural evolution is (indicated by a prime) from **E'** to **J'**. Earth-like conditions (**F'**) seem to be rare in the universe; we have as yet no indication of events **G'** and after having occurred elsewhere. The oval encompassing the orders of nature does not constitute the "Whole" of existence or a physical boundary which we can view from the "outside" (hence the line is dotted). We make guesses about the characters of things from among them as one kind of them.

The physical order is not, as many seem to think, easy to define, as we noted in Chapter 3 and shall see in Chapter 5. We may take what I will call the broad sense of "physical" to mean objects that are spacetime-occupying and energy-possessing (at least, we will see, above the Planck scale). But that is insufficient as a characterization of the objects of physics, for it applies equally to the objects of chemistry, the Earth sciences, and biology. I will mean by "the physical order" something narrower, namely the objects of high-energy physics, the domain of reality for quantum, relativistic, cosmological theories, and the laws of thermodynamics, or what is called "fundamental" physics. The physical is *the smallest components and widest environments of spacetime-energetic systems.* By "smallest" I mean sub-atomic, by "widest" spacetime and the

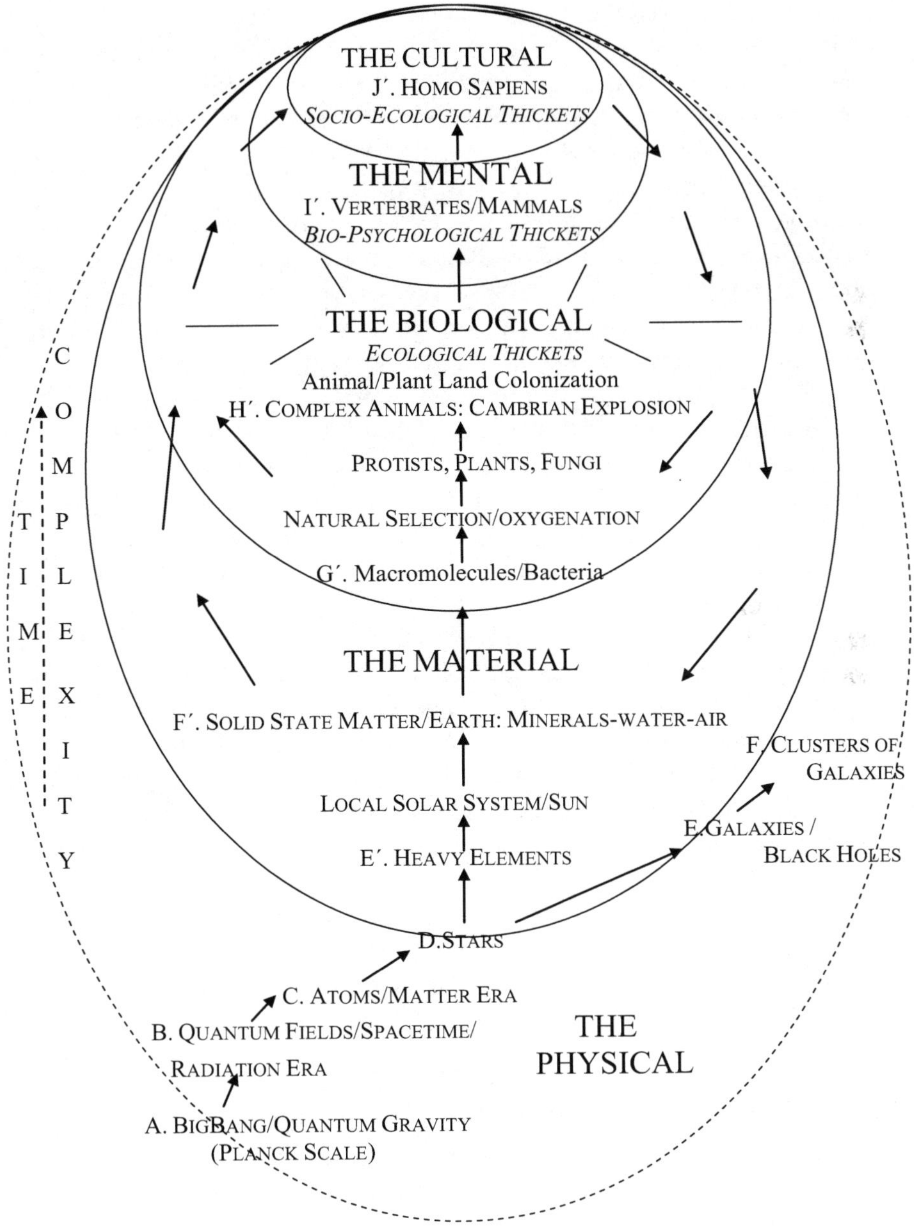

Figure 4.2. The Orders of Nature

gravitational fields that determine it. One might say the physical is the order at which spacetime, through the general theory of relativity, and elementary material particles, through quantum field theory, emerge from energy fields. We will see that fields and their quanta, the later being the quasi-individual entities in fields, along with ensembles of the latter, constitute the *systems* of the physical stratum, spacetime the physical *structure* of those systems, and the fundamental forces and thermodynamics the physical *processes* of those systems.

The order which directly emerges from the physical I call the material. This refers to non-living matter with identifiable parts, in the form of atoms (or ions), and the material entities—individuals and ensembles of individuals—which result from their combination. These emerge from the fields and particles of high-energy physics. Thus solid-state physics, astronomy and astrophysics, chemistry, the Earth sciences, and engineering all have a place here. Chemistry in particular is the study of the properties of types of matter and the specifically chemical reactions they undergo. Unlike physics, there is a lower bound to the scale of chemical objects, a smallest unit, the atom or ion, and a set of fundamental natural kinds, the elements. Taken together these sciences deal with microscopic and macroscopic material systems, which manifest emergent properties at many sublevels within the order. I argue for the existence of irreducible properties in chemistry and material science, as well as "teleomatic" processes, which is *not* to say teleological or teleonomic processes (Mayr 1974).

With life comes a massive leap in complexity, both extensional and intensional. Life is a set of processes, hence also a state, manifested by complex material individuals, which, I will argue, cannot be understood without teleonomy. Biology has a natural smallest unit, the cell, and a set of fundamental (but complex and historically changing) natural kinds, species. Biology's individuals vary greatly in scale, although far less than those of physics or chemistry; there is a minimum scale, the bacterium or perhaps the virus, but the upper bound is strictly constrained—the largest organisms are of normal macroscopic size, or a bit bigger (a blue whale, the General Sherman Sequoia, the Malheur Forest mushroom). With biology also come larger ensembles, namely societies and ecosystems, of which organisms are components: insect colonies, coral reefs, the Amazon River basin, or the Earth's biosphere itself. The social is, in the current scheme, a specification level dependent on the biological; there are further types and degrees of sociality characteristic of different species, as we shall see.

Mind is a set of processes, or more precisely, activities of certain neurologically complex animal species, studied by psychology, psychiatry, ethology, neurology, cognitive science, and philosophy of mind. These exhibit certain intentional nervous system performances we call feelings, images, thoughts, and the processing of all three. Mind is an animal, not solely a human, phenomenon. It is supported by but not composed of biological components; not minds *per se*, but minded organisms are components of some societies and ecosystems. Mental activities are intentional, they subtend or contain intentional objects which are non-physical in the narrow sense (not spatial). At lower levels intentional events are internalist, dependent only upon neural system and soma, but at more complex levels externalist, dependent upon the organism's relation to environment (including society) as well. I will argue that intentional events and properties *can* play a causal role in organisms, thereby exhibiting teleology.

Culture, as I will understand it, is an order that arises only in human social behavior (although a number of nonhuman species are capable of some transmission of local learning across generations). Here we find multiple novel phenomena: selves, or autobiographically conscious individuals; signs; and new, non-physical entities, called "meanings" which are rule-governed sets of possibilities, in effect structures of possibilities. Only with human mentality can meanings be identified and manipulated as objects. These are emergent upon joint or social manipulation of signs, itself made possible, along with the self, by a uniquely human form of social relating in which individuals take the perspective of others. As Wimsatt has suggested, at the biological, psychological, and cultural levels, along with the embeddeness of their phenomena in societies and ecosystems, causation becomes enormously complex.

The orders of the physical, the chemical and/or material, the biological, the mental, and the cultural are characterized by an ascending order of complexity, both extensionally and intensionally. I do not mean the cultural or mental *per se* is more complex than the biological, or that the unique emergent properties of a higher level are individually more complex than the system of components they arise from. I mean rather that there are systems functioning at a higher level which are more complex than systems functioning *only* at lower levels, because the former are dependent on, and include, functioning at those lower levels. The former have at least as much—or typically more—quantitative complexity than lower functioning systems, *as well as* an additional, emergent level of function enabled by greater organizational complexity than lower levels. A carbon atom functions physically and materially, and is more complex than its physical electrons, both extensionally and intensionally. Cells, which function biologically as well as materially, are more complex extensionally and intensionally than the chemical subsystems they contain. Animals with centralized neurological control systems capable of minded learning are more complex than plants or protists. And humans, whose brains are the most extensionally complex material systems we know, and are capable of behavior requiring psychiatric, cultural, and historical as well as biological and ethological explanations, are more complex than nonhuman animals.

Such a scheme of hierarchical emergence may offend in two directions. The reasons that it offends physicalists, reductionist and nonreductionist, have been described in the preceding chapter. But in the opposing direction, inclusion of the mental and, even more, the cultural, in nature will raise concerns and hackles. In the local or ordinal naturalism I am pursuing, the aim is to understand natural orders and their relations. Whatever else the mental and cultural orders are or entail, and in whatever more inclusive metaphysical order one might want to locate them, from a local, robust viewpoint it is the case that: a) we only seem to know of minds that arise in biological systems and are dependent upon and interactive with them; b) the only cultural order we know is a set of communicative signs produced by the biological creatures possessing the most sophisticated minds and brains, in the process of their social interaction. A local naturalism seeks to describe what is true of minds and culture in relation to the other orders of nature. Nothing said in this naturalism will deny *a priori* that minds and cultures have other properties, or function in other orders, that go unrecognized by my local naturalism. I have already accepted property-pluralism. The material exhibits properties not found in the physical, the biological exhibits properties not found in the material. So the inclusion of the mental and cultural in nature does *not* mean the properties of mind and culture will be physical, material, or biological. Nor will natural science techniques have a privilege over psychological, social, or cultural methods of investigation *in* the mental and cultural orders. The former remain relevant, since the mental and cultural are dependent upon the physical, material, and biological, but diminishingly adequate regarding the emergent properties of minded and cultured beings.

In conclusion, the division of nature into a complexity hierarchy of strata is *itself* a robust hypothesis, attested by several powerful facts that deserve to be repeated. First, the physical, material, biological, mental, and cultural orders of nature exhibit increasing complexity. Second, rational inquiry has found it necessary to develop a series of disciplines, with special methods and concepts, that largely correspond to this series of orders. It is not for no reason that the history of human knowledge has generated multiple sciences. Most remarkable, it is only in the last eighty years that science has come upon strong evidence of the metaphysically compelling fact that our hierarchy of complexity roughly matches the temporal evolution of nature, from the Big Bang to the formation of stars, to the development of heavy elements, geodynamics, life, vertebrates, and minds complex enough to manipulate signs. Nothing here implies that such development has been teleological or purposive; it simply seems to be true of reality's history.

Together these three considerations imply that a central feature of nature is stratified dependence and emergence. In the following chapters we will try to show there are plausible accounts of each of these domains of reality as orders of nature coherent with this claim. Each chapter presents a distinct module of my naturalism as a whole, and will have its own virtues or vices as an account of the basic science and key philosophical issues of its domain.

Part II

The Orders of Nature

5

The Physical Order

Even when reading self-proclaimed physicalists one rarely finds much effort devoted to defining the physical. One symptom of this failure is that philosophers and physicists often treat the "physical" and the "material" roughly as synonyms, although for different reasons. Philosophers do so because their interest lies in providing an antonym for the allegedly more problematic "mental," a context in which the distinction between the physical and material seems unimportant. Physicists do so because quantum mechanics and relativity blur the distinction between energy and matter at extreme scales: the energetic quantum vacuum creates particles, and matter can be converted into energy by Einstein's famous equation $E = mc^2$. But we should be cautious about inferring from convertibility that the two are the same. As one philosopher of physics puts it, Euros and dollars are inter-convertible, but that does not mean Euros *are* dollars or vice versa.[1]

There is ample justification to save the term "material" for a more narrow set of beings than the "physical." Whatever else it refers to, "material" indicates ponderable matter with non-zero rest mass, the kinds of things studied by classical solid-state physics, astronomy, chemistry, and the Earth sciences, and *not* fields, waves, and mass-less particles like photons. In the everyday world where quantum and relativistic effects are negligible, matter certainly behaves very differently from, say, electromagnetic radiation. And as of this writing it appears that the about 70 percent of the mass-energy content of the universe is mostly dark or unobservable energy. Matter appears to be a minority interest in the universe.

A more precise distinction of physical and material requires that we clarify the ontology of the former. But that necessitates a survey of four crucial areas of recent physics, three of them grounded in the revolutions of the most amazing thirty years in the history of physics, 1900 to 1930. Rather than discussing philosophical forays into these issues we will spend some time gaining clarity about the science. This will occasionally require some mathematics, for there is no getting around the fact that mathematics is the language of the ultimate truth-claims of fundamental physics. The good news is that some of that knowledge can be gleaned without being able to *do* the math. What we want is rather to see *what the math is doing*, e.g., which quantity depends on which, what happens when one variable gets very small or very large, etc. I will attempt to segregate the mathematics into paragraphs or endnotes that may be skipped without losing the qualitative point. Our reward for this labor will be that, by the time we turn to metaphysical analysis in the final section, we will be able to make some tentative guesses about the physical.

I. Where and When: The Relativistic Revolution

While it is quantum theory that is regarded as the most radical, nonclassical revolution wrought by twentieth-century physics, relativity is no less complex and difficult. Relativity asks us to imagine a very nonintuitive picture of the world. Still, Einstein's 1905 breakthrough in the special theory of relativity (STR) was conceptually rather simple. If, as the nineteenth century believed, there is a universal medium or "ether" in which electromagnetic waves (including light) propagate, then the Earth should drag some ether with it, creating a difference between the observed speed of light in different directions near the Earth relative to the Earth's direction of orbit. But the famous Michelson and Morley experiment found none. Others tried to explain this negative finding in ways that were close to the eventual solution; Fitzgerald and Lorentz suggested that at speeds near that of light, distance or size measurements alter, and they produced the correct mathematics describing the alteration. Einstein reasoned from more abstract grounds. He thought the absence of ether rational, since all inertial reference frames (with zero acceleration, zero net force) should be equivalent under the laws of physics. That is, lacking a rigid universal framework for all motion, no reference frame should be able to "know" whether it is moving or not. This must apply to light too. But then the question was, what to hold fixed and what to allow to vary, spatio-temporal measurement or the rate of information transfer and causal influence? Against all (Galilean-Newtonian) common physical sense, Einstein assumed that the speed of light ($c = 3 \times 10^8$ m/sec) is an *absolute*, that it is the greatest speed by which information or causal influence can travel and will be measured *the same in every inertial frame*. This is utterly counter-intuitive. It means that an observer rushing toward a light source at 90 percent the speed of light and another speeding away from it just as fast will measure the speed of the light coming toward them from that single source as exactly the same! To make up for light's absolute value: a) temporal simultaneity must be relative to local reference frame; b) at high velocity length must contract, time dilate, and mass increase (according to $\sqrt{1 - v^2/c^2}$ as Fitzgerald and Lorentz had suggested); c) spatial and temporal measurements must be inter-dependent; and d) energy and mass must be convertible (by $E = mc^2$). All these become variable to make up for the observational constancy of c.

The remarkable fact is that Einstein was right. Spacetime measurements are relative to frame. But all those measurements obey a transformation law *between* or operating *on* the reference frames. "Relativity" is somewhat misleadingly named; it is just as much a principle of invariance, albeit at a higher level. The event or spacetime interval or spacetime separation of events is the objective fact; all measurements *of* it are relative to frame and equally valid; then at a higher level, the relations among all those measurements of it satisfy an invariant rule. This means space and time are interdependent; there is no space or time *per se*, but only spacetime intervals. Every frame perceives light as moving at c. Each reference frame fails to perceive or measure any relativistic change in itself, any change to its "proper" time or "rest" length, since its clocks and rulers all share the same fate. But the reference frame and any objects to which it is rigidly attached will be measured by *any other* reference frame with respect to which it is moving very fast (i.e., a significant percentage of the speed of light) to be higher in mass, slower in time, and shorter in length along the direction of movement.

It is important to remember that this is a not an epistemic effect, a claim merely about observations. It has nothing to do with the consciousness of the measurers, since machines would come up with the same relative measures. Macroscopic clocks that have been carried on commercial airliners are slower. As Rindler puts it, length contraction "is real in every sense. A moving rod is *really* short! It could *really* be pushed into a hole at rest in the lab into which it would not fit if it were not moving and shrunk" (Rindler 2001, p. 62, his emphasis).[2] But even this statement can be misleading. The point is, the hole and the rod each have as many different spacetime measurements as there are reference frames moving relative to them with sufficiently high speeds. Length and time are quantities in relation to other bodies constituting reference frames, and there is no hidden, absolute but unknown length in comparison to which the relative lengths are merely epistemic or an artifact of measurement. The rod shrinks in the only way shrinking is real: relative to some reference frame. Spacetime values are *real but relative*, or, taking a phrase from Buchler, "objectively relative" (Buchler 1955, p. 128).

The best way to understand relativity is to picture the mathematics. Imagine a coordinate system (see Figure 5.1), with time as the vertical axis T and one dimension of spatial distance as the horizontal axis X. Now imagine a line segment, like a simple rod QR, somewhere in that space. This represents an event or process happening in spacetime. Now imagine superimposing on this simple picture a new set of axes, T' and X'. This is a new frame of reference. QR is now describable by two different sets of coordinates, from (T_a, X_a) to (T_b, X_b) and from (T'_a, X'_a) to (T'_b, X'_b). We could do this indefinitely many more times; each reference frame, with its own skewed axes will measure the interval or receive from it a projection that is different from all others. You can think of this as a single object throwing shadows of different sizes and shapes on differently angled surfaces. Relativity says that every one is equally valid. The interval QR is real but its length, mass, and speed are relative to reference frame. STR then gives us a statement of spacetime separation or event interval that is invariant across all reference frames ($ds^2 = c^2dt^2 - dx^2$, or with three spatial dimensions, $ds^2 = \eta_{ab}dx^a dx^b = c^2dt^2 - dx^2 - dy^2 - dz^2$). This mathematically represents the objective reality of the interval which could be measured by, or projected onto, indefinitely many reference frames.

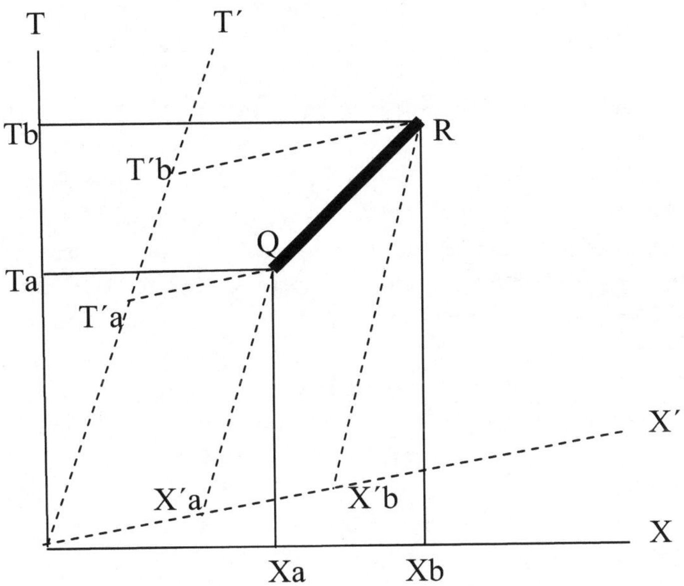

Figure 5.1. Projections of Spacetime Event (QR) on Reference Frames

Special relativity made the above argument only for frames devoid of acceleration, that is, inertial frames where net force equals zero. Einstein's 1915 general theory of relativity (GTR) generalized these results to all frames of reference, including those with acceleration, hence gravitational force too. There were three conceptual steps involved, which have been clarified by John Stachel (Stachel 1989, 1993). First was the Equivalence Principle, Einstein's proposal of the identity of inertial and gravitational mass, hence acceleration and movement in a gravitational field. Newton's laws of motion and gravitation had ascribed to bodies a mass, although the mass in each case performed a different function. The inertial mass of a body measured its resistance to any force or change in motion (in $F = ma$), while the gravitational mass of a body measured its contribution to the total gravitational force acting between two bodies (in $F_g = G\, m_1 m_2 / r^2$). These two masses had always been known to be factually equivalent, but there was no obvious reason why they ought to be; they are traits of the body which stipulate its responsiveness to two different forces. Einstein explained the mass equivalence by proposing that inertial and gravitational forces are mathematically and physically equivalent. Second, in working on the problem of the spacetime implications of a rotating disk (hence the geometry of rotating galaxies), Einstein recognized that spacetime need not be Euclidean. If this is so, then the spacetime structure in which objects move under gravity would be geometrically equivalent to an object moving in the line of least distance or time—its "straightest" line—in a *curved* space, like on the surface of a sphere. (Therefore he needed to represent spacetime in a four-dimensional Riemannian space by a metric tensor g_{uv}.) Third and most hard-won was the realization that generalizing relativity required *general covariance*, that the description of a physical phenomenon be invariant regardless of *any* change in coordinate system or reference frame.[3] The result was Einstein's field equations.

Here we need to, as with STR, picture the mathematics, although we can only picture a small part of the mathematics because far more is required for GTR. The mathematical representation of GTR is provided by tensors, like g_{ij}, which can be expressed as matrices that represent the spacetime interval or separation (the $ds^2 = c^2dt^2 - dx^2 - dy^2 - dz^2$), which is the reality we are trying to analyze.[4] The reason is so we can do tensor calculus to determine the rates of change of stresses or pressures over a curved surface represented by vectors with four components at each point, where several different quantities are all changing at once from point to point. A smooth, locally flat manifold is any space, of any number of dimensions, on which we can do calculus (i.e., it has no holes in it). A Riemannian space is such a many-dimensional manifold where the space is globally curved but flat in the small neighborhood around any point—just as the street outside may be flat, but it is a tiny part of a very big sphere, the Earth—so, locally Euclidean. Einstein required a four-dimensional Riemannian space (actually, pseudo-Riemannian). We can picture this space as the surface of a sphere (see Figure 5.2). At any point on the sphere, we can define a vector in the "tangent space," a flat plane tangent to the sphere at that point. To get the event interval (now, in GTR, a Riemannian interval given by $ds^2 = g_{ab}\, dx^a\, dx^b = -c^2\, d\tau^2$) we need to be able to compare vectors at two different points (and planes) on the sphere. In curved spacetime the measurements in different

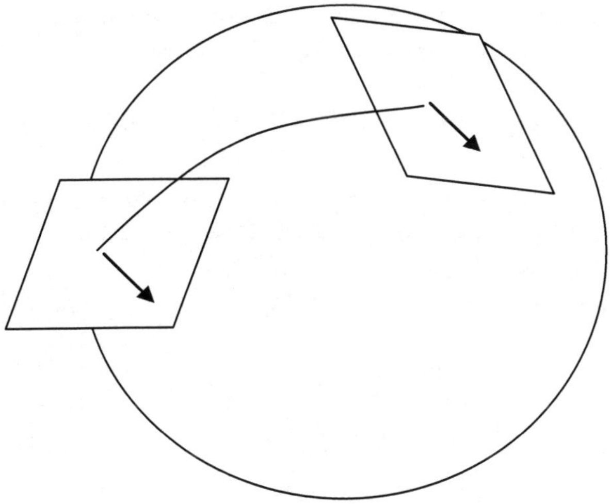

Figure 5.2. Affine Connection Between Tangent Vector Planes

neighborhoods might not be equivalent, so to compare them we must move or "parallel transport" one vector to the other's location. The path of the transport gives a geodesic (or straightest line) that would be traced by a free-falling test particle between the points or vector origins. This path describes the *affine connection* between the tangent planes at the two points. The affine connections between tangent vector spaces provide the structure of the spacetime, the connections between places, which will allow us to measure spacetime. The structure is described by *the metric tensor.*

The metric tensor is the star of the show. The Riemannian manifold is just a smooth mathematical construct, without any "metric" or means of assigning quantities to intervals. What individuates points, making them and their space *physical* rather than just mathematical, is the relation of two structures given by the metric tensor: on the one hand, the chrono-geometric (time-space) structure of the manifold, and on the other, the affine inertio-gravitational (energy and matter) field which dictates the paths of test particles. The metric tensor, which distributes the chrono-geometric structure, varies with the pressure generated by the inertio-gravitational field energy; as the latter increases, local spacetime contracts.

In GTR both of these are coded by a metric tensor field. The point is that the metric tensor represents both, for we cannot have one without the other. The chrono-geometric structure, hence all spacetime measurement, is dependent on the inertio-gravitational field. Mathematically, this results in the famous "Einstein's equation" or "the field equations." They can be stated in many ways. The simple form (if we set the speed of light $c = 8\pi G = 1$, unit of our choosing) is $G_{ab} = T_{ab}$, which equates (a 4 × 4 matrix of) gravitational force at a point in question, on the left, to (a 4 × 4 matrix of) stress or pressure and energy density at that point, on the right (mass is here being taken as energy density, according to $E = mc^2$).[5]

What is the meaning of all this? The *physical* meaning of Einstein's equation in *our* actual spacetime is rather simple: given a small ball of test particles at rest in a four-dimensional spacetime, as it falls in a gravitational field its volume will shrink at a rate proportional to its volume times the sum of the energy density plus the combined pressures in the x, y, and z directions, all at the ball's center (Baez and Bunn 2006, p. 5). Gravitation shrinks things. The equation has a *cosmological* meaning that is not simple, for its solutions stipulate what kinds of spacetimes or universes could exist. In this role Einstein's equation is really an equation *of* ten non-linear equations each of which gives a family of possible "spacetimes." Our actual universe seems to be given by the Friedman-Lemaitre-Robertson-Walker (FLRW) set of solutions for a homogenous, isotropic (same in all directions), expanding universe. But the *metaphysical* message of GTR is clear and forceful. It is not that our spacetime is curved; mass-energy curves local spacetime, but background spacetime against which galaxies move appears remarkably flat. Nor is it that spacetime is no longer a framework for all events; it still is. Nor is spacetime merely the relations among bodies, a notion offered centuries ago by Leibniz in his argument with Newton, and last century by Ernst Mach (Mach later changed his view). Schwarzschild early on discovered a solution of Einstein's equation for a static vacuum. So there is still spacetime without any ponderable matter around.[6] The fundamental point of GTR is rather that spacetime is *dynamic*, *local*, and *causally interacts* with whatever is in it. It is not independent of what is happening. Spacetime still provides the stage for the actors, but now in a poor repertory company where the actors double as stagehands and carpenters, remaking the stage on which they act, without which they could not be actors at all.

II. What: The Quantum Revolution

If relativity asks us to imagine the world in a weird way, quantum theory asks us to deal with the world without any imaginable picture at all. The road to quantum mechanics (QM, or more precisely NRQM, nonrelativistic quantum mechanics) started when Planck recognized in 1900 that at the subatomic level, action—energy integrated over time—comes in discrete quantities, multiples of $h = 6.6 \times 10^{-34}$ Joules-seconds (or more conveniently of $\hbar$, or $h/2\pi$), rather than varying continuously. In 1905 Einstein discovered, or reliably posited, the photon as a "particle" of light. In 1913 Bohr theorized quantized atomic energy levels as changes in electron orbit. Then in 1924, DeBroglie proposed (in his dissertation!) that if electromagnetic waves have quantized energy, particles can be characterized by wave features like frequency and wavelength. This broke the dam, and from 1925 through 1932, quantum mechanics poured out: in 1925 Heisenberg, Jordan and Born developed "matrix mechanics;" Schrödinger independently formulated "wave mechanics," including his wave equation (SWE) in 1926, which he later showed to be equivalent to matrix mechanics; in the same year Pauli published his exclusion principle for fermions and Born his statistical interpretation of the SWE; 1927 brought both Heisenberg's uncertainty principle and Bohr's complementarity interpretation of QM; Dirac formulated the special relativistic version of QM in 1928 and important parts of the eventual standard mathematical formalism in 1930, which was completed in 1932 by Von Neumann.

Understanding quantum mechanics—in the sense of grasping its novel claims about the world, not achieving a fully satisfying interpretation of it—requires understanding a canonical set of experimental phenomena.[7] Imagine electrons being shot from a single source at a screen armed with a detector that registers arriving electrons with audible clicks (see Figure 5.3). Between source and screen is interposed a barrier with two slits a tiny horizontal

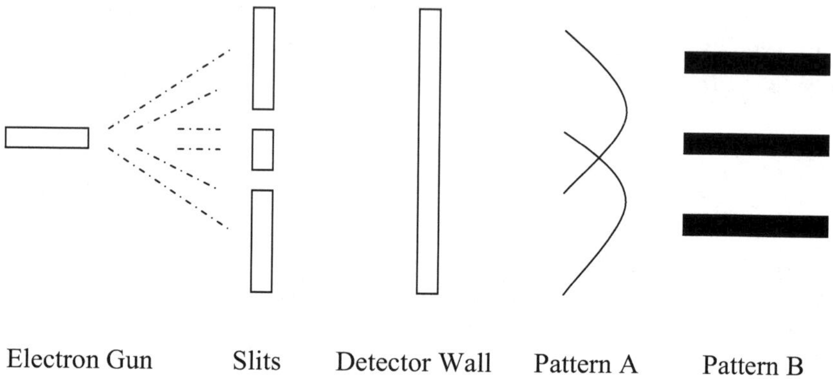

Figure 5.3. The Double Slit Experiment

distance apart, slits that can be opened or closed separately, so that when open the electrons can only get to the screen by passing through the slit(s). If electrons were particles we would expect them to accumulate like a hail of shotgun pellets on the screen opposite an open slit. That is in fact what they do if one slit is open. Then, if both are open, we would naturally expect two circles of shot opposite each slit, with a few in between (Pattern A).

But that is *not* what happens. The two slits instead produce a series of vertical lines horizontally arranged across the screen (Pattern B). That is the diffraction pattern which, it was already known, is produced by interfering electromagnetic waves (two crests or two troughs combine to make a hit, contributing to a diffraction line, but one crest hitting one trough cancels out, making no hit). At the same time, the detector shows that a reduction in the volume of electrons produces fewer hits, but each click remains equal in volume, indicating the electron hits the screen as a discrete particle. The point is, electrons—and, it turns out, photons—behave in some ways like waves and in others like particles. This is the *wave-particle duality*.

It gets worse. If we try to narrow the slits, or place detectors at the slits to see which slit any particle-wave passes through, the interference pattern disappears and we get the shotgun pattern (A)! As Feynman put it, nature appears to have arranged itself precisely in such a way as to prevent our specifying the particle's precise state and location. This "prevention" was quantified by Heisenberg's uncertainty principle. There are two pairs of properties—position (x) and momentum (p), and energy (E) and time (t)—for which any attempt to narrow the interval of one, to make it *more* precise, widens the interval of the other, making it *less* precise. Their intervals are inversely proportional: $\Delta x \cdot \Delta p \geq \hbar$ and $\Delta E \cdot \Delta t \geq \hbar$. It is as if observation squeezed a wave oscillation's width, making the peak higher, or pressed down its height, widening it out.

And worse still. The SWE accurately tells us all we can know about the state of a quantum system. Its mathematical result is itself a continuous wave function. But we are asking it discrete questions, like, is the particle in the space Δx, or is its momentum in the range Δp? What, then, is the physical reality being described by the SWE? Born answered that the wave function does not represent the particle; it represents the *probability* of the particle being characterized by the value(s) of the observable(s) at the time or place given. That is, it gives us very good estimates of the probability of a single event, or statements of frequency for a set of events.

Now, statistical indeterminacy is not new. Many phenomena in physical science are statistical. Boltzmann's nineteenth-century statistical mechanics was indeterministic in the sense that we can't count gas atoms in a macroscopic container, but we can count atoms in very small containers, then extrapolate, assuming an equilibrium. We assume that the individual objects really do have precise position and momentum at the same moment, but we can't know what they are. Their indeterminacy is *epistemic*, a matter of our observational limitations. But the indeterminism of a quantum phenomenon is different. It exhibits *objective* chance; we could not more finely characterize its state, no matter how much we might know, because the state *does not possess* finer characters.[8] The entity we are studying does not have a sharp-valued property independent of our measurement, nor does it simultaneously have sharp position and momentum. And that means *it does not have a trajectory at all.*[9]

This leads us to the last and most troubling nonclassical features of quantum mechanics: entanglement, or non-locality, and the measurement problem. But to describe them we have to look at the SWE and it is useful to know a little about the mathematical formalism of QM. The fundamental concept in quantum mechanics is that of a *state* (as opposed to a classical trajectory or path). The state of a particle or system of particles summarizes all knowledge of its varying measurable properties. The mathematical equation that describes the rate of change of the state is the SWE. When we solve the SWE we are trying to determine the wave function Ψ ("Psi") which has in it all the information about the state. But how that is done gets complicated.

The wave function can be viewed as the superposition of a set of waves, each of which is called an *eigenfunction*. In effect, the wavefunction is a collection of wavefunctions. The value of an eigenfunction for a given possible energy level (e.g., of an electron) is an *eigenvalue* of the entity in question. We are trying to figure out the eigenvalues of the state given the experimental or observational restrictions or conditions on the system. The mathematical formalism represents the state as a vector in a mathematically constructed space of potentially infinite dimensions (a Hilbert space), in which the number of dimensions used matches the number of observable quantities characterizing the state. The vectors are transformed by *operators* representing the quantities, like energy, momentum, or position. The numbers we want to know are the operators that multiply or transform the vector. The operators are mathematically complicated, but what matters in the end are the probable value-ranges for the operators that change the state vector's component vectors in that mathematical space.[10] Now we are ready for the SWE (more precisely, the time-dependent version):

$$i\hbar\,{}^{\partial\Psi}\!/_{\partial t} = \left[-\,\hbar^2\nabla^2\!/_{2m} + V(x, y, z)\right]\Psi$$

This is basically Hamilton's classical equation making (the left side of the equation) the total energy of a system equal to its (right side) kinetic energy plus its potential energy (the *V*).[11] The SWE says the rate of change of the wavefunction Ψ (on the left) is equal to (in terms of Planck's constant) the square of the change in kinetic energy of the state plus its potential energy (each taken in all three dimensions), all multiplied by Ψ. Each eigenfunction serving as an operator transforms the vector into multiples of itself yielding integrally-related eigenvalues of the SWE.

As noted there is nothing discontinuous about the wavefunction; it is smooth. As Born showed, the square of the absolute value of its coefficient indicates the probability of the system property in question having the eigenvalues indicated. The equation tells us *the probability that* x falls in the range Δr. But when we measure or observe the physical outcome, we get *yes or no*, *r* or not-*r*, *not* a probability. Put it this way: Harry may get home from gigs by his curfew only ¾ of the time, but on any given night it is the whole Harry that does or does not; we never observe just ¾ of Harry coming through the door. As Von Neumann put it, there are two different kinds of state-evolution in QM, the continuous evolution described by SWE, and the discontinuous collapse of that continuous wave into a definite value during measurement. How to understand the relation of these two is the *measurement problem*. Can we regard the measured value as having been true of the phenomenon before, or independently of, our measurement of it?

This is the problem that Schrödinger illustrated with his famous cat example. Suppose we attach a quantum mechanical experiment to a device that will drop a cyanide capsule into a box with a live cat if the particle has a certain property (e.g., it goes through slit A instead of B or its electromagnetic field is polarized or oriented in one direction rather than another). Suppose we have no other way to observe the outcome except by opening the box to see if the cat is alive or dead. Now, the question, does the system have or not have property x before our measurement?, becomes the question: is the cat alive or dead before we open the box, or is it the opening of the box which *makes* the cat alive or dead, the cat being, before we open it, in some bizarre mixed state of alive/dead? Schrödinger and Einstein both thought the latter obviously absurd.

Some say that the source of the problem is just that quantum mechanics deals with such tiny entities that any energy employed by the experimenter or apparatus must affect them. If we are studying bowling pins with a stream of bowling balls we shouldn't be surprised that every pin we study is lying down. Some writers have made much of this, connecting to broader claims about the symbiosis of subjective and objective factors in quantum mechanics, the mind of the experimenter playing a causal role in the experiment, etc. But there are many parts of physics, or science, where we know that our observation influences the phenomenon on perfectly traditional classical grounds. What is perplexing about quantum behavior is not that our measurements change the value of the property they measure, but that it is not clear that *any* sharp value of the property is there independent of our measurement.

The same problem is exhibited by the phenomenon of *entanglement.* We construct an apparatus in which training a laser on a crystal causes two photons to be shot from a central point in opposite directions at opposing equidistant films polarized in the same direction. Now, the likelihood of finding different polarization in either photon—or in other versions of the experiment, different spin—should be, by classical rules, purely random. But we discover that, in the case of the photons produced together, they will *always* be found to have opposite polarization (e.g., one blocked by and one passing through a horizontally polarized film). Hence, once we measure one, we know the value of the other without measuring it. Because of the measurement problem we cannot say the photons had this polarization prior to measurement. The experiment can be arranged so that there is no possible communication between the first particle's measurement and the second particle's measurement (i.e., where any influence would have to travel faster than the speed of light). Thus it is as if the two spacetime distinct objects have been "mixed" or entangled, so that whatever in the future happens to one—i.e., our measurement—happens to the other as well.

All this inspired Einstein to his most careful attack on quantum mechanical theory. Einstein accepted that QM's predictions were empirically right, but believed that its account of the reality producing observables must be incomplete, there must be a "hidden variable" at work that we do not yet know. If not, if QM is complete, he reasoned in a famous paper, then there could be "non-local" action at a distance, which was absurd (Einstein, Podolsky, Rosen, 1935). Today this issue has been more or less resolved due to John Bell's remarkable 1964 thought experiment showing that this disagreement had testable consequences, followed by extensions by Clauser, Horne, Shimony, and Holt in 1969, and finally the work of a number of experimenters, notably Aspect in 1982. These seem to confirm the results of entanglement. It thus appears quantum reality *is non-local*, allowing "action at a distance." (This is one of the problems in trying to integrate QM and GTR, for the later is an entirely local theory.)

The problem of interpreting the ontological meaning of QM is famously unresolved. Working quantum mechanists rarely trouble themselves with the question; others adhere, up to a point, to one or another interpretation. One of the more well-known is Bohr's ("Copenhagen") interpretation, which like positivism emphasized the instrumental nature of the wavefunction, that quantum entities need not be held to *be* particles or waves, but only to appear so when measured, and that we cannot expect to discern mechanisms underlying the discontinuity between classical and quantum worlds. Another is Hugh Everett's "many worlds" interpretation, which held that all superposed states of the SWE *do* occur, each in a branching alternate world.

Today a widely held explanation is decoherence theory (Zurek 2002). At any moment, a quantum system's "pure state" is a superposition of multiple eigenstates which each have the same *phase* (making them integral multiples). The identity of phase is what produces the superposition, entanglement, and wave-like interference effects; different phases cancel these effects (like wave crests and troughs). We observe these nonclassical effects because the objects are so incredibly small compared to their distance from nearby objects, that they are in effect isolated, closed systems. The point is that the more interactions the particle undergoes, the more phase cancellations occur, effectively washing out any interference effects; the multiple eigenstates make less and less of a contribution, leaving only *one* eigenstate, like thermodynamic randomness which leaves an average temperature because the "tails" of the Gaussian (normal) curve cancel each other. *Now* the system is in a classical statistical "mixed state," not a pure quantum state. All our measurement apparatuses are of course macroscopic objects. So interaction with the measuring device washes out quantum effects.[12] The environment "turns off" the quantum effects. Whether this interpretation will survive or some hidden variable will be discovered, perhaps by research in quantum gravity (QG), remains unknown. Whatever the future holds, physics will likely not return to a classical answer, but turn to some more complex account of which classical and quantum are limit cases.

QM was soon integrated with special relativistic effects by Dirac (RQM)—which implied all particles must have oppositely charged antiparticles—then incorporated into successor quantum field theories (QFT) like quantum electrodynamics (QED), the quantum theory of electromagnetism. In midcentury microphysics discovered more and more "elementary" particles, threatening a kind of theoretical chaos. But in the 1960s and '70s, a number of breakthroughs led to a complex mathematical simplification of this chaos, called the *standard model* (SM), which remains our account of microphysics to this day. Particularly important were: Weinberg's, then Salam's and Glashow's discovery that electromagnetism and the weak nuclear force could be treated as one unified force mediated by W± and Z^0 particles; Gell-man's proposal that protons and neutrons were composed by sets of quarks confined together by gluons, the carrier of the strong or "color" force, or quantum chromodynamics (QCD); and lastly the proposal of the Higgs Boson as the source of the mass of all other particles.

The SM lists the subatomic entities of the universe, which interact via four fundamental forces: the weak nuclear, the strong nuclear, electromagnetism, and gravitation. It includes both standard particles and their oppositely charged antiparticles. The forces are conceived as interactions activated by the appropriate charge on the relevant particles and mediated by another, force-carrying particle, a boson, which is exchanged, transferring energy between the particles (photons for electromagnetism, W± and Z° particles for the weak force, gluons for the strong force, and the graviton for gravity). Interaction is *not* particles bumping into each other, but occurs via force fields generated by charges and the exchange of a particle. Particles are created out of, and annihilated into, the quantum field vacuum's non-zero energy.

We must mention some of the profound conceptual innovations incorporated in SM. In the early twentieth century the mathematician Emmy Noether had discovered that conservation laws could be represented as symmetry transformations, that is, transformations of spacetime coordinates under which phenomena remain invariant. Symmetry simply means invariance under some operation, for example, a sphere is symmetrical under rotation about its axes. Noether realized that each symmetry of a physical law generates a conserved quantity; for example, conservation of energy is invariance of the time symmetry of the laws of physics (that they do not change with respect to time). Some physical processes "break" symmetries or move from higher to lower symmetry. For example, water molecules that in the liquid state can occupy any angle with respect to another—a random state of high symmetry—in the process of phase transition, like freezing, suddenly line up at fixed angles—a state of lower symmetry. (Symmetry does not correspond to "order." Generally speaking, it corresponds to "disorder" or a lower level of structure.) The differentiation of the weak nuclear force from the electromagnetic force in the early universe was discovered to be a case of such symmetry-breaking. The idea of "spontaneous" symmetry breaking (SSB) was incorporated from superconductivity research and condensed matter physics. The notion was that laws covering a highly symmetric system of things would still be valid of a phase transition to a less symmetric state in which they acquired an additional type of structure or preferred direction. Lowered energy under certain conditions causes a spontaneous organization at a new equilibrium state, while the laws governing the earlier, symmetric state still apply, hence the symmetry remains but is "hidden." It is this that allows the electromagnetic and weak forces to emerge from one "electroweak" force.

Another innovation was gauge theory, whose profound story we cannot recount here. Suffice to say that it uses groups, mathematical sets of transformations, in application to a physical system described in a certain way. Gauge theory in particle physics is a new, and more expanded, use of an old approach present in Maxwell's electromagnetism and in STR, whereby one conceives of the properties of a phenomenon as invariances under a group of geometrical transformations of the underlying phenomenon. The gauge theories used to formulate the SM hold invariant over independent transformations of reference frame at every point in space, rather than only over one global transformation that applies everywhere at once. This notion is part of the general relativist inheritance, namely, that any true physical description and/or explanation of any phenomenon must hold independent of any particular reference frame of observation, so in this case, independent of any gauge of measurement. That means we could use a different reference frame and different ruler at every point in space. It turns out that quarks, or the fields whose manifestations are quarks, obey a particular group of transformations, thereby exhibiting the six types of quarks, hence all protons, neutrons, and other hadrons (large particles). The weak force together with electromagnetism obeys the unified group of rotations, in which postulated mass-possessing bosonic Higgs fields produce the $W\pm$, Z°, and photonic fields. Since two forces come from the same root the four forces are, in a sense, three: the strong or color force, the electroweak force, and gravitation.[13] A sought Grand Unified Theory (GUT) that would derive the strong and electroweak forces from the same root has been not been forthcoming. The rules governing the micro-world hold at a very high level of mathematical abstraction.[14]

III. How: Thermodynamics

There is another side to the physical world, which concerns *how* processes occur, hence whether they *can* occur. That is the business of thermodynamics, which focuses on why physical processes start and stop, and in what state they stop. Events happen because of rules governing how their energy evolves over time. This is the one area of physics we are studying that did not undergo a revolution during 1900–1930; its key concepts were established in the nineteenth century by Carnot, Clausius, Thompson, and eventually Maxwell and Boltzmann.

Some systems occupy states that are stable, meaning they maintain themselves rather than spontaneously evolving or degenerating. The states of most non-living systems are either at, oscillating about, or moving toward stable equilibrium. Equilibrium is not a state of zero energy, nor necessarily of low energy, but of undirected motion at which particular processes *at the relevant level of description* stop. Thermodynamic equilibrium is a simultaneous state of thermal, mechanical, and chemical equilibrium; mechanical, because net (directional or vector) force is zero; thermal, because subregions have the same mean temperature; and chemical, because there is no net change in chemical composition. An equilibrium may be static or dynamic (the latter if mechanical processes, molecular interactions and chemical reactions occur equally in opposing directions), stable (if unable spontaneously to transition to a new state) or metastable (precarious, so a minor change could end equilibrium). Equilibrium also means that the macrostates of the system are time-independent and that potential energy between specifiable components of the system (chemical or mechanical or thermal) is zero.

Thermodynamics evolved as the study of heat and its role in motion ("heat" would eventually come to mean the quantity of thermal energy exchanged due to differences in temperature). Its laws are *not* force laws, but laws governing energy's permissible transformations. The First Law of thermodynamics holds that the total energy of a closed system is constant. Energy is neither created nor destroyed; this is one of the most fundamental principles of physics. The Second Law is more complicated. Its formulation started with Sadi Carnot's 1824 recognition that thermal energy can create mechanical work, like in a steam engine, only when there is a temperature gradient (e.g., two volumes of water at different temperatures). Rudolf Clausius realized its fundamental importance. Newton's laws of motion and the First Law describe reversible processes; they would not be violated if the processes they govern were run in reverse. So if you ran a movie of such process in reverse, physics couldn't tell you it was being run in reverse, because reverse is equally plausible. But if I put a cup of hot coffee down in a cool room, heat flows out from it; the energy present in the room does *not* flow to my cup and heat it further. This is an irreversible process; it cannot be run backwards. The heat lost by the coffee is gained by the room, so total heat stays the same, by the First Law. The quantity that has increased is *entropy*, which Clausius defined as change in heat of a substance divided by its temperature. (The quantity of heat lost by the cup is equal to that gained by the room, but since the temperature of the cup is higher than the room, *H/T* goes up.) Clausius' Second Law claims the entropy of a closed system must remain the same or increase.

Later Ludwig Boltzmann, using the kinetic theory of heat as atomic motion to create statistical mechanics, developed a different way of understanding entropy as "disorder." This term can be quite misleading; an equilibrium distribution of a gas in a room would seem quite orderly. That is *not* what Boltzmann means. Entropy depends on the ratio of the number of configurations of the components (microstates) of a closed system that yields some state of the system (macrostate) to the *total* number of possible configurations of components (microstates). There are actually three levels of description here: some macrostate property of the whole system; the total number of different possible configurations of all system micro-components; and the number of different possible configurations of all system micro-components which would yield the chosen system macro-property (actually, the natural logarithm of that number). If we have a rigid container, divided into two chambers each containing an identical number of molecules of a different gas (as long as they are non-reactive), then remove the wall between the chambers, the two gases intermingle and eventually reach an equilibrium state. In that state no region of the container contains more of one gas' molecules than the other's; the probability of any molecule selected from any region of the container being one rather than the other is 50 percent. Taking the *whole container* as an isolated system, if we added up the total number of possible states of the system, each being one spatial configuration of all the gas particles, the number of possible configurations that yield a *mixture* of the two gases (with the wall removed) is much larger than the number of configurations when the two are *separated* (with the wall intact). There are more ways for the gases to be mixed than ways for the gases to be separate. Work, e.g., the imposition of the wall, is required to maintain a condition of higher structure. According to the Second Law, closed systems move toward, not away from, that macrostate which is composed by the highest number of possible microstates compared to all possible microstates of the system.

Macroscopically, entropy is given by $dS = \delta Q/T$ (change in entropy equals change in thermal energy over absolute temperature in degrees Kelvin) or, microscopically, $S = k \, lnW$ (where k is Boltzmann's constant, 1.38066×10^{-23} Joules per degree Kelvin, and lnW is the natural logarithm of the number of microstates constituting the macrostate in question). Why, for Boltzmann, do particles behave this way, moving toward higher entropy? Because the statistical behavior of enormous number of colliding molecules will irreversibly tend toward an equilibrium state that smudges out their differences. That is what it means for the system to tend toward disorder, or what Willard Gibbs called its highest level of "mixupedness." Highest entropy, equilibrium, most mixupedness, or lowest structure is the most probable state. It takes work *not* to get there.

The Second Law does not mean that all systems are currently increasing their level of disorder. It does not deny that the universe has evolved structure or disequilibrium in its 14-billion-year existence. It just means that "negentropy" or structure in a system must always be compensated by an efflux of entropy, a discharge into the environment due to the work done to create and maintain the island of order. So a system's entropy can decrease if it can "pump" entropy into its environment. We will return to this subject in the following chapter.

IV. Whence and Whither: The Cosmological Revolution

We have one more revolution to note. In 1929 Hubble discovered the red-shift of distant stars, meaning their light shifted in frequency or color the way the pitch of a train whistle drops after it passes an observer (i.e., the Doppler effect). The farther away from us, the greater the shift. This indicated that, like the train, the stars were *receding*. In 1965 Penzias and Wilson discovered an isotropic (the same in all directions) background microwave radiation throughout space, now understood to be a remnant of the hot early universe (not, we now believe, the earliest phase of the universe, but a period shortly after when the universe was in thermal equilibrium). The most reasonable way to explain this is that once upon a time all matter and energy in the universe was concentrated in a very small, very hot volume, and has been spreading out ever since, like an explosion. This is the idea of a Hot Big Bang.

The idea does *not* mean that the fabric of space is stretching; if that were true, we couldn't know it, because we and every ruler we could use would also be stretching. Nor are we on a surface of an expanding balloon, which would again imply that space, the surface of the balloon, is stretching. (The young Alvy Singer's mother in *Annie Hall* was right: "Brooklyn is not expanding!") It simply means that all galaxies are receding from each other; within galaxies gravity is strong enough to maintain distances among stars and planets. Knowledge of existing elements and subatomic matter, hence the intertwining of cosmology and microphysics, has in recent decades suggested a rather precise chronology of the early universe and its evolution.

What might be called today's "standard cosmological model," sometimes labeled ΛCDM (Lambda, for a cosmological constant, and CDM for cold dark matter), is as follows. We assume the cosmological principle, that the universe is homogeneous and isotropic, hence regions of space we cannot observe are no different from the regions we can observe. Given this, and observational evidence, our observable universe is a FLRW (Friedman-Lemaitre-Robertson-Walker) expanding universe, with very flat curvature overall, that began with rapid expansion from a condition of very small volume at very high temperature about 13.7 billion years (4.35×10^{17} seconds) ago. We must add to this two recent discoveries to be discussed below: most of the matter in the universe is cold (moving slowly) and dark (unobservable); and expansion is now accelerating, presumably due to dark energy, in effect a cosmological constant imbuing space with energy causing it to expand (we will see how in Chapter 11).

The Big Bang itself, or its earliest phase, remains mysterious. Penrose and Hawking showed that in the beginning there must be a *singularity*, a state where temperature was infinite; so, given $E = mc^2$, mass-energy was infinite; so, given GTR, the gravitational field strength was infinite; so, again due to GTR, spacetime curvature was infinite; hence the transition from any spacetime point to another would not be smooth.[15] Infinity is a bad thing in physics, unlike mathematics, for it means our predictive powers disappear; the size, age, amount of matter, and amount of energy in the universe are all believed to be finite. So our backward extrapolation seems to terminate in an exception to our physical laws. For similar reasons, the very earliest phase of the expansion after the singularity is also a black box into which we have virtually no insight. This is the Planck time, the first 10^{-43} seconds, the time it takes light to travel the Planck length ($\sqrt{G\hbar/c^3} = 1.6 \times 10^{-35}$ meters). Just before that, temperature was above 10^{33} °K with energy density per particle above 10^{120}J/m^{-3} or 10^{19}Gev. The high energy of these subatomic particles means their mass-energy was great enough that their behavior was affected by gravity, an effect otherwise negligible. This is why neither QFT nor GTR can be trusted at this scale and why a theory of QG would be necessary to understand this tiny temporal neighborhood.

After the Planck era we believe we know what must have happened. From 10^{-43} to 10^{-37} seconds gravity was already separate from the strong and electroweak forces (and perhaps since the beginning). The universe was expanding and cooling. At 10^{-37} seconds something happened that, while it cannot be said to be proved, is today widely assumed: the universe's cooling led to a massive, sudden expansion called *inflation* in which the universe went from

10^{-60} to10^{20} m^3. (We will look at inflation in more detail in Chapter 11). Shortly after inflation ended, the continued cooling may have led to the symmetry breaking of the other forces; strong from electroweak—*if* they were unified according to yet unsupported GUT—around 10^{-35} seconds, but more reliably electromagnetism from the weak force at 10^{-12} seconds.

At 10^{-6} seconds the pair annihilation of the 10^{87} quarks and their oppositely charged antiparticles took place, leaving 1 out 300,000 quarks, or 10^{81} baryons to make up the universe's matter. This is the stuff of all protons and neutrons to this day. Around 10^{-1} seconds those quarks were confined into nucleons, 38 percent neutrons and 62 percent protons, one proton/neutron per 10^9 photons/electrons/neutrinos, the last constituting the background microwave radiation of the universe. Simultaneously electrons and their antiparticles (positrons) began mutual annihilation, and continued through the universe's 10th second, releasing photons while cooling, leaving us with the electrons we have today. At 10^2, or 225 seconds, the nuclei of the first elements, hydrogen and helium isotopes, formed, establishing their 73% to 27% ratio. This was the beginning of the transition from a radiation-dominated to a matter-dominated universe. The rest of the history of the universe is continued expansion of this equally distributed energy and matter, cooling, and gravity working its magic on sufficiently low-energy matter, massing hydrogen and helium atoms enough for their gas clouds to create stars.

Now, there may seem to be a contradiction between cosmology and the Second Law. According to the latter, the universe's total entropy cannot go down. But the Big Bang theory implies very low structure, or very high entropy, at or near the origin of the universe, with lower entropy (more structure) now. The Second Law seems to imply that must be false. One way of removing the contradiction comes from Penrose. The explanation lies in the fact that gravity, as Stachel quips, is "not just another pretty force." Imagine a set of bodies evenly distributed in a finite three-dimensional space that are on average motionless with respect to each other (they could be rotating or oscillating, but not advancing). If they were gas molecules, that would be an equilibrium or least-structured-state; to change the system work would have to be done on it. But suppose now that the bodies are planets exerting gravitational force on each other. Now they are clearly *not* in equilibrium, but in a highly structured state. Something is keeping the bodies from collapsing into each other. Because gravity is a universally attractive force—unlike electromagnetism, with positive and negative charges—gravitational equilibrium would mean all of the bodies clumping together, for equilibrium is the state where no more work can be done by the force in question. The ultimate gravitational equilibrium is a black hole, in which an enormous number of microstates are summarized very easily by a few parameters. As Penrose remarks, gravity was never "thermalized," meaning that the Big Bang must have proceeded, even at moments where it was in *thermal* equilibrium, from a condition of low entropy, a structured state, with respect to gravity (Penrose 2004, p. 706 and 728). The gravitational "degrees of freedom"—the number of independent parameters that fix a system's motion or state—were not part of that equilibrium of matter, density, heat, and electromagnetism; total entropy was low *even though* a thermal equilibrium existed. The total or average entropy has been increasing ever since, even if neighborhoods, like galaxies, solar systems, and planets, develop local structure, low entropy, and so must export entropy to the rest of the universe.[16]

So what has the Big Bang left us with in its 10^{17} seconds? On the largest scale, our universe is a distribution of clusters of galaxies, the largest structures that exist. The 10^{12} nearest stars form our galaxy, which is 10^5 light years (ly) or 10^{21}m across. Estimates of the size of the entire universe vary, but the distance to the next comparable galaxy, Andromeda, is 10^{22}m or 10^6ly. The visible universe, as far off as we can detect radiation—which is to say, the farthest light could have traveled since we believe the universe began—appears to contain *10^{11} galaxies!*

One kind of object whose existence was inferred from GTR is of crucial importance for cosmology. It appears that, in addition to the source of the Big Bang, there are singularities *inside* the universe. These are the famous black holes, stars—or actually, any material objects—massive enough that their gravitation eventually causes them to shrink below their Schwartzschild radius ($r_s = 2GM/c^2$) creating a density so great that not even light can overcome their gravity in order to exit that radius or "event horizon." Hawking and Penrose showed that black hole production must be significant in any universe, either produced by the initial Big Bang, by condensation at the center of galaxies, or the eventual collapse of large stars. Black holes are very important for cosmology, since they are, in a sense, the "edge" of spacetime, locales at which spacetime, as we understand it, comes to an end, because of infinite curvature.[17]

What is the likely fate of this universe? The question remains open and controversial, but we can at least clarify the possibilities. The universe as a whole is presumably a closed system, hence must be increasing its entropy. But, as seen by Boltzmann and others in rather poignant ways in the nineteenth century, this implies that the universe must eventually experience a "heat death," meaning the death of all heat, in which galaxies will move further away from each other, hence those stars which do explode will be too far apart for effective recycling of stellar materials to fuel new stars. Eventually, after a very long time, all stars will burn out. Boltzmann committed suicide; it is claimed the unhappy prospect of heat death played a role. Will the universe lead to such an end? Many think so. The problem in answering this question is that it depends on the universe's mass and density, its age, its curvature, and its rate of expansion, and our inferences to those quantities depend on each other.

The mass density of the universe is the crucial figure. Omega (Ω) is the ratio of the actual average mass density of the universe to the critical value which would be required for a flat universe. If $\Omega = 1$ the universe is open and flat, meaning that background (intergalactic) space would be Euclidean or have no curvature, and its radius will expand to infinity at which point its velocity will be zero. If $\Omega > 1$ space has positive curvature, like the outside of a sphere, and the universe will slow and collapse back into a fireball. If $\Omega < 1$ space has negative curvature, is hyperbolic like the surface of a saddle, and there will endless expansion that does not slow to zero, hence heat death. The critical value is currently taken to be between 1.8×10^{-29} to 4.5×10^{-30} gr/cm^3, which is *very* close to an absolute vacuum (two to eight Hydrogen atoms per cubic meter). Current estimates of the actual mass density of the universe are very close to this figure. All this assumes a Hubble constant—the rate at which the velocity of "recession" of distant bodies increases with distance—of 72 km/sec per megaparsec (3.26×10^6ly or about 10^{22}m), hence a 12.5–15.7 billion year old universe. Recent supernovae observations suggest that the universe has actually accelerated its expansion in the last 5 billion years, and this can be explained by hypothesizing, in addition to unobservable matter—which could only account for .3Ω—a dark energy equivalent to .7Ω. But that means, rather remarkably, that now, at 10^{17} seconds, the universe has returned to an energy-dominated state, in which 60 percent of the universe is dark energy, 30 percent dark matter, about 10 percent invisible baryons, and only about 1 percent visible matter (Kirschner 2004, p. 254). To speak of dark energy is equivalent to ascribing energy to spacetime itself or postulating a "cosmological constant." Before the discovery of cosmic expansion Einstein had hypothesized such a constant as representing a vacuum pressure necessary to keep the universe from collapsing under the force of gravity (more on this in Chapter 11). Hubble's discovery seemed to make the idea unnecessary; Einstein regarded it as his biggest error. But the recent discovery of acceleration changes this assessment. If there is a cosmological constant, acceleration will continue long enough to cause heat death. But we must remember this is frontier science, and changes in other factors affecting the equations could alter the picture.

V. The Ontology of the Physical

We are at last ready to ask: *What is the physical?* One way of defining it would be to stipulate the properties anything physical must have, or the necessary or minimal conditions met by anything physical. A famous option is Descartes' definition of materiality (for me, physicality) as spatial extension. Are all physical beings characterized by spatial extension?

In this context it is useful to distinguish Bose-Einstein, Fermi-Dirac, and Boltzmann-Maxwell statistics, three different mathematics we use to estimate the behavior of large collections of little things. The Boltzmann-Maxwell statistics we use to describe gas molecules employs the standard Gaussian normal curve. The reason is that we assume, fairly accurately, that the motions of each particle are unrelated because at moderate energies their collisions are more or less elastic. Hence their collective motion can be considered random. The quantum world, however, operates according to *two* different sets of rules. Force-carrying Bosons, like photons, as integral spin particles obey Bose-Einstein statistics, acting more as waves than particles, although still quantized and exhibiting particulate properties. They are superposable, meaning when they combine they produce a single distinctive wave state which is identical to an endless combination of waves. The superposed waves are not "parts" of their product, which contains no trace of distinct components; bosons integrate seamlessly and do not aggregate or clump as matter does. Electrons, protons, and neutrons, on the other hand, as half-integral spin particles, obey Fermi-Dirac statistics based on the Pauli Exclusion Principle. Electrons in an atom cannot occupy the same quantum state; each possible state (determined by its four quantum numbers, see the next chapter) fills up, so other electrons must occupy a different state. It is this inability to occupy the same quantum state in an atom that generates the impenetrability of matter and its ability to aggregate rather than superpose. This makes it possible for supra-atomic material objects, including living things, to exist.

As such, waves and fields do not take up space in the sense of *excluding* other things from space. Spatial exclusiveness is characteristic of fermionic matter. Fields and waves are not spatially *bounded* either; they extend indefinitely. However, they are spacetime *located* and *extended*; waves and fields are somewhere, not everywhere, and extend over spacetime separation. (Mathematically, point-particles are extension-less, but in physics the model of the point particle is used only where useful. Electrons and quarks do in fact have size.) Current QFT presumes a four-dimensional Minkowski space. So the fields of QFT and GTR *are* spacetime occupying. Spacetime is, at this scale and higher, causally necessary, even if it is not efficiently causal itself; it is determined by gravitational fields, but it provides the frameworks in which entities can causally interact (i.e., determining what is past, present, future, near and far).

Unfortunately, while the analysis of the physical as spacetime-located and spacetime-extended holds for all scales from QFT up through the macroscopic, it will not do for smaller scales. It is apparent that neither QFT nor GTR are "final" theories, and most students of the yet unavailable QG theory assume that spacetime itself is an emergent phenomenon (Smolin 2001). At some point between the scales encountered in QFT and the far lower Planck scale continuous spacetime must cease to obtain. Thus spacetime cannot be part of a comprehensive definition of the physical, unless one allows that the QG fields from which spacetime emerges are *not* physical. That seems extravagant.

One brief point about time. Some theorists, in what is another example of unexamined reductionism, argue that time is "unreal." Now, as noted, processes obeying only conservation of energy and charge, the gravitational and electromagnetic field laws, are reversible.[18] Also, in some areas of cosmology it may be possible to represent change spatially, in such a way that *t* drops out of a mathematical representation.[19] This leads some to say that the subjective "arrow of time," our experience or feeling of the direction of change from past to future, is not characteristic of physical reality. But all this theorizing ignores that time goes wherever spacetime goes, since they are part of one unified phenomenon, and further, all processes described by the Second Law exhibit asymmetrical temporal direction as a matter of objective physical fact. Irreversibility is the physical reality of a time-direction. If time is unreal, so is the Second Law.

But there is an even more ubiquitous physical property than spacetime: energy. What is energy, or what some call "physical energy"? Classically the capacity for doing work, or exerting force over distance, energy comes in multiple kinds, e.g., kinetic, potential (whether mechanical or chemical), and mass-energy. Energy is conserved in all closed systems, and that fact is perhaps the most fundamental physical law. Further, it seems that nothing with causal relevance can fail to possess energy; energy provides causal potency. But while apparently a universal property of physical existence, energy seems definable only in terms of its forms and products. Feynman remarked that while we know the forms energy takes and the rules for its transformation and conservation, "we have no idea of what energy *is*" (Feynman 1970, I.4–2, author's emphasis). That is too extreme, although it seems we cannot define energy in terms of anything more fundamental.[20] We can say energy is a quantity, conserved in all closed system transitions, which takes on a finite series of forms and seems to grant physical efficacy.

We might then define physicality above the Planck scale as spacetime-location and spacetime-extension, and energy-possession, hence whatever possesses *spacetime energy*. While this may be the necessary condition of supra-Planck physicality in a broad sense, it is not sufficient, for it applies equally to the objects of chemistry, the Earth sciences, biology, to societies and human cultural products like musical performances and paintings. The alternative approach is to define the physical as the objects of physics. This is the approach I will take. But it has its own troubles. It is not easy to find the common thread among physics' subfields. And taking this approach has, by definition, an important consequence: whatever is uniquely the object of chemistry or geology or biology is then *not physical*. You cannot define "physical" through physics while still calling the objects of chemistry and biology "physical," unless, following an extreme form of reductionism, you make chemistry and biology *part of* physics or some ideal, future physics.

Physics is an unusual science; rather than calling other sciences "special sciences," it might make more sense to call physics *the* special science (Chapter 3, cf. note 1). Unlike chemistry or biology, it has no simplest natural kind or smallest entity for analysis. Its objects cannot be defined by scale, at either a lower or upper bound, for physics examines both the smallest and the biggest things in the universe. Physics has the subatomic realm to itself (except for nuclear chemistry); it alone posits the smallest entities (of whatever type, particles or fields), the "fundamental" forces, two of which apply at all scales (electromagnetism and gravity), and is the science of fields, waves, and electromagnetic radiation. It also has some features of stars and galaxies to itself (except for the use of chemistry in these investigations). And it is the science of spacetime. So we might say physics studies the smallest components of, and the background contexts for, all natural phenomena. But thermodynamics and solid state physics also study everything in between the smallest and largest in terms of their energy, mass, motion, charge, etc. We might then say that physics specifically studies the spacetime (e.g., location, volume, motion) and energetic traits of systems, explaining these in relation to the spacetime and energetic features of their smallest (subatomic) components and widest environment (i.e., spacetime, gravitation).

Now, as noted in Chapter 4, we can distinguish three types of natural entities or systems: individuals, ensembles, and fields. Our notion of entities comes mainly from our experience of the first, which are most robustly accessible. Individuals include, at the least, nucleons, atoms, molecules, solid state macroscopic objects, cells, organisms, planets, and stars. They exhibit: 1) *materiality*, which is to say they are constituted by fermionic particles and have chemical characteristics; 2) *numerical individuality* or *haecceity*, individual existence, being identifiable and re-identifiable as distinct from entities of the same kind (Strawson 1990, Stachel 2006);[21] 3) *space-like boundedness*, in all spatial dimensions, including "boundary coincidence"—meaning their boundaries as registered by different causal interactions coincide (Campbell 1960)—and *space-like exclusivity*, excluding other entities of the same scale from their area/volume; 4) *structure, process, and components*, as described in Chapter 4; and, 5) *time-like subsistence*, so at least some of these traits must be *relatively* subsistent, and boundary conditions must be more internally than environmentally stabilized (which does not mean independent of *all* environmental or background conditions).

Some metaphysicians will find subsistence and "ownership" of properties troubling, since they seem to imply a kind of independence that was a highly criticized aspect of Aristotelian substance. Certainly, as noted, any individual's independence is relative, meaning comparative. Individuals are material systems that are continuous and maintain at least some traits and boundaries relatively independent of environment, which means, invariant over a relatively wider set of states and environmental changes, not over *all* states and changes. "Stability" means a relatively slow process of change. And like all entities, individuals are metaphysically scalar, they obtain at a characteristic scale.

Ensembles are entities of negligible structure. They have components, and undergo processes, but they do not have stable boundaries. An ensemble's structure or boundaries are either statistical or environmentally determined or both. Ensembles are Karl Popper's "clouds" or collections of gas or of gnats, in contradistinction to "clocks," his metaphor for classically deterministic systems like the movements of planets or balls on inclined planes (Popper 1972, p. 213). They include, at the least, volumes of gases and liquids, non-individuated collections of macroscopic solids, weather systems, ecosystems, the biosphere itself, and galaxies. Ensembles are not less fundamental or important than individuals. If the universe could be called a system it would presumably be an ensemble.

As scale decreases below the atomic level, individuality, hence ensembles of individuals, become harder to find. At these levels, physical reality is first of all *fields* and their properties. Fields are distributions of some physical content, like energy, across a region of space, to whose points they assign quantities, in some cases vectors (quantities with direction). Fields possess energy and other causal properties, are spacetime located and extended but not bounded or exclusive. They have structure, unlike ensembles, but no parts. When fields are added they superpose, like waves, rather than aggregate. Field-systems are the source of both atomic individuals and the ensembles in which they generally function. A metaphysics of fields does not spell the end of a metaphysics of "entities;" fields are causal, energy-bearing, property-maintaining systems.

The underlying "stuff" of the physical (again, absent an adequate QG) is governed by QFT. Its fields are distributed over a continuum of points in four-dimensional Minkowski spacetime (the spacetime of STR, *not* GTR). The field's energy is located at every point of spacetime by a local field operator whose point vibrations, modeled on tiny harmonic oscillators, yield field quanta interpretable as particles. When no particle is present the field has a "zero-point" or vacuum energy, of $\frac{1}{2}\hbar\omega$ (where ω is angular momentum or "spin"), which is the source of "virtual" quanta. Because of Heisenberg uncertainty the points of the field fluctuate up and down in energy, canceling out over longer distances and times to leave the zero-point energy. The smaller the spacetime scale, the more violent the fluctuation. We can speak of the global quantum field filling all spacetime, but it is local fields locally interacting that create phenomena. There is a type of field for each kind of fundamental material or force-carrying particle. Interactions between particles are understood as local couplings which create (emit) or annihilate (absorb) particles, and the field is the source and sink of that creation and annihilation. For QFT, the fields themselves are the underlying realities (Cao 1997).

GTR is also a field theory. Gravity is a field force, and that field is energy-bearing. Thus it so happens that the two major roots of our physics, QFT and GTR, exhibit a striking ontological convergence on fields, discussed in detail by Cao (Cao 1997). The significance of this satisfying convergence has been marred by the self-described "crisis" that physics has faced during the second half of the twentieth century, and which became more glaring as other areas of high-energy physics were clarified in the 1970s, namely the failure to integrate QFT and GTR. The reasons for the difficulty are many, but most are due to the discontinuous nature of energy in all quantum theories, versus the continuous nature of spacetime in GTR, and the fact that QFT presumes a continuous four-dimensional fixed Minkowski space, not the dynamic spacetime of GTR. Workers have searched for an underlying theory that would generate QFT and GTR in limit cases. For a couple of decades, string theory, which makes gravity and quantum reality the result of ten-dimensional "strings" rather than point-masses, received much attention. But however fascinating its promise, string theory has one great drawback: it presupposes a fixed four-dimensional Minkowski spacetime as the environment for the strings. This is why it is, in a sense, an expansion of QFT. Theories of QG, on the other hand, try to derive both the entities of QFT and any spacetime from a deeper source. We cannot probe the complexities of QG here. But if current, widely held guesses hold true, below the Planck length (and in cosmogenesis, before the Planck time), gravitation is quantized, resembling the flux of the quantum field vacuum, and cannot support a metric field for the chrono-geometry of spacetime.

How to characterize the particles in the fields? The particles of QFT, like the spacetime points of GTR, are also not individuals, because their "individuation" is accomplished by the field—either the fields of the quantum vacuum, excitation of any singular quantum field, or the inertio-gravitational field in GTR. As Stachel puts it, whatever *haecceity* the emergent particles have is inherited from the field. Even at the NRQM level, particles cease to have a constant core of simultaneous causal traits, they are identifiable but not re-identifiable, and, depending on how one wants to deal with the measurement problem, they acquire their identifying "markers" only via environmental interaction, that is, measurement. In addition to being entangled, or non-local, and exhibiting both wave and particle characteristics, we may say that such complexes are *Heisenberg incomplete*, meaning they do not continuously exhibit conjugate properties like momentum and position, or energy content and temporal location. In the same way, spacetime points in GTR inherit their numerical particularity from their environment (Stachel 2006). Thus, as we go from the atomic to smaller scales, we move from individuals with intrinsic properties to what Cao calls "holistic," and we could call "proto" or "quasi" individuals, entities whose properties are dictated by a narrower set of environmental roles and relations.

The development of field theories has encouraged a tendency of some twentieth-century metaphysicians to retreat from an entitative ontology altogether. Certainly the search for "structure-less particles" of "absolute individuality," devoid of context-sensitivity, seems conceptually misguided, as Cao and Smolin have separately argued (Cao 1997, Smolin 1997). Simples seem doomed eventually to be found to exhibit distinctive properties or context-dependent structures, stimulating a new search for a lower level of structureless objects. In reaction against this search, some treat structures instead as the real objects of physics. This approach has been encouraged by GTR itself, symmetry-breaking in the standard model, the gauge theories of groups, the very ontology of fields, and cosmologists seeking the wave equation of the universe (Hartle and Hawking 1983)—all seem to testify to the triumph of a kind of Platonism. This has fueled a movement in philosophy of physics called "structural realism," which attempts to maintain a realist view of QFT and GTR by asserting the ultimate reality of structure (French and Ladyman 2003).

Another perspective that seems largely compatible with structural realism is expressed by philosophers, and some scientists, who specify *process* as the ultimate physical reality. This view, famously that of Bergson and Whitehead, has been revived by Bickhard, Smolin, and Stachel (Bickhard 2000, Smolin 2001, Stachel 2006). In his approach to QG Smolin writes, "relativity and quantum theory each tell us . . . no, better, they scream at us—that our world is a history of processes. . . . From this new point of view, the universe consists of a large number of events. An event may be thought of as the smallest part of a process, a smallest unity of change" (Smolin 2001, p. 53). While Smolin has couched his claim in the common reductionist language (that only the most elementary is real, all else being "illusion") his main point is that process is primary. Stachel makes the same claim about relativity, both special and general, that the dynamical four-dimensional metric process is primary; it is what determines the structure and state of any 3-space-1-time-dimension "break-up" of spacetime at and by a reference frame (Stachel 2006, p. 65–67). Indeed, among contemporary scientists and philosophers of science, the terms "structure" and "process" are not always clearly distinguished. Thus when structural realists and others make "structure" the essence of reality, they seem to mean such structures to include both orders of relations and orders of change, thereby including under their term "structure" what I would call "process." So those promoting ontologies of structure and process are often numerically identical.

However, some of those endorsing the realities of structure and process have objected to the claim that structure can be ontologically prior to entities. Cao argues, and Stachel echoes, that structures without entities are "causally inert" (Cao 2003) and lack *haecceity* or individuality (Stachel 2006). There is a difference between mathematical structures and physical structures: the physical structures can be efficient causes, and to cause they must possess energy. The point is made by the very essence of GTR, as Stachel argues, since as we saw, the chrono-geometric structure of space is dependent upon the inertio-gravitational field. It seems that both QFT and GTR require "stuff," energy-possessing fields, without which mathematical structure cannot obtain *physically*. The failure to recognize this appears connected in an ironic way to the very traditional, entitative ontologies being criticized by the proponents of structure and process, namely, the prejudice that only material individuals can be entitative or substantive. Whereas, Cao insists, fields *are* substances or entities, albeit of a special type. The point is that there is no structure in physical existence unless it is the structure *of something*, and no process in physical existence unless there is *something to undergo it*. All the more reason to retain our heuristic of ontological parity between parts or entities, structures, and processes, even if in some systems one factor can become negligible.

To summarize, I will mean by the physical *order* something much narrower than either spacetime-occupying and energy-possessing systems, narrower even than "the objects of physics." I will mean (as noted above) *that in relation to which* physics explains the spacetime-energetic properties of things, namely, the smallest components and widest environments of spacetime-energetic systems. That is the domain of reality for quantum, relativistic, and cosmological theories, and their thermodynamic behavior, or what is usually called "fundamental" physics. The physical order is the order at which spacetime (through GTR) and particles (through QFT) emerge from energy fields. Fields and their quanta, the later being the quasi-individual entities in fields, along with ensembles of the latter constitute the *systems* of the physical stratum, spacetime the *physical structure* of natural systems, and the fundamental interactions and thermodynamics the *physical processes* of natural systems. The physical is fundamental in the sense of being that on which all other natural orders are either directly or indirectly dependent. If there were no physical order, nature would not exist. The physical is also, as noted, the most pervasive in scope. But it is *not* comprehensive: it does not comprehend the novel features of the chemical, the biological, the psychological, or the cultural, nor is it determinative or explanatory for them. We cannot say that everything natural is physical, even though everything in nature must directly or indirectly depend on the physical. Physics studies the simplest of the orders of nature. If you roll me down an inclined plane, physics can explain my velocity. But not my vomiting or cursing.

6

The Achievements of Matter

What is matter? It might seem obvious. Samuel Johnson thought so. He famously tried to render the idealist philosopher Berkeley's denial of matter nonsensical by saying "I refute him thus!" while kicking a stone. He meant that we know matter exists because we bump into it, it excludes us from its space. But the reason is not that it is a continuous inelastic lump. At the microphysical level it is mostly empty space (since its atoms are), its resistance due to electromagnetic forces between its molecules.

Matter is complex. As noted, it could be microscopically defined as what is constituted by half-spin fermionic, rest-mass-possessing entities that are capable of aggregation, under which, unlike wave and field superposition, a whole contains components or parts. At all scales, what I have called *individuals* are the key residents of the material order, even if they exist within or develop from ensembles. As noted, individuals have material parts, are numerically individual (identifiable and re-identifiable as distinct from "likes"), space-like bounded and exclusive, time-like subsistent, and exhibit structure, process, and components. There are other properties that emerge with material individuals, e.g., shape, and the standard phases of solid, liquid, and gas. Chemistry is the study of the types of matter and the reactions they undergo. But material systems, both ensembles and individuals, from atoms to galaxies, are also handled by condensed-matter physics, astronomy, geology, meteorology, oceanography, and engineering.

Our goal is to analyze the development of material complexity and to understand those features of the non-living material world that are metaphysically most compelling. After describing the cosmological emergence of our material neighborhood (Section I), we will turn to the basics of chemistry (Section II). Then Section III will deal with the statistical behavior of material ensembles, the reducibility of chemistry to physics, and dynamic far-from-equilibrium systems. In the process we will see a variety of ways in which the complex order and hugely varying scales of material systems are *not* reducible to the physical, or how, in Philip Anderson's famous phrase, "more is different." In the final section we will conclude that a kind of *telic* organization—one that has nothing to do with purpose or design—is evident among a variety of complex material systems.

I. From the Big Bang to Earth

The world accessible to unaided perception, a world of relatively stable solid objects and processes offering motor resistance, is an unrepresentative sample of material reality, dependent on the special conditions of our residency. We humans live *in the middle* (see Figure 6.1). By that I mean, we live in the middle of the history of a physical universe after stars were created and before they all

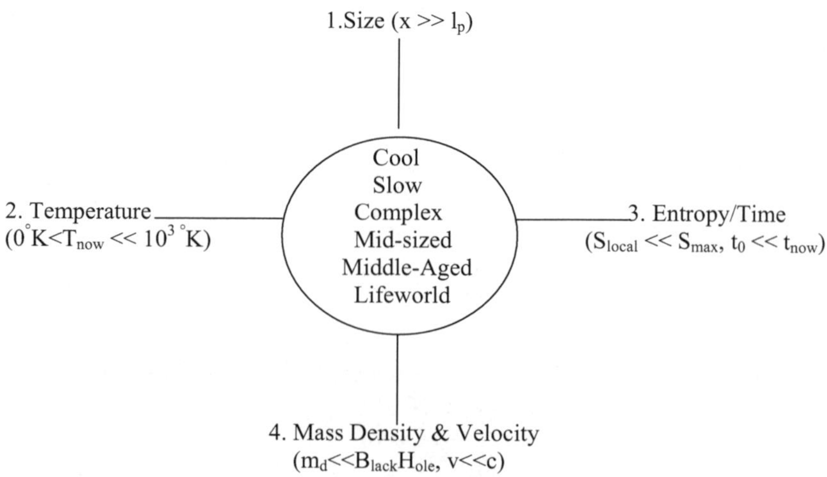

Figure 6.1. Where We Live (T – temperature, m_d – mass density, 1_p – Planck length, c – speed of light, t_0 – Big Bang, S – entropy)

burn out, at a temperature far below that of the early Big Bang yet still above zero Kelvin, in a pocket of thermodynamic disequilibrium among systems always proceeding toward equilibrium, among things of middling mass density traveling at slow speeds. Our experiential lifeworld on Earth is the artifact not only of the background laws of electromagnetism, gravity, and the strong and weak forces, but also of very particular conditions, some of which apply only to familiar macroscopic scales, some only to the particular region of material reality we occupy, and some only to our era in the development of the universe.

Taking just one of these parameters, material reality is largely an artifact of temperature. The difference between solid objects (with definite volume and shape), liquids (definite volume but shapeless), and gases (no definite volume or shape) is determined by temperature (and pressure). At sufficiently high temperature there is no condensed matter; everything becomes a gas. At far higher temperatures matter ceases to be, meaning there are no molecules or atoms, but rather highly energized leptons, like electrons, and protons and neutrons buffeted by electromagnetic radiation. Higher still there are no confined quarks, hence a quark plasma. Our metaphysics of macroscopic entities with stable boundaries is peculiar to a cool world.

One of the ways temperature determines our reality is that it acts as a *force-switch*. As noted in Chapter 5, only at temperatures near that of the Big Bang does the heat energy of elementary particles go so high that it generates gravitational force. Once cooling takes place gravitation assumes the character it has in our era, by which it keeps massive objects close to each other, and slowly brings together distant objects, but is not strong enough to override the far more powerful nuclear and electromagnetic forces that actually constitute material objects.[1] Only when mass becomes as great as that of a star does gravitation generate matter-changing processes, like nuclear fusion, or even greater in a black hole, where it can rend the fabric of spacetime. Electromagnetism is generally what holds entities, or more precisely, atoms and molecules, together in our world. Of course, without the strong or color force, there would be no nuclei in existence to work with; and without gravity to concentrate matter, there would be little for electromagnetism to do. But what we call "things" in the narrower sense exist because of electromagnetism and form discriminable macroscopic objects because of low temperature. The fact that a rock stays a rock is mainly the work of electromagnetism; this can happen only because temperatures are low.

Living in the "middle" also means we live in an *old*, or at least middle-aged universe. Imagine the universe before the formation of stars. At that time there would have been *being*, in fact, just as "much" reality as now if measured in terms of total energy or Baryon number. An observer of that earlier universe would have formulated an ontology in which there were essentially no individual stable macroscopic systems with internally dictated boundaries, only hydrogen (H) and helium (He) gas clouds animated by microwave radiation.

From our vantage point, the material universe appears as a big, cold, dark, lonely place, where life is a rare occurrence. What is less apparent is that this is the only kind of universe that *could* support life; life could not exist in a small, warm, bright, crowded universe. As R. H. Dicke argued, a universe with life in it must be at least 10 billion years old to allow some main sequence stars to have formed, matured, and exploded to seed the universe with heavy elements necessary for life, like carbon, and for terrestrial planets to congeal and cool (Dicke 1961). In an expanding universe, time correlates with size and dispersal; a 10-billion-year-old universe must be very large, its major components far apart. Hence any universe with life must be big, cold, dark, and lonely; biotic matter is only conceivable as a tiny minority of *any* universe.

The synthesis of matter by the Big Bang was very limited. All matter produced was the three lightest, simplest elements and their isotopes: hydrogen (H), in the forms of protium (H-1), deuterium (H-2), and tritium (H-3); helium (He) in the forms of He-3 and He-4; and lithium, in the form of Li-7. Even this was a long job. It took until 10^{13} seconds, 700,000 years, for the primordial soup to cool to 5,800°K, which is about the temperature of the surface of the sun. Until then all the matter was in the state of plasma, in which nuclei and electrons were free, and hence more opaque to light than is bound matter, because free electrons absorb light more completely. At 10^{13} seconds, cooling allowed electrons to bind to nuclei forming atoms of H and He, yielding an ordinary, transparent gas through which light can travel; the universe was now divided into matter and radiation, allowing the emission of Cosmic Microwave Background Radiation earlier trapped by the plasma.

At this point gravitation began to overcome expansion locally and to collect these atoms into the first macroscopic entities, gas ensembles. Molecular gas clouds of hydrogen, helium, and their isotopes, sufficiently large and dense to generate great heat, began to form stars. The heat was trapped in the envelope of the cloud, resulting in the nuclear burning of hydrogen into more helium by fusion, in the process synthesizing nuclei of heavier elements up to iron (atomic number 26). By 10^{15} seconds some stellar cycles ended in exploding novas, seeding the universe with gas and the dust of those elements, the heat of the supernovae themselves being responsible for creating all the elements heavier than iron. So, except for hydrogen, helium, and lithium, all matter we know (with human-made exceptions) emerged from the distribution of stellar material, enabling gravity to create dust, asteroids, comets, meteors, and planets.

Stars come in many types, but the most common are main sequence stars (Figure 6.2).[2] Our Sun is a mid-size main sequence star. Such stars have a life span of about 10^{10} years, during which they evolve from bright and hot to faint and cool. The star's fate

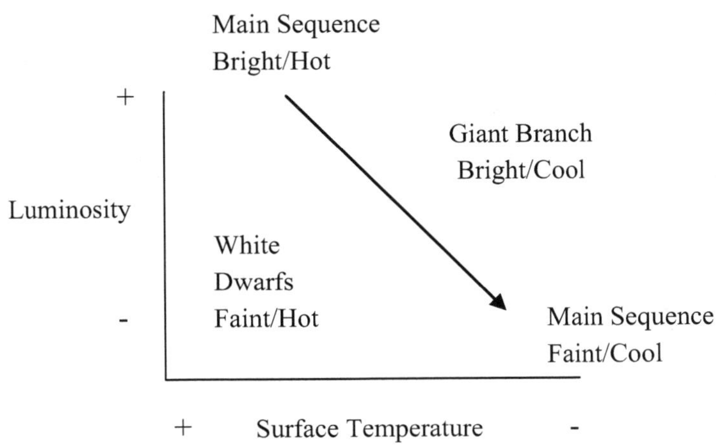

Figure 6.2. Hertzsprung-Russell Diagram

is largely determined by mass; larger stars radiate more energy and have shorter lives. Stars continually circulate some of their mass into interstellar space in the form of molecular clouds and dust. Once 12 percent of their hydrogen is burned, their core becomes unstable and begins to contract, while their gas envelope expands. If large enough, they drop off the main sequence eventually exploding as supernovae, leaving behind either a neutron star or a black hole. Stars of the Sun's mass have complicated fates, oscillating between the evolution of the main sequence and its near branches of faint/hot and bright/cool. Starting on the bright/cool branch, it loses mass to drop down into the faint/hot branch, and begins burning helium into carbon, nitrogen, and oxygen, then silicon and eventually iron. It may then return to the bright/cool branch, and go through the cycle again. When last on that branch it will eventually reach a new instability and blow off its outer layers into a ring-shaped gas and dust cloud, a planetary nebulae (about which more later). Eventually it collapses into a white dwarf, hot but faint, which slowly cools.

Stellar formation in galaxies is the first example of an "ecology," a dynamic, periodic system of energy-using individuals that recycle their waste into new growth (Smolin 1997). Galaxies recycle stellar material. Stars continue to form from enormous interstellar molecular clouds, recirculating matter while radiating heat and light, eventually seeding the universe with the products of their nucleosynthesis. Every time a star becomes a white dwarf, neutron star, or black hole, some mass is permanently taken out of galactic recycling. It is the stellar ecology that creates the qualitatively distinct material kinds that populate the universe; each star is a crock-pot version of the Big Bang, a smaller, slower, cooler kind of creation.

The rocky planets (along with rocky planetesimals, asteroids and meteors, and icy comets) are the first macroscopic solid individuals in the universe. Put it this way: rocks are *rare* in the universe. Solids are either crystalline, polycrystalline, or amorphous in structure, the former being a rigid, repetitive three-dimensional long-range structure of atoms and molecules, the second a mixture of crystals, the last really meaning *not* crystalline or lacking long-range repetitive structure. Crystals "grow" in that, once several molecules line up in a crystal structure, in a suitable environment more are electromagnetically attracted to the structure. Only with the formation of terrestrial planets do we find a host of different materials reacting with each other under varying conditions of temperature and pressure, generating chemically complex solids.

Our Sun and its system of planets are the precipitates of one nebula or cloud of gas and dust, as Kant and Laplace separately proposed long ago. (Its closest neighbor star is Proxima Centauri, 4.2ly or 4×10^{13}km away.) It is likely that as this cloud accumulated it was caused to contract by its own gravity and to rotate by a supernova in our galactic neighborhood. This collapsed into a rotating protoplanetary disk about 4.6 billion years ago whose central mass was the Sun, and whose cooling bands or rings gradually differentiated, collapsing into planets. Closer to the Sun, metallic and silicon dust accreted due to electrostatic attraction; further from the Sun, methane, ammonia, and water froze, and as it contracted, pulled in hydrogen and helium via gravity. The former became the terrestrial or rocky inner planets, the latter the jovian or gas giants. Jovians would play a major role later by gravitationally slinging asteroids and meteors into collision courses with other developing planets. Their growth was halted by strong solar winds—streams of electron and proton plasma—which blew the remaining hydrogen and helium out of the developing solar system. But the solar winds also mitigate cosmic rays from outside the solar system. The Sun remains more than 99 percent of the mass of the *entire* solar system.

The unusual character and location of the Earth are the basis for everything else on our planet: geology, along with meteorology and oceanography, is the context of life. The Earth is the right size and in the right place around the right kind of star; its orbital distance, size, and density allow rocks to be hard, air to be gravitationally retained, and water to be liquid. Before life began Earth had already become the densest planet in the solar system (5,515 kg/m^3). It is geologically dynamic or "alive." Its center is extremely hot—from gravitational compression, radioactive decay of unstable elements, and residual heat from early meteorite bombardments—and its mass contains material strata of different kinds and phases which perform an internal self-renewing cycle. Its solid inner core and molten outer core, mostly iron and nickel, are surrounded by an iron and magnesium lower mantle, the largest part of the earth (about 2,900 km of Earth's 6,400-km diameter), which conveys lighter and hotter materials toward the surface, while heavier and cooler materials sink. Earth is covered by a thin solid crust and solid upper mantle, together called the lithosphere, which ride on the liquid surface of the lower mantle, the asthenosphere. That fluid iron mantle generates the Earth's magnetic field, which provides our only protection from the solar winds that would otherwise blow off our atmosphere. These conditions enable the *rock cycle*: hot magma emerges from volcanoes and cracks in the moving crust to form igneous rock, which weathers and erodes, is compressed as it sinks into sedimentary rock, which, like subsurface heat-and-pressure formed metamorphic rock, eventually cycle low enough to be melted into magma. Furthermore, the lithosphere is divided into tectonic plates, continually moving the crustal continents and diving under each other, pressing materials into the interior. The continents merge and then disperse every 300 to 500 million years with major effects on climate (the last "supercontinent" was Pangaea 300 million years ago). Even without life, the Earth is a self-maintaining, recycling system at several interactive levels—its geological rock and tectonic cycles, which provide the gas and water vapor resulting in oceans and the atmosphere, the water cycle of evaporation and precipitation, which in turn erodes rocks, and our weather cycles of temperature, pressure, wind, and moisture.

Nasty contingencies have cropped up. There is a strong possibility that a developing planet within the Earth's orbital ring, Theia, destabilized by the Earth's growing mass, struck Earth about 4.533 billion years ago. It not only destroyed itself but ejected the Moon from the Earth and created the latter's axial tilt, the mellifluously named "obliquity of the ecliptic," which was then stabilized by the Moon's gravity. In its early history Earth was the subject of massive asteroid and meteor bombardments, which released enormous heat but also added size. The Earth has had several atmospheres in the course of its existence. The first, probably only hydrogen and helium, was blown off by solar winds and the heat of the Earth itself. Once Earth grew to 40 percent of its current size and cooled enough to maintain crust about 4.4 billion years ago, it was able to retain a second atmosphere of carbon dioxide and methane emitted from the interior by volcanoes. Growth and cooling continued, leading to rains and oceans. It was in this context that life began, eventually giving the Earth its third, oxygenated atmosphere. In perhaps five billion years the Earth may be destroyed as the Sun's outer layers expand after its core collapses. But our biosphere must end long before that. Life on a terrestrial planet near any main sequence star has a limited temporal window of perhaps five to six billion years, from when it cools sufficiently after nebular formation to the point that the star burns too hot or cools too much to support life.

II. The Kinds of Matter: Chemistry

Unlike physics but like biology, chemistry has a list of fundamental or simplest (not simple) natural kinds, the elements, and a smallest entity, the atom or ion (with some caveats we will see in the next section). Chemistry is the science of substances (*not* in the Aristotelian sense) or material kinds, particularly their propensity to undergo certain reactions, like combining or bonding. Hence chemistry can be defined as the science of the transformations of substances. Chemistry's home is the laboratory, seeking the chemical units in any mixed matter, and the properties thereof, as pioneered in the eighteenth century by Lavoisier. Since John Dalton in the early nineteenth century, we have accepted, and later known, that elements are composed of identical atoms (or ions), and that any amount of a given compound has the same ratio of component elements. Since Dimitri Mendeleev's 1869 formulation of the periodic table we have recognized the complex regularities governing the system of elements and hence their allowable transformations. Before proceeding, we need some basic information about substances, their atomic structure, and bonds.

An atom is the smallest particle of an element, with the same number of electrons and protons, hence electrically neutral. Molecules are electrically neutral complexes of bonded atoms. Ions are electrically non-neutral atoms or molecules, which have gained electrons to become negatively charged anions or lost electrons to become positive cations. Chemistry is concerned with the phenomenological (here meaning macroscopically experience-able) properties of substances, e.g., heat conductivity, electrical conductivity, malleability/brittleness, acid/base character, phase or melting-boiling-vaporizing temperatures, hardness, luster, etc. Above all, it is interested in chemical reactivity (and its antithesis, stability). Most chemists are not interested in the atomic nucleus, although there is a subfield of nuclear chemistry concerned with the effects of nuclear decay and radiation on substances.

There are 91 naturally occurring elements, currently 118 elements in all, each composed of a distinct type of atom with distinctive properties, mostly due to the way atoms generate and interact with electromagnetic forces. Each element is actually a family of isotopes with differing numbers of neutrons but an invariant *atomic number* of protons. The total number of substances, elements plus the fixed combinations of elements that are compounds, is enormous, currently around 13,000,000, and growing as humans artificially create more. These elements and compounds may be found in homogeneous mixtures (e.g., salt and water in seawater) or heterogeneous mixtures (e.g., milk and sugar in coffee), each of which are decomposable by *physical* means, like distillation or filtration, without a chemical reaction. The definition of a pure substance is that no physical process, e.g., a centrifuge, will separate out its elementary components. Substances are the kinds of materials of reality.

Elements are defined by the number of protons in their nuclei, which is normally matched by the number of neutrons, but, as noted, isotopes differ (e.g., carbon-12 has 6 protons and 6 neutrons, while carbon-14 has 8 and 6). Some elements, like the noble gases, are monatomic, meaning a single atom is already a unit of the element and exhibits its properties. Most elements are diatomic, having as their smallest unit a molecule containing two or more identical atoms, e.g., nitrogen and the halogens. Chemical compounds, stable combinations of elements, have as their smallest unit either molecules, for example H_2O, which is one molecule of three atoms, or ions. Molecules vary wildly in size; a DNA macromolecule can be of macroscopic length, and crystalline solids, due to their very stable periodic structure (e.g., a diamond) can be regarded as a single molecule (although so can their parts or decompositions).

The properties of a substance vary first of all with the electron configuration of its atoms. Electrons, or in particular the outer or valence electrons, determine the electromagnetic properties of an atom, hence its ability to bond. Electrons "orbit" the nucleus, the quotation marks indicating that, given QM, electrons do not have trajectories. The orbital is a region the electron occupies circling the nucleus at some radius, hence a probability distribution for its appearance, or an "electron density." Each orbital is characterized by four quantum numbers. The principle quantum number (*n*) is an integer designating the electron's radial distance from the nucleus, its shell, and hence energy level; the azimuthal quantum number (*l*) designates the orbital's shape or subshell; the last two numbers indicate components of the angular momentum of the electron's orbit.[3] It is this collection of four numbers to which the Pauli Exclusion applies; no atom can have more than one electron with the same four quantum numbers.

The bonds among atoms, hence their potential for chemical reactivity, are determined by the valence or outermost electron shell. Full valence shells make for stability, because more energy is needed to exchange electrons; less than full makes reaction more likely. Monatomic elements remain stable because of the forces within their atoms, but all more complex elements and compounds are stabilized by various types of inter-atomic bonds. *Ionic bonds* occur when contact causes mutual ionization through exchange of valence electrons and are common between ions of metals and non-metals, yielding brittle crystals with high melting points. *Covalent bonds* involve attraction of one atom's nucleus to another's electrons, in effect "sharing" their electrons; if the bonded atoms are asymmetrical in their electronegative charge, they are "polar," or contain two poles (a dipole). In *metallic bonds* between metals, each atom bonds with several others, permitting valence electrons to move among atoms, accounting for electrical conductivity. At a higher level of organization substances are held together by inter-molecular *Van der Waals* forces which fall off precipitously with distance and are responsible for phase: *London* or *dispersion* force interactions generated by relations between temporarily induced dipoles, without ions or permanent dipoles; *dipole-dipole* forces where there are permanent polar molecules but no ions; and *hydrogen* bonding in solutions between molecules with hydrogen atoms bonded to electronegative atoms, hence in effect a kind of dipole-dipole attraction. Last are attractions between ionic and polar molecules, and ionic bonds without the involvement of polar molecules.

Chemistry is the study of the distinctive reactivity of substances. Chemical reactions conserve composition or proportion of types of atoms and mass. There are many kinds. In *redox* (oxidation-reduction) reactions electrons are lost by one substance and gained by another. Combustion is an oxidation that reduces O_2. In *acid-base* reactions (in the Brønsted-Lowry formulation) one substance (the acid) donates a proton (a hydrogen ion) to a proton-accepting substance (the base), or alternately (in the Lewis formulation) one substance (base) donates a pair of electrons to an electron pair acceptor (acid). Substances can aggregate to create a more complex substance (*synthesis*) or decompose into component substances (*analysis*). Crucially, the presence of some substances changes the rate at which a chemical reaction among two other substances takes place, without itself undergoing the reaction, namely *catalysts*. Reactions depend on energy; some are *endergonic*, requiring additions of energy, others *exergonic* or energy-producing. This list of chemical reactions is philosophically important, for as noted above, not only chemistry but its natural kinds are defined by what counts as a *chemical* event or reaction as opposed to a *physical* event or reaction.

Take a glass of water. The smallest particle of water is a molecule of H_2O, because decomposed it would no longer be water. In the water, as in any macroscopic material object, there are on the order of 10^{23} molecules (or atoms).[4] In each H_2O molecule the hydrogen atoms are held to the oxygen by a covalent bond in which the mutual repulsion among nuclei, and between electrons, and attraction among each nucleus and each electron, generate a stable boomerang shape with electrons forming a line perpendicular to the line between nuclei (the nuclei "share" the electrons). But the intermolecular forces holding those 10^{23} water molecules to their neighbors are hydrogen bonds between the O of one molecule and an H in another (this happens to be crucial for life on Earth, as we will see in a later chapter). The water is currently in the liquid phase, because temperature and pressure occupy a middle-range with respect to its inter-molecular attractions, thus it has properties like viscosity and surface tension. If subjected to either low temperature or high pressure, or both, it would become a solid, with definite volume and shape, the molecules shifting into a crystalline organization. If subjected to high temperature or very low pressure, it would be a gas without definite shape or volume. But in all three phases it would still be the same substance, meaning it has a set of unique properties with respect to chemical reactions.

So the water—what it is and does—is a product of a unique organizational structure maintained by electromagnetic bonds between molecules with special properties at a given temperature (given favorable environmental conditions). Like any other macroscopic "thing," it is an enormous collection (10^{23}) of very tiny components (atoms or molecules) that come in relatively few kinds (10^2 elements to 10^5 substances). The tiny components are far more stable than the things made of them, the forces constituting them being under most conditions far stronger than the forces holding them in combination. That is the truth in Democritean atomism and materialism: every "thing" is made of tiny relatively indestructible bits of relatively few kinds, and the vast perceivable differences between things are largely due to complex combinations of those simpler bits. What was missing in those views—besides an understanding of chemical reactions—is the recognition that the combinations can form systems with properties that are not reducible to the tiny components, and themselves capable of downward causation on such components.

III. Is More Different?

The statistical behavior of gases and fluids, by which a system maximizes its entropy as its components randomly interact, is characteristic of the atomic and molecular ensembles which constitute most of the material universe. Schrödinger makes the point that physics and chemistry are "statistical throughout" (Schrödinger 1992). It is not the case that when we move from physics of the more simple to the more complex, e.g., the subatomic or atomic to the macroscopic, we move from the determinate to the indeterminate. On the contrary: the movement from the simple to the complex takes us from the indeterminate to the determinate. Free atoms and subatomic particles in a gas move in chaotic, unpredictable trajectories generated by heat energy, even without quantum mechanical considerations. They are, it is true, presumably determinate (above the quantum level), but we cannot know those determinations. We cannot predict the behavior of individual atoms because of their sensitivity to tiny changes in energy, especially heat; there is always epistemic indeterminacy. Schrödinger makes the telling point that the philosopher John Locke was wrong to imagine that if our senses were acute enough to see microscopic reality rather than confusing macroscopic "secondary qualities," we would know the world in deterministic splendor. This is false: if we could perceive individual atoms, our perceptual world would be sheer chaos. Inaccuracy (in effect, scale) saves us from that chaos.

We earlier saw Philip Anderson argue for the independence of solid-state physics and chemistry from elementary physics. In his response to Anderson, Weinberg admitted that even if we could produce a quantum mechanical description of all the 10^{23} atoms of a macroscopic sample of a chemical substance we would have no more understanding than we have now (Weinberg 1994, p. 62). In fact, no exact application of the Schrödinger equation to atoms heavier than hydrogen has been achieved. The complexity of component interactions makes the mathematics too difficult. (This does not make quantum mechanics ineffective or impractical; it means workers must rely on the Born-Oppenheimer approximation to infer properties of populations.) The same point was taken up by Theodor Benfey, citing the work of both Douglas Hartee and E. Bright Wilson, that "even the tabulation of an adequate description for one stationary state of an iron atom would require [a] number of [digital] entries" greater than 10^{78} (the ratio of the mass of the universe to the proton mass) (Benfey 2000, p.198; Hartee 1957, pp.16–17). Anthony Leggett points out that even with the "undreamed of" computing power to figure the Schrödinger equation for all those atoms, we would learn,

> Almost certainly, nothing. We would be faced with a pile of millions of tons of computer print-out (or, more optimistically, computer graphics) which would certainly 'in principle' contain the answer to any question we might want to ask, but in a form which would be totally useless to us without some consciously chosen principles of organization. (Leggett 1988, p. 214)

All this testifies to a metaphysically interesting fact: an individual event may have no likely outcome while many iterations of the same event may. A million particles with the same list of properties or states may well show a central tendency, that is, a normal curve in which one state is most prevalent. The larger the collection, the greater its predictability. This is the famous Law of Large Numbers of Gauss and Bernouilli: the accuracy of a measurement of a collection of N particles increases with $1/\sqrt{N}$. That is, if N is multiplied by 100, measurement error falls to 10 percent of its former value. That is because the individual motions or events are, per our presupposition, random or independent. Where we can make the idealizing assumption of the mutual independence of the components of the system, a host of motions and/or properties will cancel each other out, that is, yield a particular central value of a variable, with as many components deviating above that value as below it. This is also what thermal equilibrium means. Microscopic randomness yields macroscopic order. This is, we may say, the order *of* disorder.

The great probabilistic achievement of pre-twentieth-century physics was Boltzmann's use of statistical mechanics to explain thermodynamic processes. This has been taken by many to mean that thermodynamics was "reduced" to statistical mechanics; indeed, it has become a canonical example of an inter-theoretic reduction, along with its own subcase, the apparent reduction of temperature to mean kinetic energy of a system's components. Many philosophers of science, and scientists, commonly remark that temperature of a macroscopic system is identical to and replaceable by mean kinetic energy.

But the issue is not so clear. Von Brakel points out that temperature is only locally, not globally identical to or replaceable by mean kinetic energy (Van Brakel 1997, pp. 260–61). First, the identification does not hold for nonequilibrium states of fluids, nor for solids, plasmas, or vacuums. Second, temperature can be fundamentally defined, as Primas points out, by the Zeroth Law of Thermodynamics, which establishes transitivity among systems in thermal equilibrium. If two systems are each in equilibrium with a third, they will be in equilibrium with each other (Primas 1985). "Being in thermal equilibrium" means no change in temperature. But this definition is inherently macroscopic. Temperature can be defined as change in entropy over change in energy, $T = \delta S/\delta E$, where S and E are properties of ensembles, not individuals. Third, there is phase. The capacity of systems of a chemical substance, however pure or mixed, to occupy very different global states with distinct, causally efficacious observable properties, is not an atomic or even a molecular fact; it is an *intermolecular* fact. A water molecule cannot boil; only a collection of water molecules can. As Van Brakel puts it, ". . . even if temperature could be reduced, it doesn't yet follow that 'boiling point' can be reduced. . . . No molecular definition of what a liquid is exists; nor is there a theory in molecular or even microphysical terms which confirms how many aggregate phases there are" (Von Brakel 1997, p. 261). Equilibrium, entropy, and phase are supra-molecular traits.

This touches on the topic of multiple realizability. As we saw in the last chapter in defining entropy, a macrostate property is realized in an equivalence class of microstates or what Boltzmann called "fluxions." It is the multiple realizability of macrostates that makes probability, hence collective predictability, out of randomness. As we have seen, under multiple realizability reductionism fails: "the [relative] stability of macro-states . . . entails that the vast majority of neighboring (dynamically accessible) micro-states map into the same or (more rarely) neighboring macro-states" (Wimsatt 2007 p. 217). Stable macrostates have to be tuned so that the chaos at lower levels does not lead to deviation amplifying that would destroy macroscopic stability. While macroproperties must be sensitive to *certain kinds* of microchanges, "it is crucial that most differences [at the microlevel] do not have significant [macrolevel] effects most of the time" (Wimsatt 2007, p. 218). Phase, temperature in many cases, entropy, and equilibrium all require reference to the state of the system and/or its environment, a reference that is supplemented, but not replaced, by a microdescription.

Let us consider chemistry in particular. Chemistry is based in and exhibits a set of distinctive concepts that play a central role in its practice: *substance*, *reactivity* or *affinity*, *periodicity*, and *shape*. The concept of substance has no analogue in physics. As noted, chemistry is the study of the transformations, or dispositions to react, of distinctive substances. Substances possess characteristic properties at all supra-molecular scales (e.g., hardness, melting/boiling point, conductivity, etc.) When the properties are present, this indicates the substance is present. Above all, the chemist is interested in the stability/reactivity of the substance, its likelihood to bond with another substance to create a third (*affinity*) by the transformation of their molecules and/or atoms. Then there is periodicity, a concept that, like substance, has no analogue in physics. In the course of building up the list of elements by increasing atomic weight (which would eventually be replaced by atomic number, number of protons), it was noted that "after certain regular but varying intervals the chemical elements show an approximate repetition in their properties" (Scerri 2007, p. 16). In modern forms of the table horizontal rows present periods, vertical columns groups, proceeding from metals at the left, through transitional metals, to non-metals on the right. So the repetition of properties at horizontal intervals or periods means vertical groups tend to share properties. The arithmetical whole number relationships among the elements provide the key template for chemists in their examination of elements and substances. Last is the electronic configuration of the atoms or molecules, hence their shape. Shape matters in chemistry. Metaphysically, atoms and molecules are the smallest beings for which shape, or structured volume, is a causal reality. Smaller particles can be treated as point masses. But chemically, bonds depend on the shape of their component atoms and molecules.[5]

The fortunes of physical reductionism were fundamentally changed by the promise, expressed by Dirac, that all basic chemistry could be reduced to quantum mechanics because the latter explains electron configuration. But some recent philosophers of chemistry think otherwise. Primas has argued that electron configuration or "orbitals" cannot be derived from quantum mechanics (Primas 1998). Woolly explains, "if one starts from a description of a molecule as an isolated, dynamical system consisting of the number of electrons and nuclei implied by the stoichiometric formula [which calculates amounts of reactants and products] that interact via electromagnetic forces," that is, if we think of the molecule not as a hierarchical assemblage of atoms but a collection of subatomic particles, "one cannot even calculate the most important parameters in chemistry, namely, those that describe the molecular structure" (Woolly 1978, p. 1074). Scerri argues that the ascription of individual electron states, quantum numbers, and orbitals to subatomic particles is not justified; these are properties of the atom or molecule, which are distributed over components rather than individuated items. He concludes that electronic configurations "cannot be derived from quantum mechanics" (Scerri 1991, p. 318). Or as

Ramsey puts it: "If 'reducible' is interpreted as 'explained by the laws of physics,' shape is not reducible in any recognizable sense of 'explained' " (Ramsey 1997, p. 244). The point is that the explanation of the contribution of components to system must itself refer to the role of the system in determining that contribution.

Scerri points out that the determination of electronic configurations in terms of quantum numbers predates quantum mechanics; it is related to the old (pre-1926) quantum theory, which itself was motivated by the search for electronic configurations of the elements of the periodic table, beginning with Thomson's discovery of the electron in 1897. To this day, Scerri insists, our account of the configurations derives from three principles: Bohr's *aufbauprinzip* or constructive principle, according to which orbitals fill in ascending order outward (like the construction of a building from bottom to top); Hund's rule, which requires among other things that electrons of equal energy spread out in a given orbital; and Pauli's exclusion principle. Quantum mechanics does not explain any of these principles; its application to electronic configuration *presumes* them (Scerri p. 233ff).[6] Scerri's conclusion is that "The reduction of chemistry to quantum mechanics has neither failed completely, as some philosophers have claimed, nor has it been a complete success, as some contemporary historians have claimed" (Scerri 2007, p. 248).

Wimsatt's language can, I believe, put this more informatively. As we have seen, reductive explanation works *to the degree that* a system property or performance is aggregative, which requires that it be invariant over inter-substitution of parts, decomposition and recomposition, size scaling, and devoid of feedback (amplification or damping). These conditions hold only for some properties of the system, typically when the contribution of parts to whole is context-*in*dependent; explanation of nonaggregative properties require supplementation by either phenomenal or functional-selectionist explanation. The latter implies "downward causation," nicely described by Schröder: "Downwards causation is the influence the relatedness of the parts of a system has on the behavior of the parts . . ." (Schröder 1998, p. 447). The explanation of the electronic properties of atoms is a mixture of reductive and nonreductive explanations.

A number of writers, like those mentioned above, accept that a host of chemical explanations require reference to such influences.

Nevertheless, many who rightly argue against reduction of chemistry to physics accept ontological reductionism while rejecting explanatory reduction (Scerri and McIntyre 1997). But as argued in Chapter 3, there is no reason to accept this position.[7] The justification for positing an ontology is explanatory; we posit those entities which are necessary causally to explain the phenomenon. Furthermore, as noted, our historical tendency is to tie ontological claims (and also, epistemological realism) to material entities, hence components, ontologically ignoring processes or interactions and structures or relations (Bickhard 2000). A benzene molecule (C_6H_6) is in fact not just six carbon atoms and six hydrogen atoms; it is those twelve atoms *undergoing a process* of electromagnetic interaction and thereby occupying a particular *structure*. The structure and the process "are" as well as the components, and play a causal, hence ontological role, as properties of the system.

Recent decades have seen a wealth of scientific work on another context in which more seems to be different, namely far-from-equilibrium systems. The most common form of "order," in the generic sense, is equilibrium, the most diffuse state a system can reach. We have seen that equilibrium can be dynamic or static, and stable or unstable. Either way, equilibrium yields the "order of disorder" described above. But it has been discovered in recent decades just how many interesting states can occur in a system driven *away* from equilibrium. Interest here means "structure," rather than mere order; it means improbability or locally lowered entropy. Adding energy to the system, for example by heating, may drive the system away from equilibrium, at first simply producing more molecular chaos, until a critical temperature is reached that produces unexpected, spontaneously structured behavior.

Consider the case of Bénard cells. Convection heating of a shallow volume of liquid achieves, at the right temperature, as if in a phase transition, a very striking state in which the liquid spontaneously divides itself into small but visible convection cells—like little boxes—in which the fluid circulates locally. This is irreversible and negentropic, it moves from a state of higher to lower entropy, so must export entropy. The system achieves a locally metastable state, best understood as a plateau or hump in a potential energy graph. In Nicolis' analysis, what happens is that some critical parameter, in this case thermal energy, changes its flow to prevent the system from reaching equilibrium (Nicolis 1989). In equilibrium molecules are virtually independent, and there are no privileged spatial or time directions, so movements are reversible. Increase in energy causes perturbations that die out in asymptotic stability. But if the energy increase drives the system far-from-equilibrium, the system may seek a new *local* symmetry, a more structured spatial arrangement in which energy can be stored. In this state larger volumes of the system come to act as a unit, creating long-range correlations or interdependence. This is not due to physical forces; two distant, correlated particles do not exert force on each other. Similar phenomena include Belousov-Zhabotinsky (BZ) reactions, amazing colorful chemical reactions in solutions that repeat a series of stages over time, as a chemical "clock"; Taylor vortices, vertically differentiated ring-like flows of liquid in the space between two nested cylinders, one of which is rotating; and many different auto-catalytic chemical reactions that operate through feedback or amplification, where the product re-initiates the reaction.

In some far-from-equilibrium systems, when even more energy is pumped into the system, we get turbulence, or using the more generic current term, *chaos*. This means complex nonlinear behavior, where outcomes are not proportional to the first power of one variable (either they are proportional to some higher power or to more than one variable), typically due to amplification (or damping) that does not settle down to metastable equilibrium but exhibits some cyclic behavior that never precisely repeats. A dynamical system is chaotic if it is unpredictable because the evolution of its states are highly sensitive to differences in initial conditions so small as to be un-measurable, an extreme form of epistemic indeterminacy.

Critical point phenomena and the everyday event of phase change offer related processes in that, similar to chaos, a small change in conditions yields a nonlinear leap into an entirely different condition with novel properties, only in this case stable. This typically involves global or distant relations among its parts and novel forms of spatial structuring, which can no longer be analyzed as an association of one- and two-body micro-systems. An example is ferromagnetism. When the temperature of certain metals lowers, the metal spontaneously becomes magnetized, and that means the energy field about each magnetized component suddenly "picks" one of an infinite number of directions in which to stabilize. A system perturbed by changes in temperature or slight differential pressure seeks a stable, lowered symmetry state out of an infinite number of degenerate states. If this occurs without any asymmetrical input (e.g., a push from one direction), then the symmetry breaking is spontaneous. The linear differential equations which the system obeyed before the change, and which incorporate all degenerate solutions, are still obeyed by the new state, but those equations cannot pick out the new preferred state, or as Anderson says, "construct" it (Anderson 1972).

The name most famously associated with complex phenomena in chemistry and thermodynamics is that of the central figure of the Brussels School of thermodynamics, Ilya Prigogine. Prigogine worked on many problems in his career, and was inspired by the reading of some philosophers, particularly Bergson on time. While he admitted to great respect for the work of Boltzmann, he felt that Boltzmann had basically created "thermostatics," a nondynamic probabilistic theory of irreversible systems, contrasted with the Newtonian dynamic theory of reversible systems. What Prigogine sought instead was a nonprobabilistic dynamic theory of irreversible systems, which seemed to him *constructive*, not mere heat noise but the creation of complex, novel forms of order. His early contribution was a proposal of a "minimum entropy production" principle for nonequilibrium stationary states. He had in mind the analogy, suggested earlier by Bénard himself, between physical equilibrium and the very active yet apparently stable states of living organisms ("homeostasy"). Later, working with nonlinear far-from-equilibrium systems that did not obey the Onsager relations (certain lawful relations between the flow of energy across temperature differences and the flow of matter across pressure differences), he posited that stationary states were achieved in what was otherwise a "giant fluctuation, stabilized through matter and energy exchanges with the outer world." These were his famous "dissipative structures" (Prigogine 1977).

Prigogine observed that structured material collections, like our everyday crystalline solids, typically exhibit a high degree of interdependence or correlation across microscopically large distances. Thus the behavior of the molecules cannot be understood by one- or two-body analyses of components. Prigogine pointed out that only in reversible, nonequilibrium phenomena can perturbations be dampened to yield reproducible events, unlike say, in a frictionless pendulum, where a change in energy would produce a permanent change in motion, and the original period could never reassert itself (Prigogine and Nicolis 1989, p. 21). Analogous to Schrödinger's point about the indeterministic micro-world, an idealized deterministic (frictionless) model would, given perturbations, be *less* stable. As Schneider and Sagan suggest, nature's "abhorrence of a gradient" is such that when the tendency toward equilibrium is blocked, the system is forced to create structure (Schneider and Sagan, 2005). Just as a sailboat can tack against the wind by using the wind, chemical and stochastic systems can create new forms of complexity in and from the Second Law's flow toward higher entropy.

IV. Individuals, Ensembles, and Teleomaticity

The complexity of the material order, studied by chemistry, solid-state physics, the Earth sciences, and engineering, rests upon but goes beyond the physical order. At the level of atoms, and more distinctively the molecules formed from atoms, true individuality begins to emerge as quantum entanglement or EPR (Einstein-Podolsky-Rosen) correlations wash out. Bonds among structurally distinctive molecules lead to a far larger set of substances now capable of phase differences based on temperature and pressure. As we depart from fundamental physics into condensed matter physics and chemistry we find increasingly nonphysical properties and performances of the systems under study. Non-physical means systems cannot be regarded as homogenous spacetime entities under interaction rules; because the smallest causal entities are now complex, inhomogeneous dynamic ensembles or hierarchically composed individuals of distinctive natural kinds bearing complex chemical properties. These substances are the letters of any sentence the universe can utter.

Matter consists of individuals and ensembles of individuals; most of the matter in the universe is fluid ensembles, mainly gaseous. "Clouds" was Karl Popper's metaphor for stochastic ensembles, "clocks" for deterministic individuals. Popper claimed that clouds are more basic than clocks, indeed that, "to some degree *all clocks are clouds*" (Popper 1972, pp. 213–15). This is insightful but exaggerated. Individuals or clocks may be collections of lower level components, but a highly structured "cloud" with coincident, internally dictated boundaries across a wide set of environmental conditions is not a cloud any longer, it is a "clock" or an individual.

We may distinguish two types of complex material systems from the less complex (e.g., gas and liquid at equilibrium). There are *complexly structured individuals*, stable amorphous and especially crystalline solids. These objects self-maintain their boundaries across a wide spectrum of environmental changes, can be at internal equilibrium and in thermal-equilibrium with their environments, while maintaining a distinct chemical distribution. Then there are *complex dynamic ensembles*, like Bénard cells, Belusov-Zhabotinsky reactions, and ensembles undergoing autocatalytic chemical reactions. These are characterized by unstable structures temporarily maintained by dissipation, accepting energy from and discharging entropy into their environment, their components undergoing relatively higher numbers and rates of processes

Ernst Mayer coined the word "teleomatic" for inorganic phenomena "in which a definite end is reached strictly as a consequence of physical laws" (Mayr 1974). He did so as a foil for his notion of "teleonomic" biological phenomena, which he was trying to distinguish from the teleological. Mayr left "teleomatic" to include even the fall of an object under gravity, hence virtually all physical and chemical interactions. This is a waste of a perfectly good neologism. It makes no sense to use any version of *telic* for phenomena falling to equilibrium; their motion was not selected, there is no sense in which their eventual state or encompassing system plays a causal role, the explanation of their behavior requires only a force and/or the Second Law. We cannot say "why" a system falls to equilibrium; that is what systems do when not prevented.

But in the cases of complex structured individuals and complex dynamic ensembles we find that a nonequilibrium state of a system serves as an *attractor* which recruits materials into a complex, improbable structure. It is the system state which "does" the recruiting, rather than an external stimulus; the energy under the Bénard cells is not telic, but the *state* of the Bénard cells is. This gives a functional or selectionist explanation of what happens to a microcomponent (e.g., a water molecule): its state is what it is *because of* the macrostate and the microcomponent's role in it. Likewise, the crystal's state selects the orientation of the molecule that it recruits to add to its structure. This is what functional explanation implies, that the N-1 level event is caused to be what and when it is in part because of the state and activity of encompassing system N. There is nothing mystical, intelligent, designed, or purposeful here, just a complex whole causing components to behave in an improbable manner as they otherwise would not.

The most impressive cases of wild (non-artificial), non-living, stable, complex material systems are planets, stars, solar systems, and galaxies. Solar systems are very stable ensembles, largely maintained by inertia and gravity. They are dominated by one very active individual, their star, and other members can be minimally active systems of crystals or gases whose role in the maintenance of the system is mainly gravitational. Stars are complex, dynamic stable individuals that project energy across massive distances and create new material elements. Then there are galaxies, as noted earlier, vast ensembles recycling disarticulated components into new (stellar) growth. And with the Earth we see a dynamic material individual, with interactive geological, hydrological, and meteorological recycling. I hold that these all systems are telic, implying at least a rudimentary final cause in the sense of a "toward-which" in Aristotle's sense. Again, this does *not* imply purpose, or teleonomy, nor the representation of an end, or teleology. It means one state of the system functions as the improbable toward-which that selects behavior of microcomponents thereby sustaining complexity. Even without life, the Earth would be the most dynamically complex wild individual we currently know.

But there is yet another form of material complexity. Sometime around the universe's 10-billionth year, something even more stunning arose on the already remarkable Earth: individuals capable of articulated and improbable structure otherwise found only in crystals, while maintaining themselves in this state by ceaseless dynamism, like far-from-equilibrium systems, and recycling their own products and components, like the Earth itself. And they added a higher degree of controlled exchange with environment, a level of adaptiveness unseen in other material forms, as well as the novel ability to copy themselves. With these creatures the contribution of the material wanes, not because the systems cease to be materially complex—they are more so than any non-living system—but because they are much more besides. To such amazing material systems we now turn.

7

The Phenomena of Life

The fecundity of life on Earth is staggering. E. O. Wilson writes of a handful of forest soil,

> This unprepossessing lump contains more order and richness of structure, and particularly of history, than the entire surface of all the other (lifeless) planets. It is a miniature wilderness that would take almost forever to explore. . . . The abundance of the organisms increases downward, according to size, like layers in a pyramid. The handful . . . is home for dozens of insects, mites, nematode worms, and other small invertebrates. . . . There are also about a million fungi and ten billion bacteria. . . . Each of the species has a distinct life cycle fitted to a portion of the micro-environment in which it thrives and reproduces . . . programmed by an exact sequence of nucleotides . . . evolved as independent elements for thousands of generations. (Wilson 1983, p. 4)

Living things have a number of relatively uncontroversial characteristics. They must maintain a high level of continuous activity just to remain alive. For this they metabolize to garner energy from the environment. They are homeostatic or self-regulating, self-organizing, and even self-making, or self-remaking, since they must rebuild their own structures. They are irritable or sensitive and responsive to their environment; they sense and behave. They reproduce, a complex activity requiring the storage of information in specialized macromolecules. And over many generations they evolve.

Life brings with it not one, but several novel strata of phenomena. The most central, and that at which the present chapter will be aimed, is the living organism itself. But there are other levels or kinds of systems that come along with organisms, four of which are theoretically indispensable: *macromolecules*, *societies*, *ecosystems*, and *minds*. (I leave out the biosphere, or Earth-biosphere system, as too big to handle here.) Except for artificial polymers, the only macromolecules we know are parts of living organisms, who must manufacture them. Some organisms are social in a strong sense, their nature and function being fitted to a local society of conspecifics. All life is situated in ecosystems in the sense that ecosystem features are necessary for the organism's evolution and existence; we cannot understand the organism without understanding the environment in which it evolved and lives. And among the innumerable forms of life, complex animals have evolved who exhibit mental or intentional activities. Except for the last, which is the subject of the next chapter, we will take these levels or kinds of systems as components, traits, or contexts of our necessary focus, organisms.

Few biologists or philosophers of biology think biology can be explanatorily reduced to physics and chemistry, or that functional and telic explanations are inadmissible. In biology "reductionism" usually refers to "gene-centrism" or genic reductionism, regarding all biological phenomena as in principle explainable by the selection of genes. Some non-biologists deplore "biological determinism," especially since the 1970s rise of sociobiology, the search for the genetic roots of human social behavior, which some fear to mean "biology is destiny" (the popularized form of Freud's remark, "anatomy is destiny"). One goal of the present chapter is to demonstrate the necessity of nonreductive explanation, hence emergence, at multiple levels of complexity within biology.[1] But this also means recognizing what is *not* to be feared in finding conditions and sometimes causes of human traits and behavior in our biology. After four sections presenting the basic chemistry of life, life's history, its exhibition of sensitivity and learning, and societies and ecosystems, the final two sections will argue against genetic reductionism and for teleonomy, purpose, and value in the biological realm.

I. The Structure and Processes of Cells

Like chemistry, biology has a typology of fundamental natural kinds crucial to the discipline, namely, species. Living organisms come in species characterized by a common inherited form.[2] Also like chemistry, biology has a wonderfully clear answer to the question of its elementary units. With the ambiguous exception of viruses, everything living is composed of cells, the smallest and simplest living things are single-celled, and nothing that is not a cell or made of cells is alive.[3] Consequently, some (certainly not all!) of the properties of living organisms must also be properties of their cells.

Cells are bounded or self-contained, even if that boundary must be selectively permeable, enabling diffusion, osmosis, and active transport. They must metabolize food, which, even in the case of plants' photosynthetic harvesting of solar energy, means breaking down organic molecules for the energy necessary for growth and self-maintenance. They inevitably have a liquid interior, or cytoplasm, enabling the internal transport of materials. They must be able to move relative to an immediate environment, which even in bacteria is made possible by miniscule extensions of their cytoplasm. They must build structures out of proteins. They must produce enzymes, which lower the activation energy for all the chemical reactions presumed by the above. They must maintain constancy in key parameters of their internal state. They must be able to reproduce, hence store and transmit genetic information in chromosomes.

To do this life employs 25 different elements, although six compose over 96 percent of cells by mass and are considered essential to life as we know it—carbon (usually C-12), hydrogen (H), oxygen (O), nitrogen (N), phosphorous (P) and sulphur (S).[4] The average cell is (again by mass) 70 percent water, 6 percent monomer amino acids, monosaccharides, nucleotides, and fatty acids, 1 percent inorganic ions like sodium, chlorine, potassium and phosphates, with the remaining 23 percent organic macromolecules. These are polymers made of chains of molecules, in particular four essential to life: proteins, which are amino acids linked in peptide chains; complex carbohydrates or polysaccharides, that is, polymerized sugars; DNA or RNA, polymerized nitrogen-based nucleotides of sugar-phosphate groups; and lipids, a large set of molecules composed of varying ketoacyl and isoprene groups. The first provides the structures of the cell and the enzymes that make reactions possible; the second is the form in which energy is acquired and stored; the third is the informational template for duplicating the proteins; and the fourth serves important functions of energy storage, membrane structure, and signaling. As for the isotope carbon-12, it is a remarkable atom. The reason is that its outer valence electrons are four, but admit an additional four through covalent bonds, and are very unlikely to lose or gain electrons. It is its elaborate bonding possibilities that make carbon crucial.

Protein-constituting amino acids consist of a carbon atom that bonds a carboxyl (C-OH) group on one side to an amine group (N-H) on the other and to a side chain. The amino acids appearing in proteins are *chiral*, in particular left-handed, so that right-handed amino acids are unusable. Proteins are amino acid chains in which the carboxyl group of one amino acid is attached to the amine group of another by a peptide bond. The resulting complexity of proteins is overwhelming. As noted, a protein may have 100 amino acids, the latter coming in 20 types (our bodies make ten, but must ingest the other ten), and their sequence determines the nature of the protein. There are thus 20^n types of protein of length n. But proteins also coil, fold, and form hydrogen bonds between their amine and carboxyl groups, yielding far more structured complexity. The typical human cell, a cube .0015 cm on a side with a volume of 3.4×10^{-9} cm^3, contains almost 8 billion protein molecules organized into about 10,000 different proteins!

Metabolism is the making and breaking of chemical bonds. The energy which cells need, store, and produce is not an entity; it is the *binding* of entities, electromagnetic forces among atoms and molecules. Cells metabolize sugars, particularly glucose ($C_6H_{12}O_6$). Glucose itself is a means to an end, the construction of adenosine triphosphate (ATP). It is in the bonds of the ATP molecules that energy is stored; in them a nitrogenous adenine group is bonded to a sugar, ribose, which undergoes phosphorylation, the adding of phosphate groups (bonded oxygen and phosphorus, P) or breaking them off to form ADP (adenosine diphosphate). ATP has very high chemical potential energy, a great potential for reaction, hence its bonds can store a lot of energy. The hydrolysis of ATP—an *exergonic* or energy-exporting process in which water molecules break into one hydrogen atom and one hydroxide (OH^-) ion—releases phosphates to become ADP. This is the energy key: adding or breaking off phosphate groups from the nitrogenous adenine-sugar ribose compound. Then an active, *endergonic* or energy-importing process—relying on photosynthesis, fermentation, or respiration—is required to re-phosphorylate the ATP for another cycle.

The first step is getting the glucose, either from eating or absorbing other life forms or making it through photosynthesis. In the latter case the effect of light on the chlorophyll of photosynthetic cells makes an otherwise endergonic reaction possible, in which CO_2 or CH_4 and two molecules of H_2O undergo a redox reaction enabled by the light excitation of their electrons. Either way, glucose then undergoes glycolysis, in which it is oxidized, losing an electron, to produce pyruvate (CH_3COCO_2H), which reduces NAD+ to NADH and CO_2, and phosphorylates some ADP into ATP. This leads, in an aerobic environment, to the citric acid or Krebs cycle, and in anaerobic conditions, to fermentation. The final step, oxidative phosphorylation, contributes 90 percent of the ATP produced. Each cell makes *10 million ATP molecules per second.*

Cells live because of their ability to continue a process of chemical reaction that couples downhill *catabolism* (breaking down molecules to release energy) and uphill *anabolism* (forming molecules to store energy), hence breaking down *while* building up, so that spontaneous or free energy is never exhausted. Gibbs Free Energy (G) is the energy available in the cell/organism for work. Exergonic reactions release energy, endergonic reactions absorb it, and equilibrium is the state constituted by equal rates of forward and backward chemical reactions.[5] Life is a slowly unfolding chemical *dis*equilibrium. It is a thermodynamic anomaly, which is not to say it violates the Second Law. The cell pays the Second Law by exporting entropy into its environment.

Life also means the past actuality, and usually the future possibility, of reproduction. There is no room here for an exploration of the complexities of genetics, but we must at least define the basics. Life is only possible for systems that contain their own self-description. (John Von Neumann, formulator of the mathematics of quantum mechanics, famously developed an analysis of self-copying machines.)[6] Nucleic acids provide that description. Proteins, constituting both structures of the cell and enzymes, are constructed at ribosomes within the cell. The templates or codes that control the manufacture are nucleic acids, complex molecules linking a phosphate group, a nitrogenous base, and a sugar, either ribose in RNA or deoxyribose in DNA. A single DNA molecule can have 10,000 nucleotides. The copying mechanism is the linear sequence of base pairs (adenine-thymine and guanine-cytosine in DNA, uracil substituting for thymine in RNA) on one of the two sugar-phosphate branches of the famous twisted ladder or double helix in a lock-and-key interaction, after they split apart in *mitosis* (reproduction of somatic cells) or *meiosis* (reproduction of germ cells). The DNA directs synthesis, or transcription, of messenger RNA (mRNA), which carries the genetic code to the ribosome site, where it is translated or read by the ribosome to construct a protein sequence of transfer RNA (tRNA), each nucleotide of which is attached to a complimentary amino acid to construct a protein. The rate at which this happens is remarkable. A ribosome can add about 20 amino acids to any given protein polymer per second; a eukaryotic (nucleated) cell can have a *million* ribosomes, meaning thousands of new protein molecules are synthesized every second.

DNA codes all genetic information in eukaryotic and prokaryotic cells (some viruses, like HIV, have only RNA). *Chromosomes* are complex strings of DNA along with protein groups. Chromosomes are pairs of chromatids, and all chromosomes in the diploid cell (all cells but gametes have two sets of chromosomes, one from each parent) have a corresponding type of information. A *gene* is a sequence of DNA at a site on a chromosome which codes for a particular characteristic, whatever its value (e.g., eye color). *Alleles* are the forms the gene can take, coding for the *value* of the gene (e.g., blue or brown or green). In sexual organisms, germ cells are produced with half the chromosomes present in all other somatic cells; in fertilization, then, a full compliment of chromosomes is achieved, each character derived from one allele from each parent. In the nineteenth century Gregor Mendel discovered that different inheritance information for a trait from two parents is not mixed or averaged, but shuffled. Offspring may carry and pass on "recessive," unexpressed traits while manifesting "dominant" traits.

There are great complexities here, to some of which we will return. But it is true of all living organisms and cells that: they contain a nucleotide code capable, given the requisite cellular/organismic machinery and environmental conditions, of producing copies of itself and inserting the code therein; features of the organism's structure, and some of its behavior, have particular sections of that code as their necessary condition; and their genes, while being subject to mutation and sexual reshuffling, are capable of faithful reproduction over many generations, hence constituting the difference between one species and another. It is legitimate to consider the accomplishments of nucleic acids in living organisms a kind of "proto-semiosis."[7] These activities of the living cell amount to a translation or transcription, a "reading" of the nucleotide sequence by the ribosome and its reproduction in a sequence of amino acids.

Organisms accomplish the above tasks with very complex organization. As seen in Chapter 3, it is not even possible to decompose the organism into a single exhaustive set of parts. Anatomically, the organism is composed of organ systems, then organs, then tissue types, then cells. But the boundaries of this decomposition do not coincide with those of decompositions based on cell types, developmental gradients, chemical substances, types of biochemical reactions, and cybernetic or information-command flow. And there is "interactional complexity" *between* these perspectival decompositions (Wimsatt 2007, pp. 182–86). The degree of complexity of living things far outstrips the complexity of nonliving systems. But how did such things get here?

II. The History of Life

The Earth collapsed from its protoplanetary ring about 4.54 mya. It is generally believed that sometime after 3.85 billion years ago, still in the extreme Hadean geological eon, when the massive "Late Heavy" meteorite bombardment of the Earth ended, and before 3.4 billion years ago, the age of our oldest fossils, life arose or arrived on Earth. How, we do not know. There is currently no plausible way to extend the processes of chemical self-organization to make probable the emergence of a self-regulating, self-duplicating single cell. Miller's and Urey's famous experiment at the University of Chicago in 1953 demonstrated how amino acids could spontaneously assemble under early Earth conditions in a soup of the requisite molecules. Redox reactions on atmospheric elements like methane and ammonia created organic compounds like formaldehyde (H_2CO) and hydrogen cyanide (HCN), which under heat and sunlight produced amino acids like ribose, glycine, and acetaldehyde.

There have been questions about Miller's and Urey's reconstruction of the early atmosphere, but we may take from their work the plausibility that amino acids could indeed have formed under early Earth surface conditions.

The problem is at the next level, the chemical evolution of nucleic acids and proteins. We do not know how these massive and complex macromolecules could have formed, nor how a process of chemical evolution could have led to the arising of some community of the tiny chemical factories we call living cells, all able to make proteins and RNA and DNA in a highly regular and controlled way.[8] A typical protein may have 100 amino acids in its polymer chain. The number of different possible 100-chained groups of the 20 different amino acids is 20^{100}, a number far larger than the number of elementary particles in the universe (10^{81}). But only a very small number of those combinations constitute *workable* proteins for a living cell. Not only the Earth, but the universe—today at 10^{17} seconds—is not old enough to make it likely that those few combinations would have been achieved randomly.[9] Genes present a similar improbability, as figured by Barrow and Tipler. Using an average gene length of 1800 nucleotide bases, the odds of assembling a single gene randomly are one in 4.3×10^{109}. There are estimated 10^{27} bacteria on Earth today. Assuming a reproduction time of one hour, there have been no more than 10^{40} bacteria on Earth in its history. Given about 10^{7} nucleotide bases per bacterium, the number of nucleotide combinations that could have randomly been tried on Earth since life began is 10^{47}, or 10^{62} too few (Barrow and Tipler 1986, p. 565).

Nobody knows where the chemical generation of life first happened, whether in primal soup tidal pools, on the ocean floor around hot vents, below the Earth's surface, or on a faraway planets conveyed by meteors or cosmic dust to Earth. Opinions differ as to which came first, nucleic acids or proteins. It may be that the first living cell was the product of the merger or "symbiogenesis" of discrete pre-living chemical factories each having only some of life's functions, alternately labeled "protobionts," "protocells," or "autocells." Just as significant as our inability to understand life's origin, nobody knows why it is no longer happening; as far as we can tell, on the contemporary Earth life only comes from other life, not from chemical evolution. This may mean that after life's origin the immediate precursors of living cells were used up or consumed by the new cells. Life may have eliminated the last rungs of the ladder that led to it. The emergence of life is more mysterious than any other emergence in the physical register. It is the largest "explanatory gap," with the possible exception of the origin of the universe itself.[10]

However life began on Earth, this much seems clear: in the beginning was the bacterium. The current typology of life divides it into three domains: unicellular prokaryotes or bacteria, unicellular bacteria-like archae, and finally all eukaryotic organisms, from protists to plants, fungi, and animals. Prokaryotes, single-celled but without a nucleus, are the most abundant form of life on earth. Archae are a fascinating sidekick to bacteria, in form much the same but inhabiting very harsh environments like hot deep sea vents, able to metabolize salt, sulphur, or methane. For almost *two billion years* all life on earth was bacteria and archae. Among them we find precursors of all later kingdoms: animal-like heterotrophs ingesting other living organisms; fungus-like decomposers of organic material; and autotrophs ("self-feeders"), plant-like photosynthesizers and chemotrophs gaining energy from direct oxidation of hydrogen sulfide, ammonia, and iron ions. The needs of such organisms are simple: water, carbon, an energy source, and in some cases air. Bacteria have been essential for all complex life in three ways. They recycled CO_2 into oxygen, utterly changing the Earth's atmosphere. They fixed atmospheric nitrogen into usable organic molecules, crucial for plant and animal life. And all more complex organisms have bacteria in and on them, in enormous numbers, and many of these perform necessary functions for their hosts.

Later came single-celled seagoing eukaryotes, probably evolved from symbiotic groups of prokaryotes. Eukaryotes have nuclei and other organelles, compartmentalization of function being their great achievement. For *a billion more years* unicellular eukaryotes were the most complex form of life. Thus for the first three billion years of life on Earth, all life was single-celled and mostly—as far as we currently know—in the sea. (This was not surprising, given that until 500 million years ago, when the ancient single continent Gondwana formed, there wasn't much land without ice.) Only one billion years ago did multicellular eukaryotes emerge in the sea from clumps of algae. Some absorbed nutrients from the aqueous solutions in which they lived, some ingested other organisms, others were photosynthetic (green algae). From these evolved protists, including mildew, amoeba, protozoa, diatoms, and those eventual heroes of nineteenth-century biology, the slime molds. These led to the first true animals, multicellular heterotrophs with strong protein structures, who emerged in the sea as tiny plankton-like, flagella-propelled creatures only 700 million years ago (mya).

Then at 565 million years ago came the Cambrian Explosion, a sudden efflorescence of large numbers of diverse and complex species: worms, arthropods (e.g., crabs), mollusks (e.g., clams), echinoderms (sea stars and urchins), and some vertebrates, including, a bit later, fish. On land, heretofore home only to bacteria, algae and fungi developed, and plants evolved from algae managed to take root as mosses and ferns by 440 mya. Thus began the great colonization of the land by photosynthetic autotrophs, providing the ground floor, along with decomposing fungi, of almost all land ecosystems, and in the bargain recycling more CO_2 into oxygen.[11] Within 70 million years amphibians and insects joined the party, reptiles and seeded plants around 370–360 mya. Little mammals split from the reptiles at 340 mya to ply the forest floor for insects. Giant herbivores roamed Earth's grasslands by 245 mya, to be followed by the dinosaurs. Birds split from the reptiles about 200 mya. Some plants started flowering at 120 mya. The disappearance of dinosaurs at 65 mya triggered a mammalian explosion and the foresting of the previously savannah-like temperate zones (trees had been held in check by the dinosaur mega-herbivores). Shortly thereafter pro-simians began the road to primates, figuring out that the tree canopy had food and less danger than the forest floor.

None of this development should be seen as linear. Earth's geography and climate have been cyclically changing. Due to movements of the plates of the Earth's crust, for more than three billion years the continents have been joining and dispersing every 300 to 500 million years (the last supercontinent being Pangaea, 300 million years ago). This, along with the wobble of the Earth's axis, and their joint impact on sea levels and currents, have led to oscillations of world temperature, from "snowball earth" conditions with ice at the equator to "greenhouse" conditions with no polar ice caps. There have been repeated ice ages, periods with at least polar ice. Our current Pliocene-Quaternary ice age, beginning two-and-a-half million years ago, has seen at least five glacial expansions or "pulses." We are now in an interglacial period (the Holocene) following the Wurm/Wisconsin glacial pulse which ended 12,000 years ago.[12] Climate change has been the context for the growths and extinctions of thousands of continental species. There have been at least five catastrophic extinction events, the worst at the Permian-Triassic boundary 251 mya, but the most famous and recent being the asteroid that struck near Chicxulub, Mexico, 65 mya, extinguishing the dinosaurs. Life has waned in its diversity, geography, and sheer volume again and again over its history on Earth, rebounding each time.

Central responsibility for life's diversity lies above all with a novel form of evolution. In 1859 Charles Darwin (and, independently, Alfred Russell Wallace) proposed that species were not fixed, but evolved, and offered *natural selection* as a key explanation of adaptation and evolution. The idea of adaptation was not new; in the eighteenth century William Paley had noted creatures' remarkable "fit" with their environments, which he explained by divine design. Nor was evolution: Lamarck had explained it as the inheritance of "acquired characteristics," successful learned habits and their morphological consequences. In contrast Darwin employed natural selection of inherited chance variations to explain both (Depew and Weber 1996, p. 6). The core of his notion was: a) given variation of some trait in a population of a species, b) if it is heritable, and c) if the environment will not permit all population members to reproduce to their full capacity ("differential reproduction"), then d) some variants will be "naturally selected" by the environment to increase in the population, and e) if the process continues long enough under the right conditions, may result in a new species. Natural selection is not a march toward improvement. It acts negatively, culling organisms from the reproductive pool, leaving those variations that produce more fertile offspring.[13] The result, however, is the creation of novelty.

In the late 1930s through the 1940s the "Modern Evolutionary" or "Neo-Darwinian Synthesis" was formed by R. A. Fischer, Sewall Wright, J. B. S. Haldane, and Theodosius Dobzhansky, famously encapsulated in Julian Huxley's 1942 *Evolution: The Modern Synthesis*. It combined Darwinism with Mendel's earlier, neglected account of heredity, and August Weismann's claim, versus Larmarck, that information can only pass from the organism's genotype (genome) to phenotype (morphology and behavior), in terms of "population genetics" or studies of the frequency of traits in populations.[14] Darwinism was then understood in the following way: a) populations exhibit variations of phenotypic traits due to genetic mutations or copying errors in gene replication, b) as heritable genetic mutations, c) some mutations reach expression and enhance adaptive fitness for the organism's phenotype in its current environment, hence improve its reproductive success, d) increasing the frequency of such mutations in the local population, and e) given Ernst Mayr's definition of a species as a closed reproductive pool, if the population is reproductively isolated, speciation could result.

We should note that, even for classical neo-Darwinians, evolution has several possible causes besides natural selection. Most evolutionary changes in species are due to *genetic drift*, or change in allele frequencies due to random spread of a set of mutations without environmental selection. There is simple *gene flow*, or migration of a population with certain genetic characteristics into a new area. And not all selection is "natural." "Selection" refers to the non-random picking out of some organisms for higher probability of spreading their genes into the next generation relative to conspecifics. Darwin had taken the term from human breeders of domesticated animals. Natural selection proper is selection of individuals by environment. Darwin himself noted as well *sexual selection*, which explains traits in terms of the competition among members of a species to reproduce, e.g., among males for female receptivity. Sexual selection is thus a mediated form of natural selection, in which potential mates operate as part of the environment which selects organisms. *Kin selection* is a different type of phenomenon, the selection of features of individuals for benefit to their relatives rather than themselves, for example, parental protection of offspring. Then there is *group selection*, where the genetic predispositions of individuals are selected only in the context of other conspecifics, who benefit from those individual predispositions. This is controversial; many doubt it exists. Speciation also depends on *mutation rate*; evolution cannot occur where there are either too few mutations, so no variations to select, or too many, which typically means organisms cannot survive or reproduce at all. Mutation rates change due to factors not well understood. Lastly, *symbiogenesis*, the merger of two symbiotic organisms where one genome acquires the other genome wholesale, may play a large role in evolution that is distinct from mutation (Margulis and Sagan 2008).

To this day there is great controversy over Darwin's theory. Even among biologists some argue that natural selection plays less of a role in evolution than the Modern Synthesis, and perhaps Darwin, believed.[15] Arguments over the unit of natural selection, that is, *what gets selected*, the gene or the organism or some more "extended" version of the organism, remain. But in its basic claims Darwin's theory of evolution remains a crucial part of the explanatory framework of contemporary biology and a definitive achievement of modern science.

III. Types of Sensation, Action, and Learning

Unicellular organisms, including bacteria, are irritable and capable of simple physical responses to stimulation. This irritability is due to the structure of the cell membrane. Under the modern *fluid mosaic model*, the plasma membrane of the cell is understood as a lipid (composed of fats and fatty acid soluble molecules) central core flanked by proteins. But these proteins are intermittent nodules, not continuously spread. They are capable of responding to particular types of stimulation, e.g., photo, chemical, mechanical, etc. Each stimulus causes chemical and/or electrical potentials in the relevant proteins. The interior of the cell contains other proteins that are capable of contraction in response to that transmitted stimulus. If enough of them contract, the cell lurches toward or away from the external stimulus. These potentials have limited effect on neighboring cells, and the transmission of the stimulus dies out at the gaps. In multicellular animals and plants this type of response can be very highly organized, so that cells transmit a similar influence along the body. Some simple organisms release chemicals into the body's circulation in response to stimuli, a primitive version of the endocrine system.

Neurons are an animal invention. They channel excitation along pathways, permitting discriminations of types of sensory input. In evolutionary terms the neuron is a cell with the same kind of stimulus response described above which has changed in order to generate a continuous action potential along its length in response to a particular kind of stimulus. Neurons are remarkable cells with a characteristic star-like body surrounded by tentacles (dendrites), and a long tail (axon). They produce a chemical-electrical charge that travels from the cell body down the axon. At the gap between its axon and the dendrites of the next neuron (the synapse) the action potential causes the release of chemicals (neurotransmitters) to cross the synapse and begin the action-potential in the next neuron. Neurons can be inhibitory or excitatory. In the course of evolution, nerve cells become linked by *interneurons* so that a stimulus at one site, for example the body surface or sense organs, can continuously and with a consistent effect be transmitted to a distant "target." This enables body length to increase.

Two other changes were essential for neurological evolution. First, the surface responsiveness of the plasma membranes of all other cells must be *inhibited*. This is crucial to specialization; not only must a dedicated type of cell or organ evolve, but other cells that earlier performed this function must be quieted. Second, motor function must also be specialized, because the neurons terminate somewhere, and if the organism is to respond to specific information, the terminus must have the power of specific action. Muscle cells have to evolve and become coordinated. All this was necessary for the advantages of neurology to be selected.

Neurons evolved already in simple hydras, corals, and anemones, exhibiting *net innervation*, networks of nerve cells. But further sophistication began with bilateral body plans. Already in the well-studied flatworm complexity reaches the point where multiple lines of neurons must communicate with each other. Such crossings, or *ganglia*, make possible a new function, the processing of multiple sources of information. Without central processing, multiple messages and cross-talk among nerve pathways would lead to an internal Tower of Babel. This centralization is the phylogenetic forerunner of encephalization. Neural centralization at a certain level of intensity is a *brain*.

Nematodes have one of the simplest brains, composed of two ganglia and about 300 neurons. But removal of their brain does not render them incapable of survival. Crustaceans, spiders, flies, and hymenoptera—ants, bees, and wasps—have brains, and in addition spatially separated ganglia. Hymenoptera are capable of complex, nuanced social behavior. Cephalopods, that is squid and octopus, have the largest centralized brain and most sophisticated nervous systems of any invertebrate. But it is vertebrates that developed the most complex and powerful brains; among them we find the characteristic tripartite bulging, and eventual folding over, of the superior end of the spinal cord into hind-, mid-, and forebrain. Only in mammals does the forebrain or cerebrum become the dominant structure of the brain, and develop a series of highly structured cortical or outer layers, culminating in the neocortex. In larger mammals the neocortex develops a wrinkled, rather than smooth, appearance, due the presence of *sulci* or folds, providing far more surface area. But we cannot say only these structures enable advanced cognition; there is evidence that evolution has plied other brain structures for advanced learning off the primate and cetacean branches of the evolutionary tree—or bush—among birds, particularly corvids or crows (Kirsch et al., 2008).[16]

The acquisition of more and more complex information processing would do little good without greater complexity of action or movement. The most primitive form of organismic movements are *taxes*, lurching movements of cells, often translation from a stimulus. Taxes are whole body movements or changes in orientation. *Tropism* in plants is directional growth of stem or roots toward resources, as well as the opening and closing of flowers (nastic movements). Animal *reflexes* (e.g., blinking and swallowing) are more sophisticated. They are innate, automatic body-part responses to stimuli that utilize the immediate connection of receptor and effector neurons with minimal processing in between. *Instinct* is more complex still, an innate fixed action pattern, a coordinated series of acts for which the stimulus only needs to be present to begin, not to guide or maintain, the action. Instincts are also deeply proprioceptive, reflective of internal conditions, and display the organization of drive, goal, and satiation.

Now, none of these types of action in response to stimuli involve *learning*. Learning in the broadest sense can be defined as the acquisition of a behavioral pattern, in effect, acquisition of a type of selective response. Many kinds are distinguished: *habituation* and *sensitization* or, respectively, lessening and enhancing of response by repetition of a stimulus; *accustomation* or habit formation; *imprinting* or mimicry, which may be instinctively motivated but the content is learned; *classical conditioning* associates a response (e.g. salivating) with a neutral stimulus (a bell) by reward or punishment (ringing bell while feeding); *trial-and-error-learning*, or *operant conditioning*, where reward or punishment following spontaneous responses to a problem causes retention of the solution; *insight*, in which novel spontaneous actions suddenly are recognized to show a solution; and *discursive* learning or reasoning. The bounds of these can be defined somewhat differently, hence there is disagreement on their phylogenetic application. Habituation can be found in very simple organisms (e.g., hydra), and discursive learning only in humans. Lorenz argues that associative and trial-and-error learning require a centralized brain (Lorenz 1973). We must postpone a discussion of advanced animal learning until the following chapter.

The point for now is that, like Aristotle's tripartite notion of the soul (*psyche*) as a life principle of plants and animals as well as humans, we find a pluralistic continuum, not a dualistic opposition, in the phylogenetic development of sensation, action, and learning. They are features of all life, not of "mind," unless one wants to, like Peirce and Whitehead, ascribe mentality to all living things. If that seems extravagant—and it does—then we must say that by themselves sensation, response, taxes, and reflexes are not mental phenomena. They are life functions.

IV. Ecosystems and Societies

At the level of life we must make a new distinction in the process of emergent evolution, which will apply to more complex orders. For the evolution of organisms has involved as well the development of larger types of systems. Each organism must be a member of a *species*, a local *population* of that species, and some may as well be members of a local *society* of conspecifics. As a reproductive entity, it is a member of a *lineage*, or temporal line of ancestors and descendents (although it may fail to have the latter). It is a member of an *ecosystem* of conspecifics, other species, and terrain, aquatic or atmospheric systems, and ultimately the Earth's *biosphere* as well. Particularly important for our purposes are ecosystems and societies.

Organisms exist in ensembles of both conspecifics and heterospecifics. The point is not that they exist as water does in a cloud or must eat or live on other organisms. It is more fundamental: *life itself is ecosystemic*. Life presupposes the presence and/or past action of large groups of other organisms. All breathing organisms rely on the oxygen produced by cyanobacteria and plants, almost all organisms need the nitrogen fixed by bacteria, and the carbon and other elements left behind by dead organisms in soil. The vast majority of organisms could not live if their environments did not include, and were not maintained in their current state by, great numbers and variety of species. A change in the distribution or habits of one local species can affect other ecosystem species.[17] More intimately, many organisms exist in local "autocatalytic" feedback loops where each promotes the other, like the bladderwort plant, the periphyton algae that grows on it, and the zooplankton which, drawn by the algae, is food for the plant (Ulanowicz 2009, pp. 65–68). Indeed, natural selection implies the organism has been designed, at least in part, *for and by* their ecosystem, which defines its *niche* (although some organisms, like beavers, partly design their own ecosystem). The point is just that natural selection works in and through ecosystems.

While ecosystems are crucial to all life, some species develop a novel mode of conspecific organization: they are *social*. What I mean by "society" is more than either colonies of organisms, like a coral reef, intra-specific communication, or the offspring-parent pair. Societies are systems with their own structures and processes. A society assigns *roles* to individuals and subgroups, requiring interlocking practices or institutions, and such roles are an essential—if changeable—condition of the individual's existence. Where the properties of a society are nonaggregative (in Wimsatt's sense), where they cannot be explained by the aggregation of properties of asocial individuals, then the explanations will require reference to the social whole. Communication, while not only characteristic of social species, becomes particularly important for them, and may well be a driving force in the development of complex neurology and learning, as Cheney and Seyfarth have argued for primate evolution (Cheney and Seyfarth 2007). As one of the early emergentist thinkers W. M. Wheeler argued, "The social is a correlate as well as an emergent of all life in the sense in which [C. Lloyd] Morgan speaks of the mind as being both a correlate and an emergent of life" (Blitz 1992, p. 122). While elegant, this is too extravagant; we would rather say that the social is emergent upon certain types of animal species where membership in a local conspecific group is definitive of individual behavior and where distinctive social membership and role matter and can be selected for.

Here we have the striking fact that, while among mammals and birds we often think of sociality as corresponding to intelligence—sometimes called the "social intelligence" hypothesis—some of the most social creatures are invertebrates. Ants, bees, wasps, and termites are *eusocial*, living almost like extended organisms. They have a strict division of labor—not merely sexual—live in nests or hives of hundreds of members, and raise offspring communally. Their social behavior is genetically determined and instinctual. They have a uniquely cooperative breeding situation; all members are produced by a single queen, with a small number of male consorts, other members being infertile. Thus all members are siblings and very similar genetically. This is in contrast to the societies of, say, a wolf pack or a troop of primates, in which individuals live in relatively small groups and constantly interact, communicate, compete for resources or mates, and negotiate the individual's place in a social hierarchy via contingent learning. Both cases nevertheless involve communication. Organisms send signals necessary for the coordination of the individual's behavior with other social members. We will later see that humans, while genetically more similar to other mammals, are as social or more so than the eusocial insects, albeit in a distinctive way.

V. Reductionism and Emergence in Selection

There is an issue in contemporary evolutionary theory as to what is the "unit of selection." Candidates include, most prominently, the individual organism and the gene or some ensemble of genes, but also species, the gametes or germ-cells (for sexually reproducing species), or some groupings of organisms. Most evolutionists still consider the individual to be the unit, some defend group selection in special cases, but others have followed W. D. Hamilton, George Williams, and most famously Richard Dawkins in genic selection-ism. It is conceivable that different kinds of selection can operate at different levels and times. It is also possible that the debate hangs not only on biological fact but on basic ambiguities in what the debate is about. That is, what gets selected, what determines selection, what is selected for, what adapts, and what benefits from selection may well have different answers (Mayr 1997, Lloyd 2001).

There will be no attempt to adjudicate the controversy here. My only interest lies in what is typically called "reductionism" in biology, i.e., genic or gene-centered reductionism. Arguably this view was enabled by the success of population genetics in the wake of the modern evolutionary synthesis and culminated in the Human Genome Project. But its most famous theoretical expression remains Dawkins' 1976 *The Selfish Gene*. His theory was driven by rigorous presumptions that Weismannism and the Central Dogma rule out *all* Larmarckian-like effects, and that kin selection, the existence of junk DNA, and parasitism disprove group selection and are better explained by genic than by organismic selection.[18] Dawkins imagined the development in the primordial soup of a plurality of primitive individual entities in competition to reproduce. The entities are "replicators," nucleic acid strands that copy themselves. (Dawkins' view is wedded to the nucleotides-first theory of life's origin). Sophisticated replicators evolved which could code for the construction of a "vehicle," first cells and eventually multicellular bodies, which enhanced the ability of that DNA to copy itself into future generations. He claims that all biodiversity results from the "selfish" gene.

We should note that Dawkins qualified his view in several ways. He admitted that genes must be defined as the smallest bit of chromosomal information that can be selected, so selection *defines* what counts as a gene. Genes are selected indirectly "by proxy" and as part of "assemblages" rather than as individuals. Dawkins allowed that the organismic perspective is "equally" valid and important. He also proposed the "extended phenotype," the notion that the organism's phenotype can include its environmental interactions, even other organisms (e.g., in parasitism). Note that the apparent holism of this last point is just the result of his genic reductionism: once the gene is the central causal agent, the organism or "vehicle" is epiphenomenal, and can be made to include parts of the environment. The gene, Dawkins says, is like the center of a field of nested concentric circles representing its influence; the organism is the closest ring, environmental effects are further out, and the fields of all nearby genes overlap. Any phenotypic trait-point in the field is a joint product of the intersection of all local fields at that point, whether they are housed in its organism or not.

There are problems with genic selection. While he endorsed thinking of natural selection in terms of populations, Mayr criticized population genetics as "bean bag genetics," meaning its mathematical treatment required the assumption that a gene's impact on phenotype was as isolable as individual beans in a bag. Genic selection, he wrote,

> is completely impractical; a gene is never visible to natural selection . . . it is always in the context with other genes, and the interaction with those other genes make a particular gene either more favorable or less favorable . . . people like Dawkins . . . are evidently wrong. In the 30's and 40's, it was widely accepted that genes were the target of selection . . . but now we know . . . it is really the whole genotype of the individual, not the gene. (Mayr and Brockman, 2001)

Mayr's remarks have been amplified by the last twenty years of empirical biology. Some of this ferment was ironically due to the Human Genome Project itself. Completed in 2003, it concluded the remarkable fact that the human genome possesses about 23,000 protein-coding genes, not many more than the mouse's genome and only twice that of a flatworm! This means one-to-one mapping of genes to phenotypic traits cannot be the standard form of expression.[19] Then there has been the rebirth of embryology, hence the study of the complex route from gene to developmental expression of gene, which has led to the field of "evo-devo," the study of developmental processes in evolution, and "eco-evo-devo," which includes the environment's role. Students of complexity in biology have pointed out that order in evolving biotic systems comes not only from natural selection, but from self-organizing and autocatalytic processes in organisms and ecosystems (Kauffman 1995, Ulanowicz 2009). That is, we cannot attribute all order in the biosphere to natural selection or reduce selection to gene selection. Even the mid-twenetieth-century orthodoxy that genes remain unaffected by anything but exotic outside influences (e.g., radiation) has been significantly altered by recognition of the roles of development and environment in gene expression.

All my somatic (versus germ) cells possess my entire genome, the same genome that was contained in the fertilized egg that was my beginning. My blastocyst, a few days after fertilization, contained an inner mass of embryonic *stem cells* which eventually differentiated into all the specialized cells of my tissues. What made one become a lung cell versus a liver cell is *gene regulation*; certain genes in each cell were activated and others suppressed by methylation (adding methyl groups) which tightly aggregate proteins preventing further gene transcription (Gilbert and Eppel 2009, pp. 12, 43). A variety of complications affect gene expression, including: *pleiotropy*, or the effect of a single gene on multiple characteristics; *epistasis* or one gene controlling the expression of another gene, particularly the role of *Hox genes* which switch other genes on and off; *linked genes* that tend to be inherited together; *polygenic* or multifactorial inheritances in which a group of genes act together, making expression a matter of degree or continuous variation rather than a binary on/off; and *phenotypic plasticity*, where cues or environmental conditions alter the developing phenotype. Jablonka and Lamb point out how porous all this makes the Weismann barrier (Jablonka and Lamb 2005). For genes reside in and are completely dependent upon cells. Nucleic acids are not alive; cells control the reproduction, maintenance, and repair of chromosomes through enzymes, in effect "editing" their DNA. Stressful environmental conditions lead to changes in both sexual selection and repair processes. The same is true of mutation rates, both globally—mutation rates across the genome may increase in stressful times—and in cases of "induced local hypermutation," where particular areas of the chromosomes undergo differentially changed mutation rates. Stress produced hormones can turn genes on and off. So soma can and does causally affect genes.

Lastly, *epigenetic assimilation*, in which environmental factors and behavior that combine to impact gene expression can eventually affect the genes themselves, has made for a limited return of Larmarckism. The idea is not new. Two famous examples were the "Baldwin Effect," proposed by James Mark Baldwin in 1896, and the theory of epigenetic mechanisms developed by Conrad Waddington in the 1950s and '60s. Baldwin's argument concerned the heritability of behavior and was quite simple (see Weber 2007.) Assume some stable selection pressure in an environment that will sort out non-heritable responses to the pressure. Some organisms discover a "useful trick" or behavioral response that can be nongenetically transmitted via imitation. The trick will be selected and passed on without genetic transmission. Then, whatever mutations arise that make the trick easier to perform or transmit will likely be naturally selected. A phenotypic trait has acted as an attractor for genomic change over time to support it.

Waddington, the father of epigenetics, proposed a model of the "epigenetic landscape," a two-dimensional surface with various canals or gullies, down which a poised ball—the species—might roll, signifying evolutionary change. The key is that at any given moment there are many unexpressed but potentially adaptive genetic mutations in an organism. Mutations already expressed in the phenotype have been *canalized*, stabilized by repeated selection, making deep ruts in the landscape, but a substantial amount of genetic variation is present without reaching the threshold of expression that would allow selection for or against. Now imagine that in a population of ostriches with soft-skinned bellies the young suffer from injuries to skin. Some of these however are genetically gifted *to be able to* calcify quickly. Stress on the ostrich population will increase the percentage of the quick-calcifying members in the population over time. After enough time, this alternate canal of thickened skin will grow deep enough that the evolutionary ball will roll down it. Waddington argued that environmental selection pressures provoke *different level changes* in individuals and populations in a step-wise fashion, based on the costs of the means of adaptation, with the cheaper strategies exhausted before more costly strategies take effect. Behavioral change or learning is the cheapest and most immediate response to pressure, followed by physiological changes in individual members (i.e., epigenetic), and lastly change to the genome, the most costly kind of biological change.

Classical ecological processes may also select different reproductive strategies of survival. E. O. Wilson and Robert MacArthur showed that the ecological formula for an environment's "carrying capacity" for a species population affords organisms two distinct strategies (MacArthur and Wilson 2001).[20] In unstable environments where population is far below carrying capacity it may be more adaptive for a species to reproduce less robust organisms in high numbers, or "r-selection" (*r* for rate of reproduction). In stable environments where a population nears environmental carrying capacity (or *K*), hence more competition for resources, "*K*-selection" produces fewer but more robust offspring. Some species are characteristically *r*- or *K*-selected, others can oscillate between the two given environmental change. Related to this is Geist's distinction of "dispersal" versus "maintenance" phenotypes (Geist 1978, pp. 13–16, 113–41). Here the question is not reproduction rate in response to environmental change, but the sensitivity of unexpressed genetic endowments to the opening up of new environments. Near carrying capacity a maintenance phenotype is encouraged, a syndrome focused on efficiency of resource acquisition in a condition of intra-specific competition (e.g., smaller size, especially of low-priority organs, and behavioral consistency). But when the population experiences a breakthrough to a new environment, a dispersal or colonizing phenotype is encouraged with a premium on robust, innovative, individuals able to overcome formerly beneficial aversions and adapt to novelty in a condition of relative abundance.

Given all this, what could genic selection mean? It could only mean that *assemblages* of genes are *indirectly* selected as a *necessary but insufficient* condition of directly selected organismic phenotypic traits (of morphology and behavior, however extended). As some put it, natural selection of genetic information can explain the *survival* of the fittest but not the *arrival* of the fittest. The arrival may be governed by self-organization processes and complex plastic development tied to environmental conditions before there is a developed organism for environment to select. None of this implies that the organism's habitual behavior, or its possible effect on its cells' protein structures, chemically alters the gametic genes that will instruct protein manufacture in its offspring. It remains true that genetic information provides *sufficient* conditions for *some* phenotypic traits, and *necessary* conditions or the structural underpinnings of *all* morphology and behavior, including the organism's plasticity. But it does mean the fear of biological, and genetic, "determinism" is overblown. The notion that "biology is destiny" is abhorrent to many humanists and social scientists. It need not be. Biology *is* indeed destiny, but destiny is *vague*. Biology is never irrelevant to our fate, since we *are* biological. But its role is distributed through the complex circuit of genes, soma, developmental processes, behavior, environment, learning, society, and culture, whose multidirectional causal arrows make a simplistic determinism impossible.

VI. Autonomy and Teleonomy

Organisms are individuals, even if extended and sharing their bodies with symbiotic parasites. But life is not an individual; it is a state or more broadly a *process* maintained by the activity of uniquely complex systems. Sixty seconds after death there will be virtually no structural or componential distinction between my body then and one minute earlier, but certain processes will have ceased. There are today disagreements over the definition of human death, and the relation of clinical or medical to legal death. But wherever we draw the line between life and death, it is a line drawn across a continuum of activity. Life ends when a process ends.

Certainly we have listed many of life's activities in the present chapter. But how do we philosophically characterize the forms of activity that distinguish a living from a non-living material or physical system? Nonreductive biological theorists of numerous stripes have suggested various characterizations, all accepting a special emergent property of biological systems. Some believe life employs signs or intentionality (biosemiotics), or some primitive form of mentality (panpsychism). The present theory does not in principle reject such possibilities. However, the ordinal perspective must warn against reading properties of mind into life. Life is distinctive as it stands. My aim in this section is just to describe, not explain, the minimal characterizations most essential to life. These will be *autonomy* and *teleonomy*. Here it is instructive to turn to two naturalistic philosophers, Hans Jonas and Holmes Rolston.

In *The Phenomenon of Life* Jonas' main purpose was to oppose the tendency of all modern metaphysics to regard either matter or mind as the basis of reality. For life cannot be derived from either or both. Living things, he claimed, are uniquely characterized by "needful freedom." This means first that they manifest a novel form of freedom with respect to the environment, possessing a degree of autonomy foreign to inorganic existence. Jonas writes, "Profound singleness and heterogeneousness within a universe of homogeneously interrelated existence mark the selfhood of organism" (Jonas 1966, p. 83). The existence of living things is based on their ability to maintain boundaries from the environment and control their internal distinctiveness. "The point of life itself," he writes, is "its being self-centered individuality, being for itself and in contraposition to all the rest of the world, with an essential boundary dividing inside and outside . . ." (Jonas 1966, p. 79). This is connected to autonomy; there must be *identity*, or else there is no "auto" to be the "nomos," the law or governor, of its activity. The organism remains a unity, with an inside and outside, and the difference *matters*. This self-segregation from environment is negentropic. Segregation maintains identity against the tendency of the cell or organism as a physical system toward equilibrium with its inorganic environment.

This means, for Jonas, living things have *form*. The relation of their structure to their components is more independent than that of non-living things, for the structure remains the same through a virtually complete exchange of token molecular components. The form or characteristic set of structuring activities which accomplishes this is constituted by something utterly foreign to non-living matter: a complex macromolecule that functions as a code to dictate the continual reconstruction of the individual. The organism is a not a form; it is a member of a natural kind. But it is *formal*, it contains or is characterized by a form and shares most of it with others of its natural kind. Cells are the first things that, as far as we know, are literally constructed according to an existing detailed plan. Cells and organisms are *designed*. This does not mean designed in every way, nor that there is a designer. But there is design nonetheless.

We may compare cells to the classic case of inorganic structured growth. Crystalline solids can "grow" in the sense of attracting new molecules from the environment that automatically adopt a particular spatial orientation and fit into the crystal's molecular structure. Such growth is accretion. The crystal does not reorganize itself internally, rebuilding itself and treating new materials as mere matter for its structure. In the crystal, the components-plus-structure, or components-fitted-into-structure, acts electromagnetically as a unit. In the cell components are continually thrown out, put in one structure, taken out and put in another structure, all according to an internal recipe. Cells and organisms are self-organizing and -reorganizing (Kauffman 1993).

Here we can use Maturana's and Varela's notion of "autopoiesis" or auto-making. They meant the term to signify "the central feature of the organization of the living, which is autonomy . . . what takes place in the dynamics of the autonomy proper to living systems" (Maturana and Varela 1972, p. xvii).[21] But we cannot take this to mean "*self*-making," for we need to avoid the implication of the kind of agency or mentality characteristic of either intelligent animals or humans. By auto-making I mean that life is the state of: a) a particular bounded material entity; b) manifested in ceaseless activity (homeostasis, growth, and behavior); c) in order to achieve and maintain a complex, largely fixed and inherited typical structure (much of it common to a species); d) as part of a lineage (life only comes from life); e) in material and energetic exchange with an environment.[22] The *prima facie* evidence for life is an entity engaged in complex auto-maintaining activity; death is cessation of that activity and consequent loss of boundary with environment.

This autonomy has nothing to do with *independence*. Jonas writes, "The privilege of freedom carries the burden of need and means precarious being . . . living substance . . . has taken itself out of the general integration of things in the physical context, set itself over against the world, and introduced the tension of 'to be or not to be' into the neutral assuredness of existence . . ." (Jonas 1966, p. 4). Living things are *more* dependent than their non-living precursors; they remain invariant over only a tiny bit of their phase space. Simply, organisms *need* and *die*. Precisely because they are so distinct, they are precarious. It is need which generates "biological time," the "internal direction toward the next impending phase . . ." (Jonas 1966, p. 86). Unfulfilled need means death, the loss of inner activity leading to merger with the environment. Eventual death is the price of life. Life is "hazardous being" because it forms an identity that can and will end. Death is complete, rather than controlled, openness to environment.

The only way for this hierarchical form and identity to be maintained is through ceaseless activity or *dynamism*. There is no comparison between the temporal demands on living and non-living macroscopic objects. The needs of the organism are constant; something always has to be gotten, escaped or avoided, attended to, or processed, usually many such things at the same time. Activity never stops short of death. The organism's complete dependence on chemically rare macromolecules means it must always be a process of manufacturing its components. The speed at which its reactions take place is remarkable. Every cell is running in place. Jonas writes, "The organism has to keep going, because to be going is its very existence . . ." (Jonas 1966, p. 126).

This remarkable intensity of activity requires a novel form of control. Colin Pittendrigh coined the term *teleonomic* for "A physiological process or a behavior that owes its goal-directedness to the operation of a program . . . [hence] depend[s] on the existence of some endpoint or goal which is foreseen in the program regulating the behavior" (Pittendrigh 1958). As Mayr wrote, in biology we do not merely say the Wood Thrush migrates in the fall and *thereby* escapes the northern winter. Such a statement misses a crucial bit of information: the Wood Thrush migrates *in order to* escape the winter (Mayr 1974, p. 152). That is why the behavior has been naturally selected and maintained. We saw in Chapter 3 that functional explanations are necessary to biology, they are "teleological"—we can now say teleonomic—and have an ineliminable reference to purpose. Some trait is naturally selected because, Wimsatt remarks, it "promotes the attainment" of some "intensionally defined classes of state-descriptions of the system and its environment." He concludes, "Only by denying that evolution has occurred as a result of selection processes . . . could teleology be eliminated from biology" (Wimsatt 1972, pp. 66–67). No reference to intent is implied; as Dewey would put it, the Wood Thrush's behavior exhibits an end, not an "end-in-view" (Dewey 1958). This is *biotic agency*, distinct from either intentional or *mental agency*, or *autobiographical agency* or selfhood.

Purposes are ends and imply value, that is, good and bad (which is *not* to say a moral sense). There is value at the least wherever there is *evaluation*.[23] The naturalist and ecological philosopher

Holmes Rolston has developed an account of the role of value in living things (Rolston 1988). First and most obviously, organisms value themselves, their states, and sometimes their offspring. They act so as to maximize their survival and/or reproductive chances. They value preferred states of their own organism, pursued homeostatically. They have *norms*; they are living or dying, healthy or sick, succeeding or failing. Second, they value parts and states of the environment; they turn toward the sun, stretch their roots for water, explore for resources, chase prey, seek secure locations. This value is *relative, but objective*, not subjective: the valued things have a value for that organism or species due to its genetically programmed nature as evolved in that environment. The presence of certain carbohydrate-hungry beings means a potato has food value. Seed-bearing plants develop fruits because their seeds are scattered in the feces of animals *because* the fruits have value for the animals who eat them. Third, value *circulates* through an ecosystem. When a predator eats its prey, or a herbivore eats vegetation, they are "capturing the value" of the organism eaten. Fourth, *the species* evaluates, not as a collective agent, but as a "form" that presses its case genetically. The developing bird is *supposed to* have two wings. Rolston thus rightly characterizes the organism as a "valuational system." Part of what makes it survive is that it evaluates in the ways it does; it is genetically coded to make those evaluations and embody certain norms. Valuing is one of its crucial activities. The valuational system itself evolves, in that populations can be selected by environment because of their distinctive preferences.

This leads Rolston to take one more step, also endorsed by Jonas. Rolston writes: "Life flows on. Life *ought* to flow on. Few can specify how we make that descriptive-prescriptive jump, but here, where biology and history draw so close to ethics, it is made easier than anywhere else. Fact and fact-to-be-desired join in 'the ought-ness *is* the is'" (Rolston 1989, p.69, my emphasis). Living things bring something new into the universe: a material being that strives, makes a demand, asserts an "ought" or a value to be achieved or captured, for which live/dead, healthy/sick, live-offspring/dead-offspring, alternative binary states matter intensely. The question as to whether there is value, or are values, in nature, has a clear answer: yes, because living things and species value states and things. (A different question is, can one infer from the valuings performed by and embodied in an organism that what they value *is* valuable *per se*, that is, *should be valued by us*? We shall turn to that thornier issue in our final chapter.)

Biology thus gives us clear examples of *final causes*. Non-living encompassing systems or environments select components or traits because of their function. Within organisms there must also be *teleomatic* material processes as described in Chapter 6. Such physical and chemical events are selected, meaning functionally explained by the action of an encompassing system, but not selected "for." "Selection for" implies purpose, and purpose only comes with *teleonomy*. With life functional explanation, the causation of N's properties by N+1, becomes "final" or "purposive." Note this refers not only to natural selection, selection of organism, its phenotype, or genetic code by environment over time. When any encompassing biological system—society, organism, organ system, organ, tissue, cell—restricts the output of a component system, dictating any properties or behavior of the component, this too is selection. Lastly, because the organism participates in an environmental ecology and species evolution, its form has actually been selected over time for that ecology, hence *its form serves a final cause*, adaptive fit, given a genetic inheritance.

One may say there is no reason to reintroduce final causes here. Aristotle thought a ball rolling down an inclined plane exhibited final causation, moving to rest in its "natural place" on the surface of the Earth. We moderns say that is not telic (and we are right). So isn't the wren eating a seed doing the same mechanical thing? No. For even putting aside its greater complexity, there remains this: if the ball doesn't roll, it is still the same material object, but if the bird doesn't eat, it *dies*. The dead bird obeys the rules of chemistry and physics as before, and as the ball does. But it is no longer alive. And it was life, after all, that was being investigated here.

8

Mind and the Hard Problems

It is on the shoals of mind that, most non- or anti-naturalists believe, naturalism founders, or at least it must work to earn its pay. Most philosophers of mind are all too familiar with "the" hard problem, as David Chalmers called it, of how a brain, which is a spacetime extended, massive entity, can make or interact with nonphysical conscious events, like thoughts, which have no mass or spatial extension (Chalmers 1995). This is the problem that bedeviled Cartesian dualism, and despite nearly everyone's rejection of Descartes it remains with us: how could material objects give rise to immaterial (i.e., intentional) acts and properties? But a second problem, the problem of mental causation is, if not equally hard, awfully close.[1] Its vexation has been made clear by Jaegwon Kim (Kim 2000). Kim admits that, like everyone else, he believes mental events can cause both other mental events and physical events, e.g., our actions and the neural events that are their proximate causes. But how they could is a puzzle if we assume the physical world is *causally closed*, meaning that only physical events and properties can cause physical events and properties. Kim shows that several apparent ways around this problem fall into the same ditch, named "redundancy." For if we suppose that mental events are "supervenient upon" or dependent on neural events, it seems we must accept that the mental event is caused by its underlying brain event, for without that brain event or its equivalent, the mental event could not occur. We thus already have a cause of each mental state without referring to antecedent mental states as causes. This holds not only for mental-to-physical, but for mental-to-mental causation as well. Given dependence on neural events, if one mental event is to cause another it would have to cause that other's brain or central nervous system (CNS) state, which already has a cause in the preceding CNS state. Mental causation seems to be superfluous, or alternately, the neurological cause "excludes" or screens off the mental cause.

These hard problems are two aspects of *one* problem: how mental events or properties can interact with material or physical events and properties, like those of the brain, whether that interaction means "arise from" or "cause," in either direction. They are hard because they are metaphysical. After all, consciousness is not intrinsically puzzling: we all experience it, it is describable, and it is open to scientific study. There is no mystery in the "neural correlates of consciousness" either; we are nowhere near a comprehensive account, but nevertheless we can state some simple correlations when we can isolate them. What is mysterious is the metaphysics of consciousness *in light of* physicalism, the ontological status of consciousness and its contents in the context of what are considered physical realities on which consciousness depends and about which we contemporary inquirers think we are ontologically more secure.

But this conception may be part of the problem. There is a vast amount of sophisticated work in contemporary philosophy of mind and cognitive science, to which the following can do no justice. Much of that work is constrained, however, by a set of assumptions that may make its difficulties more intractable: the restriction of "mind" to human mind; the notion that nature is divisible, at most, into the physical and the mental; the belief in the causal closure of the physical world; and the notion that mental content is "supervenient" either on internal physical states ("internalism") or external physical states ("externalism"). Before setting out we need to consider these in order.

First, to put it bluntly, there is no good reason to deny mind to nonhuman animals wherever they exhibit behavior *and* neurological structures and processes similar to those which indicate or manifest mind in ourselves. This means there are likely minds without the capacity for language, abstract conceptual thought, or discursive learning. I will offer no proof of this beyond the widespread convictions of those who study animal behavior for a living. The study of mental activity by rights should be an offshoot of ethology, and what is called "philosophy of mind" should, by dint of actual practice, be called "philosophy of *human* mind." At its narrowest, the metaphysical problem of philosophy of mind lies in the relation between, say, an animal's feeling of pain and the electro-chemical behavior of its CNS. The narrowest part of the river is usually the best place to ford. Consequently, the present chapter is not about human mind; it is about mind in general, which is to say, *animal mind*. We will use contemporary theories of human mind as clues to the genus of which it is a species.

Second, the impressive metaphysical generalization of running the scope of a presumably "physical" reality from dark energy through crystals to bacteria to animals and right up to the doorstep of human propositional attitudes (e.g., beliefs or desires), makes it seem there is only one, enormous "explanatory gap" in nature. Now, in Chapters 3 and 5 we have seen how unclear our usual notions of the physical are. Nevertheless, we can accept that the contents of intentional states are neither spatially extended nor objects of physics, hence not physical in that broad sense. But admitting this need not commit us to property-dualism. As Steven Horst puts it, there are "explanatory gaps all the way down" (Horst 2007).[2] Once we recognize that there are many places in natural science at which reduction fails, where novel properties appear, the "anomalousness" of the mental shrinks. Horst rightly asks, if physicalist reductionism is inadequate to biology, why should we expect it to work in the investigation of mind? Our problem is the relation of the mental, not to the physical, but to *the biological*. Central nervous systems are living systems of cells exhibiting teleonomic behavior. The problem is not, how do intentional properties prevail in a physical world, but how do they prevail in a living biological system in interaction with a physico-material-bio-social world?

Third, the commonly held thesis that the physical world is causally closed is neither obviously true nor even clear. The question is, with respect to *what kinds of causality*, is the physical supposed to be closed? As noted in Chapter 3, if "physical" means the objects of physics, causal closure would mean that the unique objects of the other natural sciences cannot make a causal difference to the objects of physics, which seems absurd. As seen in Chapter 4, today many philosophers conceive causality as efficient causality, which restricts causality to events. If causality is efficient and physical, the causal closure of the physical is true by definition. But the natural sciences regard many different things, processes, and structures as making causal contributions in very different ways. I will accept that changes in non-space-occupying complexes cannot *efficiently* cause changes in space-occupying complexes. The semantic content of the idea of the number 2, a mental image of a tree, or felt sadness cannot by itself push or pull a neuron or an electrical field. The question is, is there another way mental states and contents can make a non-redundant causal difference to subsequent neural events?

Fourth is the debate over internalism versus externalism of mental content, usually cast in terms of supervenience. Internalism claims mental content supervenes on properties of the individual alone, e.g., its brain states. Given the definition of supervenience (see Chapter 3) where the individual's "physical" properties (e.g., brain states) show no difference, its mental contents can show no difference either. Externalism claims that mental content supervenes on the organism's relation to its environment. Its famous thought experiments, originally by Putnam and Burge, argue that two creatures could be in precisely identical physical (i.e., brain) states but different mental states, that is, be intending different meanings. This has the attractive feature of corresponding to externalism of meaning, the view that meanings are objective and not subjective, as philosophy of language in the twentiety century came increasingly to hold.

But even if externalism is true, this cannot mean that *no* equivalence classes of CNS states are necessary for a mind to entertain a meaning. That a mental state is realizable in multiple CNS states does not mean it is realizable with either *no* or just *any* CNS state. For that would mean the fact of my entertaining mental content has no dependence on my CNS. Any naturalism, internalist or externalist, must make plausible the dependence of mental acts and objects on neuronal activity. As we will see in the following chapter, meanings can be independent of particular minds, but *having* or *intending* a meaning is a mental state. That some mental contents depend upon environment (hence are external or "wide" in content) rather than only on brain or somatic events (hence internal or "narrow" in content), makes no difference to the need for a complex brain. It may well be that more simple contents, e.g., the feelings of pain or hunger, depend on CNS/soma alone but once we reach the level of a mind that maps its somatic state onto a perceptual account of its environment, mental contents depend on environment, and possibly social communication, as well.

In any case, as noted in Chapter 4, "supervenience on brain state" is too specific a kind of dependence for these relations. Rockwell points out that many mental states depend on non-brain CNS and endocrine activity (Rockwell 2007). Hormones released by the brain affect mind and behavior, glands outside the brain affect brain processes, and studies in the last decade have indicated a communicative role for glial cells in the brain. Thus information processing may be chemo-molecular as well as neuro-electrical. And of course brain, CNS, and endocrine systems themselves depend on, and are affected by, other somatic and soma-environment interactions. But if so, this means the biological and/or environmental substrates of a mental event are often too complex to know whether two of them *are* identical. If we cannot precisely delimit a subvenient state, then we cannot tell when it is identical to another, hence whether supervenience holds or not. In what follows it will be enough to say mental states are dependent on neural states.

I will argue that mind is a suite of intentional activities, states, and properties emergent upon (at least) embodied neuro-electrical-chemical (NEC) processes in a wide variety of encephalized animals capable of complex learning (including humans), and that mental states and contents sometimes make a causal difference to subsequent neural states and behavior. The first two sections set the stage by locating mind in an evolutionary context and by defining mind and consciousness using the work of neurologist Antonio Damasio. Section III then argues that the character of intentional objects is a semantics emergent from the reading of certain NEC states by others. Emergence has been exploited by other theories of mind (e.g., Sperry 1976, Clayton 2006). The notion is that somatically-embedded central nervous systems can generate phenomena not attributable to their biological components or component processes. Let me be clear: if "solving" the hard problem means showing how an intentional content can be derived from the properties of individual NEC events alone, it is no more a "solvable" problem than how natural selection could be derived from the interaction of chemical substances alone. Emergence accepts that complex organizations of components can yield properties not possessed by components. In Section IV I will suggest how there can be non-redundant mental causation in a way similar to philosopher Fred Dretske's notion of "structural" causation. The overall point is that our pluralistic theory of nature makes a reasonable response to the hard problems possible.

I. What Good Is a Mind?

We must start with the most basic question: are there mental acts and objects that are non-identical to their NEC processes and properties, and which might potentially affect the latter? There are those who either deny the existence of mental acts and objects that are not identical to brain states, or if allowed, deny they can make a causal difference to brain or behavior. That would mean that, to the extent mental phenomena exist, they are an ineffectual evolutionary "spandrel" or free-rider.

Mind is first of all a biological phenomenon; that is, as far as we know now, the only minds belong to life. There seem to be minds which fall within the purview of the field of zoology and ethology, namely nonhuman animal minds (Griffin 2001, Cheney and Seyfarth 2007). As noted, if there are human mental acts and objects non-identical to brain processes and properties, there is no good reason to deny the same to those species where we find behavior resembling minded behavior in us, and a substantial version of the neural and morphological hardware we believe makes such activity possible for us. Those conditions are *prima facie* good reasons to suspect the presence of consciousness or mind. And we do find such organisms.

As noted in the preceding chapter, in complex animals dedicated nerve cells provide sensitivity and information processing. Some animals have very simple receptor-effector neuronal connections, others have nets of neurons, and others have single or multiple ganglia, the beginning of encephalization. Crustaceans, spiders, and insects—notably ants, wasps, bees and termites—have brains, and are capable of reflexes, instinct, and complex, nuanced social behavior. Vertebrates, with long bodies integrated by a central nervous system, have complex brains of increasing size and structure characterized by a tripartite bulging of the spinal cord into hind-, mid-, and forebrain. It is only in mammals that the forebrain or cerebrum becomes the dominant structure of the brain, and develops a series of highly structured cortical or outer layers, culminating in the neocortex. In larger mammals the cortex develops the *sulci* or folds that provide far more surface area. But as noted, there is evidence that there can be alternate paths of the evolution of complex intelligence quite different from that of mammals (Kirsch et al. 2008). The claim that a particular neural structure is a *necessary* condition for a given cognitive capacity may be untenable, because other structures may, given an alternate phylogenetic inheritance and set of environmental pressures, come to enable that capacity. Thus, where on the evolutionary scale to claim that mind or intentionality first emerges is very difficult to say. If one were to guess at the criteria for picking out mind, it would presumably be a combination of relative brain to body size, some equivalence class of neural structures, *and* the presence of the higher forms of learning, especially trial-and-error and insight learning (not even to mention discursive learning).

Fortunately, for present purposes it is less important to know *who* had the first mind than to consider what nonhuman minds are able to do. To answer this question we can operate at higher levels, leading to reliable if conservative guesses. Given the neural and behavioral criteria mentioned above, there is no good reason to deny mind to primates and cetaceans, or having done so, to large mammals and some birds. On the supposition that a sizable brain is a requirement, intentionality of some kind *may* be characteristic of all vertebrates and cephalopods, or even arachnids, crustaceans, and some insects.[3] We don't know. One reason not to extend "mind" too far is to avoid the dualist tendency to let mind co-opt the functions of life. All living things have sensitivity and reflex behavior, and many have instincts, but we have no reason to say that intentionality is required for these capacities.

Regarding mental content we will have to trade on a distinction that will only be justified later. It seems best to employ a tentative, minimal classification of mental states and contents. There are at least three kind of such states and contents: *feelings*, e.g., of pain, heat and cold, bodily state and orientation, appetites, drives, and simple emotions like fear, excitation, and anger; *images* of body and environment that provide a continuous mapping of both in an ongoing causal relationship; and *ideas*, by which I mean non-analogical representations, what some call "thoughts."

The processing of all these, which some call "thinking," I will simply call *mental processing*, by which I do not mean to imply a computational viewpoint.[4] The point of this categorization is that finding intentionality in an organism means at least finding that it experiences feelings and/or images.

What might a mind be good for? One possibility is that mental activity is a sophisticated device for stimulus-response coordination that goes beyond what is supplied by non-intentional neural connections, that it develops "between" and "around"—in the sense of informing and contextualizing—receptors or stimulus capacity and effectors or response capacity, all three being neural developments. Dealing with the world through feeling and imaging may provide survival opportunities that dealing with the world without intentionality does not. To interpose between receptor and effector not merely a processing device, but one with the capacity for holding images and feelings as guidance for response, especially long and complicated responses requiring continual adjustment or extended investigation of alternate possible actions, may make possible kinds of learning, insight into environmental possibilities, and control of response that enhance organisms' abilities to succeed. The ability to act on feeling and image implies a greater mediacy and flexibility of the organism-environment interaction.

Animal learning was explored by one of the major ethologists of the twentieth century, Konrad Lorenz (Lorenz 1973). All organisms have the capacity for short-term or instantaneous acquisition of information through irritability. Simple organisms can become habituated or sensitized through repeated stimuli. In some cases sensation triggers an "innate releasing mechanism." The organism must normally inhibit action patterns not adaptive to most situations, so the selected action arises from cessation of inhibition. Lorenz analyzes instinct as a "species specific drive action" built out of the combination of: a) a fixed action pattern of spontaneous appetitive behavior searching for key stimuli; b) which, when hitting on the right stimulus, triggers an innate releasing mechanism; c) that enacts another fixed action pattern which achieves what is indicated by the key stimulus, terminating (a) and (b).

Lorenz makes the point that animal behavior is typically a built-up hierarchy of such patterns, always based in short-term information gain. Where adaptive, such action is learning, in a broad sense. But Lorenz insists that learning *proper*—beyond sensitization, habituation, accustomation or habit formation, and imprinting—requires stored information based on success or failure. The most important form of learning is trial-and-success/error learning. Lorenz believes it occurs only among encephalized organisms, among which he includes arachnids, insects, crustaceans, cephalopods, and vertebrates. In trial-and-error we find the acquisition of spontaneous novel forms of action in response to a problem from which the solution behavior is selectively retained. This involves a feedback loop by which success or failure strengthens or weakens appetitive behavior and sensitivity to key stimuli. It works like induction, through the sorting of "innate working hypotheses." Vertebrates and particularly mammals and birds add novel forms of learning through the visual representation of space, which allows both insight and exploratory behavior or play and curiosity, along with voluntary behavior and mimicry. We may guess, then, that mind enables the coordination of sophisticated responses to specialized sensation, internal or external, that it permits trial-and-error/success learning, hence acquisition of adaptive behavior patterns, allowing enhanced selective response. The flower turns toward the sun, the protist withdraws from touch or heat, thereby sensing and responding, but without the use of a mind. The neurologically complex animal can do more: it can feel hunger, construct a perceptual schema in which distal objects are indicators of food, maintain felt motivation and perceptual image as a guide to action, and connect past to present in retained memories of successful, and unsuccessful, actions.

What then would it mean to deny mind's presence or causal role among even the more sophisticated nonhuman animals? It would mean believing that mental or intentional activities, hence at least the experience or awareness of feelings and images play *no* causal role in the guidance of conduct. It would mean all conduct is accomplished by the neural connections of stimulus and response without mental states or content, that is, without the feeling of fear, the pangs of hunger, the image of a predator, playing a causal role. That would mean that evolution has gone to great lengths to construct a brain sophisticated enough not only to link sensory nerves to motor nerves, but to construct feelings and images, for no behavioral payoff whatsoever. That would be a lot to swallow. The development of the capacity to feel, to perceive, to represent the internal and external environment must have had huge costs. Such brains are big and complex and took hundreds of millions of years to develop (after, say, the Cambrian Explosion). They require, in the most sophisticated animal species, serious infantile incapacity and maternal vulnerability, a kind of *neoteny* that leaves room for substantial developmental and learning. Consciousness, which we know can be lost, is energy-intensive to maintain. Intentionality has been paid for in energy, time, and blood. Any brain, including the human brain, has lots to do besides generating consciousness. If mental acts are "epiphenomenal" then evolution has gone to a lot of trouble for something that makes no difference to adaptive fit. Now, evolution has no scruples about waste. Still, *prima facie*, the fact of brains capable of costly intentionality seems evidence for believing that mind and consciousness provided a novel way of responding to the world with naturally selectable advantages. The question is: "How?"

II. Intentionality, Consciousness, and Mind

We cannot begin to address the hard problems without clarifying the relevant terms, "mind" and "consciousness."[5] To do so we turn to neuropathologist Antonio Damasio's account, for it is in the medical examination of neurological disorder that many clues to mental organization lie (Damasio 2000). The bare bones of Damasio's view are as follows. Everything we call consciousness and/or mind is an evolutionary extension of an animal's monitoring of its own body's internal states in relation to environmental changes, to augment the organism's automatic forms of self-regulation and behavior control. It rests on modifications of drives, the capacity to experience pleasure and pain, and other homeostatic mechanisms of internal hormonal signaling and immune reflexes. Consciousness for Damasio is cognition, knowing, in the sense that to be conscious is to "know" something, although not necessarily in a discursive sense. He specifically denies the dependence of human consciousness on language acquisition.

Damasio finds three different levels or types of consciousness in human functioning. First is a minimal or "proto" consciousness that grows out of and accompanies the automatic neurological and chemical monitoring of the body's internal state. Pain, pleasure, and emotional states occur here as feelings. While consciousness is essentially cognitive, for Damasio it is always dependent on feeling and emotion, since basic emotions arise from organismic self-regulation to provide "biasing" devices (e.g., fear). When consciousness loses its relation to emotional functioning, will becomes negligible. He implies that this level of consciousness occurs in nonhumans. Let us pause to understand a bit better this most basic level of mind or consciousness, which, if he is right, was presumably the first intentionality to appear on Earth.

In Franz Brentano's classic notion of intentionality, a mental act by definition contains "something as object within itself." The act of perception contains the perceived, the desire the desired, etc. (Brentano 1973, pp. 88–89). What is contained is an "intentional inexistence," of course, not a physical object. This notion of intentionality is typically parsed today as "aboutness," a property possessed by mental states and cultural signs (e.g., words, sentences, pictures). This leads some philosophers to deny that consciousness of "qualia," felt sensory qualities like redness or pain, is intentional. For "aboutness" seems a matter of *representation*, and awareness of qualia does not seem representational. But I suggest that the preposition is a bit too strong. Some intentional states are indeed "about" something, but many are not. Intentionality in general is not *aboutness*, but *ofness*. Intentionality, which, I suggest, phylogenetically begins with feeling, is a state that is *of* something, not necessarily *about* something. The feeling of pain is a feeling *of* pain, not, like a perceptual image or a belief may be, about it. Feeling is intentional even if it is not representational.

Phylogenetically, feeling presumably began with experiences of internal bodily states like pain, heat, cold, hunger, perhaps spatial orientation, or some combination thereof. Perhaps later came the basic emotions, like fear and anger, or desire. For Damasio, emotions derive from monitoring of internal state and bias representations of actual or possible actions as "somatic markers" of options, making decisions possible. The connection accounts for the dual meanings in many languages of "feeling" (e.g., *sentir*, *gefühl*) as somatic and emotional.[6] Feeling is a state in which the organism experiences the state it is in, such that, the feeling and the state are indistinguishable referentially.[7]

Second for Damasio is "core consciousness," a second-order mapping of the feelings of proto-consciousness in relation to objects and processes perceived in the environment. The crucial achievement here is the construction of an integrated, ongoing representation of world and soma involving perception, memory, and emotion, all linked to action. It provides "images," although not only visual.[8] Images are normally *of* or *about* their causes. For Damasio images are "dispositional topographic representations" of the body state in causal relation to perceived environmental states or objects. The feelings enabled at this level are sophisticated emotions occurring in the second-order causal maps of the organism as modified by its internal processing of representations of the environment-body state. Damasio suggests the assemblies, or recruitments, of areas of the brain in these activities is spatially analogous, that the topographical pattern in/on the brain itself models aspects of the content thereby cognized. This is essentially the construction of objectivity in the organism's mental-action space.

Last of Damasio's levels is extended or enhanced consciousness. This is a third-order representation *of* the processing of the first two levels *as* an activity of the agent, as "owned." Core consciousness's representation becomes an autobiographical narrative centered on the self. It is this which humans usually call consciousness, in Damasio's terms, the "self-in-the-act-of-knowing." Here is where language, inference or reasoning, and greatly enhanced imagination reside. (We will discuss uniquely human mind in the following chapter.)

With this clarification we can offer a definition of mind. I suggest that mind is a suite of intentional activities, canonically listed among humans as perception (somatically external and internal), memory, emotion, will, thought, imagination, and consciousness, *and* the activity of processing or interrelating their contents. *A* mind is an integrated *subset* of those activities performed by an organism; not all of them are required for a mind to be active or present (i.e., nonhuman minds have only some of these abilities.) These activities are intrinsically intentional; for every intentional act there must be an intentional object, e.g., a visual act "contains" or targets a visual image, remembering contains or targets a memory, etc. To use a linguistic analogy from J. L. Austin, the canonical "illocutionary acts" that constitute mind—perceiving, remembering, imagining, etc.—are intrinsically intentional (Austin 1975). If they do not have such content—a percept, memory, or emotion—then they just are *not* those acts, but rather some other level of act (e.g., for Austin a locutionary or phatic act). Without the content we could speak of the neurological act but not its *mental* act. The brain has multiple functions, only some of which are to perform mental activities, to construct and process feelings, images, and ideas.

Consciousness is best thought of as *one* of the activities of mind, but a central and necessary one, because it is the subjective awareness of the other activities and their objects. (I take "awareness," "subjectivity," and "consciousness" to be synonymous.) Consciousness by itself has no special connotation of self-consciousness. It can have degrees in the sense of the classic medical indicators of consciousness (wakefulness, emotion, attention, purposiveness), dreaming, trance-states, etc. There are forms or levels of consciousness in Damasio's sense, types of subjective awareness, which implies there are different kinds of "subjects."

It may go unnoticed that Damasio's account makes consciousness and mind coextensive. When an asognosic or epileptic patient, during a loss of extended consciousness, perceives, acts, and nonverbally expresses emotional preferences, what do we call that? We can either say there is *brain functioning* but no mental functioning at all, or mental functioning going on *without consciousness*, or mental functioning going on with *consciousness of an abnormal kind*. Damasio's response is the last. The reasons for this are instructive.

Imagine four organisms interacting with their respective environments: one with some neural development but presumably no mentality or consciousness at all (perhaps a nematode); one with proto-consciousness alone (maybe a crab); one with core consciousness (maybe a deer); and one a human being in the midst of an asognosic or epileptic episode while negotiating a crowded lobby. The question is, regarding each of the organisms in question, which seems more right to say: a) it senses and does something based in neurology but without any intentionality or mentality at all; b) during its sensation and response, its neurology constructs feelings/images, hence mentality, but without consciousness of them; or, c) its neurology constructs the feelings/images with a particular level or kind of consciousness? This is in effect to ask: in which case(s) does the ascription of unconscious *mental* activity—not merely non-mental CNS activity, but CNS-supported mental activity—make sense? More simply: could these organisms be feeling pain or fear, seeing a predator, or perceiving the lobby, *without knowing it*?

My guess is that the answer is *yes* for the human and *no* for the nonhumans. The reason is that only one of Damasio's levels of consciousness involves self-consciousness, while the others are "access consciousness" or awareness of objects, even internal ones, without self. In the case of the human with brain injury or disease, it makes sense to say the mental activities of perception, short-term memory, emotion, etc. are proceeding, but without being attached to the "I know that." For the other animals, even the deer, it seems there is no dividing line between seeing a predator and *knowing that* it sees it, hence we can't make sense of their unconscious seeing. In the human case we can because there is an additional kind of consciousness which can be turned off or disconnected. In short, it seems that the mental activities of perception, memory, emotion, etc., may indeed always be conscious in the sense of "proto" or "core" consciousness. This means that *completely* unconscious activity is just non-mental brain or somatic activity, like my brain's electrical control of my heart rhythm. What we usually mean by "unconscious" human mental events is "unknown to *autobiographical* consciousness."

To summarize, our hypothesis posits five relevant levels of organismic capacity. First, sensation, response, and reflex can occur in living organisms without neural, encephalic or mental development. Second, encephalized organisms can process information through their CNS, generating and coordinating NEC patterns. These processes are in themselves non-mental and unconscious; those most crucial to life involve homeostatic regulation and the provision of reflexes and stereotypical response patterns. The simplest brains presumably have no other function. Third, some more complex brains (whether beginning with crustaceans, arthropods, and hymenoptera, or with cephalopods and vertebrates), permit complex trial-and-error learning and construct the intentional activities we call mind. The simplest level of mental activity is probably constituted by feelings of internal state (from pain and hunger to, perhaps, basic emotions like fear). These provide the animal with ongoing motivation over indirect, lengthy action sequences, memory of which they may retain. Fourth, more complex animals (perhaps mammals and birds, perhaps many vertebrates) have in addition the ability continuously to perceive images of the objective world and their body, connecting both with feelings and emotions to produce an ongoing mapping of one onto the other. This provides a powerful capacity for monitoring of internal and external reality. Fifth, humans seem alone in having an autobiographical consciousness that ascribes ownership of the first two forms of consciousness to a self *as* object, which is somehow connected with enhanced imagination, linguistic ability, and reasoning (to be explored in the next chapter).

III. Mind as the Self-Reading of Brain Activity

Now, how philosophically to understand the possibility of such intentional activity? I suggest we think of mental content as the semantics produced by the neuro-electro-chemical syntax of the living CNS. This is itself based on analogy; I am analogizing the *relation of* the content of the experience we call mental—the intentional content of a feeling, image, or idea—to the bio-physical events from which they arise, *to the relation of* the semantics of a sign or string of signs to the syntax, or graphic/phonic rules of formation, of that sign or string. Given the language of earlier chapters, whatever else mental events depend on, they depend on neurological events of an organism. If anything is a complex, self-organizing teleonomic system, the central nervous system of a neutrally complex animal is. Its cells function to produce a dynamic system of momentary electrical-chemical pulses forming temporary patterns, inside background patterns that are renewed continuously during waking states.

What then *is* a feeling, the most basic of mental contents? The NEC states of the CNS are so constructed that a change in neural signals is felt *as* the *qualē* we call "cold," or "pain," or "hunger," or spatial disequilibrium. Cold is how that state feels; having that state feels cold. The state is an active construction by the CNS, as Humphrey has it (Humphrey 2006, p. 81). The feeling is the feeling *of* (not about) the NEC patterns, given background neural and endocrine activity, long-term dispositions, and organismic learning.

But what does it mean to say the pattern "feels" a certain way? It means the CNS represents the pattern as a feeling to itself. This is the emergent property of "being mental." *Mentality is how the relevant neuro-electrical-chemical patterns feel to a living central nervous system sophisticated enough to produce and feel them.* There is no way to reduce this feeling to something else, meaning, to construct it out of a physical, material, or biological description. Pain, cold, hunger, fear—these are just how the NEC patterns feel to a CNS-endowed creature than can feel in the context of its interactive role in the soma-environment transaction. Mind is the activity of intending these semantic contents, dispositionally and regularly generated by electrical-chemical energy transfers of a special society of living cells; it *is* those transfers in a type (not token) sense *as* experienced.

To accomplish this, one must imagine that one NEC pattern is registering a somatic-sensory state and another is "reading" it (Humphrey 2006, pp. 87ff.). My hypothesis is that mind and/or consciousness is how the neural-endocrine system reads itself, the product of *auto-lectitious* embodied neurology. The CNS makes and reads what it makes; the contents of mind/consciousness are the semantics of that self-reading activity. The neural system supports an ability to read a continual succession of distinctive, specially constructed neural patterns *as* feelings, images, and (later in phylogeny) ideas. This is depicted below (Figure 8.1). At the simplest level these are feelings (1), but at more complex levels the CNS creates states that register the former in relation to *images* which are of their causes (2), and in the human case the CNS creates another process that reads the former as owned by a historical self (3). In the case of primitive feelings, in a primitive organism, the self-reading is probably internalist. But in more complex cases, in addition to the first-order reading, body-environment stimuli are read by core consciousness as environmentally related, as well as communicative behavior with conspecifics, hence externalism of content arises.

Loosely speaking, if the greatest jump in informational complexity in the universe in the chapters up until now has been the dynamic teleonomy of un-minded living organisms, itself based in a kind of genetic "self-reading" of a nucleotide code, then what we now see is a different kind of self-reading, in which a centralized

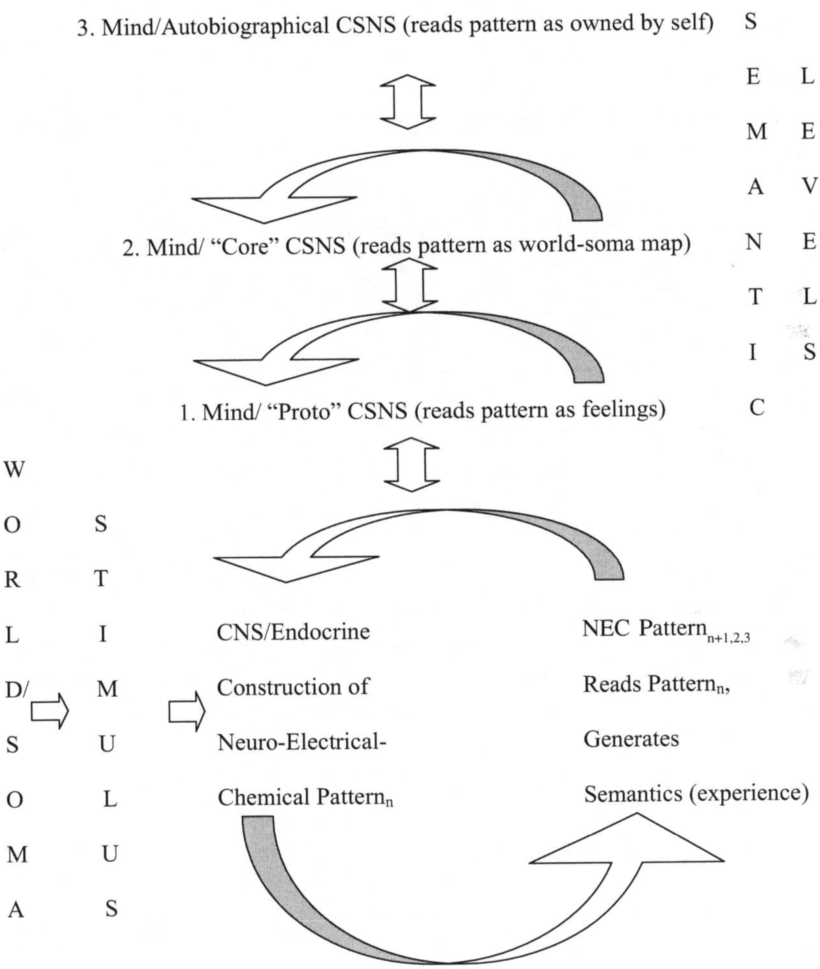

Figure 8.1. Mind as Self-Reading of the Central Nervous System (CSNS – consciousness; CNS – central nervous system; NEC – neuro-electro-chemical)

nervous-and-endocrine system constructs a novel representational system in the organism and reads it. Thus the neural process, whereby one NEC pattern evolves into another, does something unprecedented in nature. By one pattern reading another, it performs a semantics which plays a new role in the management of behavior.

IV. Mental Causation

The problem of mental causation is how the semantic information of a mental content could causally affect the firing of a neuron or release of a chemical. As noted earlier, the thesis of the causal closure of the physical world is too strong a claim. Nevertheless I too accept that semantic content of the idea of the number 2, a mental image of a tree, or feeling of hunger cannot by itself push or pull a biological or chemical system. The physical world may well be closed under *efficient* causation. But there may be another way mental contents can make a non-redundant causal difference. Note also that probably most of our mental states make no difference to what comes next neurally or behaviorally. To claim the reality of mental causation is merely to say that there are *some cases* in which mental events make a causal difference to other mental and brain events.[9]

My hypothesis is simple: where there is mental causation, the organismic NEC pattern of the moment has been produced by the CNS out of all its possible states *at least in part because* it yields or is necessary for a mental state and its content to arise. This means that the CNS, which is itself a complex living system, has selectively produced the NEC state because of what the state yields or "codes" for. "Selectivity" here implies not environmental selection of heritable mutations, nor an intentional process, but the production of a particular component state or event by a biological system because of the former's function in that system. The CNS has produced a neural state because of its mental content or "supervenient" mental state. That is the core idea.

This hypothesis assumes three things which would have to occur to make such a phenomenon happen, realistically, in my view. First, such coding would have to have been established prior to the causal instance. The CNS has had to have learned to respond to some neural state by producing a subsequent state because of what the later state codes for. Presumably for many of the mental states of neurologically complex animals learning plays the largest role, albeit on the basis of some innate linkages. Second, this presumes downward causation in the operation of the CNS. By that I do *not* mean a neural state being caused by a mental state (that would be begging the question), but the firing of some pattern of neurons being caused by some property (e.g., an action program or learned behavior) of the CNS/organism as a system. Third, the ability to control or organize a sequence of neural states through or by their mental content, where the content plays a role without which the NEC sequence would not occur, must have bestowed naturally selectable advantages on such organisms in the past. I am suggesting that the most complex material systems we know in the universe, the living central nervous systems of neurologically sophisticated animals, sometimes selectively produce an NEC pattern because it yields or codes for an intentional event with intentional content.

Let us put this in the usual diagrammatic terms, recognizing that they are idealized abstractions from complex embodied neural and mental processes arising in an interaction between an organism, with its learning history, and its environment (Figure 8.2) Let M's stand for mental states and N's for some NEC patterns produced in the CNS, their subscripts indicating their place in a temporal series. We assume N_x is necessary for M_x to occur. These subscripted states may refer to equivalence classes of states, i.e., N_x may be not one but a class of states that have the same relevant properties. So M_x may be "multiply realizable" by an N_x class of states. (Multiple realizability does not affect the exclusion problem.)

Now the question is, how can, say, M_2 play a causal role in, make a non-redundant causal difference to, the sequence $M_1/N_1 \rightarrow M_2/N_2 \rightarrow M_3/N_3$? It could if the brain, in response to M_1/N_1, produced N_2 *because* N_2 yields or codes for M_2. All else being equal, N_2 would then not have occurred if it did not yield M_2, and not occurring, could not cause either N_3 or M_3/N_3. The intentional object or semantic content of the mental state would thus be a causal difference-making part of the ensemble M_2/N_2. Whether we face a mental-to-mental or a mental-to-physical case

$$\frac{M_1}{N_1} \longrightarrow \frac{M_2}{N_2} \longrightarrow \frac{M_3}{N_3}$$

Figure 8.2

makes no difference, since we already assumed M_x cannot be present without N_x. This is to say the CNS must have learned upon "reading" one of its states N_1 to produce a brain state N_2 that codes for a mental event M_2, after which N_2 causes the production of N_3. It doesn't matter that the M's are not "physical." What matters is that the brain can produce N's that have associated M-properties, can recognize the M-properties, and can learn to produce some phases in a sequence of N's because of those M's. This would mean that minded organisms have gained *the capacity to organize behavior through M-content*; a series of M's becomes the way the brain organizes a sequence of N-states, hence behavior

This can be depicted by a modified version of the above diagram (see Figure 8.3). The CNS produces N's (solid vertical arrows) which yield M's (dotted upward vertical arrows). Each N acts as a triggering cause of successive N's (solid horizontal arrows). But on the basis of prior learning by the CNS, M_2 has become a necessary condition (dashed downward vertical arrow) for the production of N_2. All M's and N's are produced by the relevant CNS systems, which downwardly cause N_2. Note that this might be far more complex still: these phases may be embedded in a larger process in which the N's coordinate a series of other CNS states, particularly effector or behavior states, to which M_2 would thereby be causally relevant.

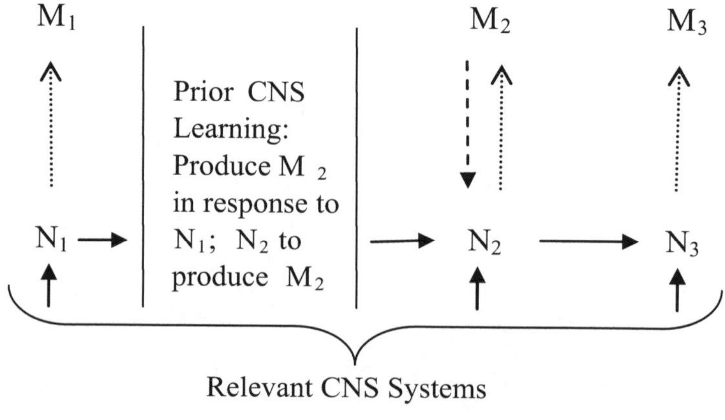

Figure 8.3

One might object that the effect of M_2 is still redundant, that N_1 remains "the" cause of N_2. To be sure, if it were true that N_1 were causally sufficient to produce N_2 without M_2 playing a role, then N_1 would indeed be *the* cause and any causal role of M_2 would be redundant. Certainly no M_x could make a causal difference to future mental or physical states unless it itself arises from and depends on some N_x. But if the brain would not respond to N_1 by producing N_2 *unless* it were also true that N_2 yields M_2—hence potentially to the N's which follow it—then M_2 would be making a non-redundant difference to the occurrence of N_2, and N_1 would be *a*, rather than *the*, cause of N_2, a necessary but insufficient condition for making N_2 occur.

Making M_2 a difference-making part of the "ensemble M_2/N_2" may appear to invoke a "dual-aspect" theory in which mental and neural ("physical") states are coupled as parallel processes. The similarity here is that both M_2 and N_2 play a causal role. But in the current model the M's and N's are not causally segregated into parallel processes. M_x is nomologically dependent on N_x, and N_x can be considered a cause of M_x in the sense that M_x cannot be present without it. But N_2 is also being caused by M_2 in the sense that N_2 would not occur without its link to M_2. It would also be true to say that the central nervous system causes both M_x and N_x in response to some internal or external stimulus.

As noted my hypothesis does imply downward causation on the part of the CNS. But no more so than what biologists commonly assert. The amoeba selectively responds to a chemical gradient, and the cell, whose proteins are produced by its chromosomes, selects which of those chromosomes to repair. The homeostatic self-regulation of an organism leads organ systems to produce a response to a departure from homeostasis (e.g., sweating in response to a rise in temperature) because that response will ameliorate the departure. The presence of an infectious agent in a human organism leads to the production of white blood cells that attack the agent; the agent is a cause in the sense that the white-cell production would not have occurred without it, but so is the organisms' bone marrow, the organization of its immune system, its current chemical sensitivity, and the learning history on which the latter is dependent. Just as the immune system learns selectively to respond to an agent by producing a type of counteragent, where there is mental causation the nervous system's production of N_2 as its response to N_1 is dependent upon N_2's yielding or coding for M_2, meaning M_2 has made a causal difference without which, all other things being equal, N_2 would not have arisen.

My hypothesis bears similarity to Fred Dretske's theory that "structuring causes" can explain how a meaning can make a difference to a physical process (Dretske 1988, pp. 41–43). Suppose that in a thermostat a flexible metal strip's movement C "triggers" a furnace shut-down M because of the strip's behavior in response to a rise in room temperature F. C is the "triggering cause" of M. But what caused C *to cause* M, what established their connection, is the engineer's design, the "structuring cause." The structuring cause recruited C as an "indicator" of F, allowing C's *representation of* F to become causal. We could also say the structuring cause assigns to C the function of representing F. Notice that in this case, C's representation of F is a material or physical relation established by the configuration of the thermostat.

Turning to ethology, Dretske points out that behavior is a process, not mere events of bodily movement. Among rats in a maze, associative learning reinforces the behavior of finding the reward, not specific bodily motions or pacing, which vary between trials. The rat does not learn to reproduce a particular set of muscular movements in a particular order and rhythm; the rat learns how to get to the reward. Structuring causes explain holistic behavior, not just movement. A rat may then be trained to press a bar M that releases food F upon hearing a tone C. C is thereby recruited as an F-indicator. It "acquires the function of indicating" F (Dretske 1988, pp. 84). *That* it represents F plays a causal role in making behavior M happen. C has a causal role in virtue of the fact that it represents F. Notice that in this case C's representation of F is something the rat has learned and neurologically stored. What has been stored is both the significance of C and how to get to F. As Dretske puts it, "Learning of this sort mobilized information-carrying structures for control duties *in virtue of the information they carry*" (Dretske 1988, p. 99, his emphasis).

My suggestion is different from Dretske's, but it is similar in asserting that: a) there is a way that a meaning or representation can make a causal difference in a succession of physical states; b) this is made possible by a prior set-up, or "structuring cause;"

and more generally, c) the place to look for a model to explain how mental content can make a difference in a succession of physical states is the biology of animal learning. In the current case, M_2 (or any M) is or includes a meaning, that is, it has intentional or semantic content (e.g., the idea of the number 2, the visual image of a predator, a feeling of hunger). It arises from or is produced by N_2. The "structuring cause" is now twofold: a) the CNS has recognized or registered the fact that N_2 yields or codes for M_2; and b) the CNS has learned to produce M_2 in response to N_1. Thus N_2 has been recruited as an M_2-indicator, and M_2 has become the preferred response to N_1. Now, given the "triggering cause" N_1, the CNS produces N_2 because it yields M_2. All neural states are produced by the CNS. Where the CNS generates neural state N_x because N_x has semantic content M_x, it has used M_x, a semantics, as the means by which it organizes at least some members of a chain of neural states. If so, M_x has made a causal difference to what neural state is generated. Mental content has become a difference-making part of the causal event.

Several criticisms have been leveled at Dretske's model, only some of which would apply to my hypothesis.[10] It has been objected that it does not give the kind of immediate mental causation we intuitively find in our experience, and that the mental cause remains superfluous to the chain of "physical" or neuro-biological causes, so the mental is still not a cause.

We must admit that causal relations among neural states, mental contents, and acts, however theorized, must be largely invisible to personal experience. Neural processes, like many biological processes, are far too fast and complex for us to expect them to match our intuition. There are at least three different levels of relevant causality that would have to be involved in mental causation (even ignoring interactions with environment at the level of the whole organism, i.e., its "acts"). First, the CNS, as a living (organ) system of living systems (neurons), is teleonomic and capable of downward causation upon individual cellular events. By "teleonomic" I mean the system is genetically designed to adjust to a variety of environmental and somatic changes in such a way as to maintain or achieve some state. Second, some CNS states have associated intentional properties (the M's, above), however wide the set of CNS or organism-environment interactions on which they depend.

Our question is whether these can play a causal role. Third, in my hypothesis, part of what the CNS does in its downward causation on its individual states (point one), is sometimes to produce NEC states because of the M-properties arising from them (point two). This recruiting of NEC states must be mostly learned. That is not surprising. The CNS, particularly the brain, is *par excellence* that biological system which is genetically programmed to learn.

Of course, presumably most mental states are not causal at all, that is, accompany neural states without making a causal difference, just as most of my neural events do not generate mental states in the first place. And there are mental states that do not intend somatic or environmental events at all, e.g., fantasy and dream states. The reality of mental causation only requires that sometimes, in various, occasional action sequences, there be moments where a mental state plays a causal role for some short period of time, thereby changing the neural-behavioral events that would obtain without it. Some neural-behavioral processes that occur with non-causal mental accompaniment in a minded organism's contemporary behavior may have required contributions from mental causation at some earlier point in its history of learning before later habituation. How all these different kinds of processes actually intertwine might be one of the most complex issues in science, what Wimsatt called the bio-psychological "causal thicket."

We are now ready to define teleological—*not* teleonomic—behavior as behavior in whose guidance a mental content makes a causal difference. If a feeling or image or idea makes a difference to what the organism does, then we have teleological behavior. This must be the case in many nonhuman animals. In humans some of these cases are desire-and-belief states, but in nonhumans we cannot be so sanguine about the ascription of beliefs (as we shall see). Once we have accepted nonhuman mind and mental causation, it becomes impossible to deny that, say, the deer's visual, auditory, and olfactory images of food, predator, or mate, and its fear or desire play a causal role in guiding the behavior that ensues. How far down the phylogenetic scale teleological causation goes is unclear. But if this account is valid, we can say that in at least some cases embodied and centralized neural teleonomy can create teleology. This should not be surprising. For minds, their properties, and their causal role are parts of nature.

9

Meanings of the Cultural Mind

It would be against the spirit of the current work to say with Alexander Pope that the study of man [sic] is the proper study for mankind. But it would be true to say it is the most controversial. One measure of the causal thicket of human experience and behavior is the number of different disciplines studying humanity—history, economics, political science, psychology, psychiatry, sociology, biology and the medical sciences, cognitive science, anthropology, linguistics, art and literary history, cultural studies, and large parts of philosophy—all with multiple ways of understanding us. Moreover, in the modern era the parties bent on inflating humanity's self-conception, and those bent on deflating it, have been in constant struggle. Mind, consciousness, soul, reason, free will, morality, justice, art, technology, God's likeness—all have been invoked as the unique character of the human, then from opposing quarters either deconstructed or ascribed to nonhumans. As long as zoology and ethology continue to advance, we will ever find animal species to be more intelligent and remarkable than we knew in the past. Pursuing these advances, some greet any claim to human uniqueness as a remnant of a Victorian mentality. In the opposite corner any discussion of human animality is taken to be threatening.

But if ideology recedes enough for us to credit the evidence of both science and our senses, there is a clear answer to the question: Are we like or unlike animals, are we animals or are we "exceptions" in nature? That answer is: *yes*. We are like and unlike. We are exceptional animals. Aristotle knew this 2,300 years ago. Since Darwin we have known we are a species of primate, and share traits with our primate ancestors as well as other animal lineages. It is a category mistake to treat "human" versus "animal," or the related "culture" versus "nature," as concepts of equal level, rather than as species and genus. And we can accept with contemporary ecological ethics and animal welfare theorists that views restricting moral value to the human domain betray an unthinking, and historically narrow, prejudice. But once we abandon any psychological or ideological need to justify a human superiority to nature, it remains impossible not to notice how remarkably distinctive we are. The complexity of our brain, the depth of our cognition, the inventiveness of our communication, and the extent of our artificial constructions swamp those of nonhumans. If we were to discover an ape or cetacean cooking, dressing up, recording stories in pictograms, or worshipping an icon of King Kong or Flipper, would we have to revise our understanding of apes or cetaceans? If so, then the absence of such a discovery is likewise significant. The human difference is real, even if our understanding of it will continue to evolve, or as Peirce would say, "converge" for an indefinitely long time. We cannot avoid guessing about that difference, for we cannot conceive humanity without differentiating it from other living things, and we cannot live as humans without a conception of humanity and the typical capacities we expect a human to have.

The problem with the question of *how* we are distinct is troubling, not because of ideological concerns, but because it is hard to answer. Indeed, the question itself needs clarification. Every biological species is different from every other. To ask for the *specifica differentia* of a species is usually to ask how it differs from other species of its genus. But when we ask about the human difference we are seeking how we differ from *all* other Earthly species, because the difference between us and all others seems a different order of difference. This does not mean the difference between a human and a chimpanzee is "greater" than the difference between a chimpanzee and a protist. It also does not mean there is *one* difference. We are distinct in many ways; compared to other primates, for example, we sweat a lot and have smaller canines.[1] What we seek in the following are the core determinants of our most remarkable mental and/or behavioral differences from all other species. But even here there are many candidates for "the" core human difference, currently lying at the crossroads of anthropology, paleontology, primatology, neurology, cognitive science, developmental psychology, linguistics, and philosophy of mind. We cannot begin to do justice to this discussion. It would, however, be particularly striking to find a candidate core human feature that cuts across and plays a role in the genesis or expression of *many* of the candidate features, if such can be found. Does some human capacity link to or enable many of our distinctive mental, behavioral, and linguistic traits, forming a thread that runs through or enables the whole nexus?

Some contemporary research suggests there is. The capacity is called *mind-reading*, the ability to recognize and attribute mental states to conspecifics, which in the human case seems to develop through the phenomenon of *joint attention* between infant and caregiver. Interestingly, it is tantamount to what the American philosopher George Herbert Mead long ago called the ability to take the perspective of another. This is to say that a unique kind of social relating is central to our uniquely human mental capacities. It is a prime candidate for *a*, if not *the*, core human difference. My proposal will be that Mead was essentially right, with some modifications.

After a survey of human origins, Section II will present my argument for Mead's claim, and Section III a metaphysical discussion of *meanings*, which I claim humans alone cognize and manipulate, followed by a discussion of our uniquely cultural nature in Section IV. But before we go any further we must remind ourselves, as they say, of where we come from.

I. The Evolution of Homo Sapiens

Paleoanthropology is an evolving science (pun intended). Not only is our understanding of the evolution of human beings and their ancestor species rudimentary, even our picture of the chronology and identification of those precursors is sketchy and subject to dispute. To cite a novel proto-human fossil at a time and place tells us only that by that point in the past a primate has differentiated, or branched off, from its ancestral lineage. How long it had existed before that, how long after, whether it evolved further, where else it existed (if anywhere), whether it is a new species, a local subspecies or population, typical of its species/population or an outlier on the normal curve of developmental variation—all this requires complex interpretation and investigation. There are plausible alternate hypotheses of the lineage of Homo sapiens and the stages of their evolution. I will discuss only the broadest parts of the standard current account.

The first clarification is that we did *not* evolve from apes. We are, indeed, primates. Primates fall into four categories: *prosimians* (e.g., lemurs, tarsiers, bush babies); *monkeys* of the new world (e.g., spider, capuchin and howler monkeys, and marmosets) and the old world (including baboons and macaques); *apes,* both "lesser" apes (e.g., gibbons) and "great" apes (gorillas, chimps, bonobos, and orangutans); and *humans.*[2] Homo sapiens evolved directly from an earlier species of the genus Homo, which evolved from yet earlier Homo species, which evolved from earlier non-Homo "hominin" species, which evolved from an unknown primate precursor we shared with chimpanzees 6 to 8 million years ago, which evolved from an unknown common ancestor of all the lineages of existing great apes long before that.[3] There are 6 million years of evolution separating us from apes. Many hominin species stretch over those millions of years; the past is littered with them. Homo sapiens shared the Earth with several, some only recently dying out.[4] Many of them are not our ancestors; we evolved from some and not others. But all of them, with whom we did or would have shared more than with any existing nonhuman animal, are extinct. Only we Homo sapiens are left.

Furthermore, while hominins are, and evolved from, primates it is misleading to think of humans solely as a primate product. Long before Homo sapiens arrived on the scene that scene had changed from the one occupied by primates. While we evolved from hominins who evolved *from* primates, we did not evolve *like* other primates. It is the environment which does the natural selecting. The environment that played a central role in selecting Homo sapiens' distinctive traits was unlike the environment in which any ape has ever lived. We evolved more *like* large mammalian herbivores and omnivores because we shared their environment, even though that environment was, in our case, operating on the genotype and phenotype of a primate.[5]

In the most common current understanding, there are four broad stages in this evolution. They are: 1) pre-Homo hominins that branched off from ape ancestors, for example Australopithecus afarensis and others (cf. note 7); 2) followed by the earliest Homo species, one example of which is Homo habilis or "handy man," the tool-user; 3) then a more advanced Homo stage, one of which is Homo erectus; and finally, 4) Homo neanderthalensis and we Homo sapiens. Except for Homo sapiens, each of these is really one type of fossil remains from one region or place which represents a family of similar and similarly aged remains from elsewhere. How close we are to them, especially to the Neanderthals, remains controversial.[6]

Australopithecus (and whatever precursors it had) evolved from a common ancestor of humans, chimpanzees and bonobos in northeast Africa at 6 to 5 million years ago (mya), early in the Pliocene geological epoch.[7] At some point reaching 70 pounds with an average 500cm^3 brain Australopithecus had left forest for savannah, meaning grasslands with some trees, achieving bipedalism. The next stage, the origin of our genus Homo, came in the yet drier environment of steppe, grasslands without trees, around 2.5 mya in the late Pliocene, where Homo habilis began using stone tools (making this the beginning of the Paleolithic cultural age), with brains at 550cm^3. We and our ancestors are the only primates to have made the move to a treeless environment. By about 1.8 mya at the time of the first major glaciations, the start of the Pleistocene, a new Homo species arose, erectus or ergaster. Its development was remarkable. Over its eventually more than million-year sojourn it achieved a 1250cm^3 brain, double the size of habilis'. It is believed to be the first Homo species to spread across Africa-Eurasia, some thinking it migrated or expanded from Africa, others that it may have emerged in multiple regions. (The debate over the "Out of Africa" hypothesis usually refers to the much later migration/expansion of Homo sapiens.) Here we find systematic communal hunting, including stalking and use of spears, seasonal tracking of game migrations, bifacial stone tools, and the control of fire.[8] Erectus mostly died out with the massive Riss/Illinoian glaciations around 200,000 ya, although some remained in the lowland tropics of South-East Asia until 50,000 ya and on South Pacific islands until a mere 15–25,000 ya.[9]

Homo neanderthalensis emerged sometime around half a million years ago, before 200–250,000 ya when we Homo sapiens arose. Brain growth continued to reach about 1500cm^3 for both around 80–100,000 ya. Neanderthals were physically more "robust" or massive, we more "gracile," and had a slightly *larger* brain mass-to-body mass ratio than we do. They were more prevalent in wooded and wetter northern Eurasia, we in the savannah-steppe of northeast Africa, but we both traveled and sometimes jointly inhabited regions until the Neanderthals died out about 25,000 ya (Finlayson 2009). Here multiple controversies loom. First, recent work has raised the serious possibility that we two were one species: it is evident that Neanderthals shared the FOXP2 gene associated with spoken language, and Neanderthal DNA has been discovered in living Eurasians, implying interbreeding at some point. Second, and much older, is the question whether modern Homo sapiens arose from a founding population of around 10,000 individuals in east Africa, then replaced erectus and Neanderthals around the world—the "Out of Africa" hypothesis—or evolved from erectus or some similar precursor in multiple locales—the "Multiregional" hypothesis. Also unclear is when Homo sapiens achieved behavioral modernity, the full panoply of language, creation of ornaments and artifacts, burial rites, etc. Jared Diamond called this the "Great Leap Forward," pegging it to the upper Paleolithic, about 50,000 ya (Diamond 1999). But many believe human language must have evolved over a long period, starting perhaps as early as the first appearance of Homo sapiens (and perhaps neanderthalensis).

At any rate, over a couple of hundred thousand years Homo sapiens eventually replaced erectus through the Middle East, northwest into Europe, and across Asia, including Indonesia and Australia, and the Neanderthals of Eurasia, by 50–20,000 ya. Rather than actual conflict and competition between hominin communities, this may have been the result of the continual climate oscillations in the form of at least five glacial expansions or pulses over the last 2.5 million years, some of which may have favored our tool kit and skill set over those of erectus and neanderthalensis (Finlayson 2009). After the extinction of the Neanderthals we crossed the land bridge into the Americas for good (hominins may have reached it earlier but temporarily) around 12,000 BCE, our "Clovis point" spearheads probably causing the extinction of the American megafauna (Flannery 2001, ch.13–15). Shortly afterward the Fertile Crescent branch of the family haltingly began to domesticate plants and animals, eventually to remake our, to that point, 200,000-year-old hunter-gatherer lifestyle.

We may summarize some key developments. Brain growth reached its apogee 80,000 ya. Homo sapiens (and Neanderthals) possess the largest brain for body size of any animal in Earth's history, with a particularly well-endowed cerebral cortex and neocortex. The Homo sapiens brain contains 10^{12} neurons, each neuron is connected with up to 1,000 others, together capable of producing 10^{130} different electrical patterns, making it the most complex material-physical system we know. Artifactual culture and ritual were present by 50,000 ya, but there is some evidence as early as 164,000 ya, so at some point in this interval we have what is called fully "modern behavior." We do not have direct evidence of complex languages before this period, because speech leaves no archeological trace. There is some evidence that regions of the cerebral cortex, Wernicke's area and Broca's area, as well as the FOXP2 gene, which are responsible for linguistic ability, evolved with Homo sapiens 200,000 years ago (and perhaps earlier with Neanderthals). In Homo sapiens we find the descent of the larynx, unique among existing mammals, creating a larger pharyngeal space for the tongue to be used to create multiple consonant sounds. Some argue this must have evolved gradually over 200,000 years, while others think language could be an evolutionary "spandrel," a by-product that was not selected but arose from other selections and only later came into service, or an "exaptation," selected for one purpose but only later achieving its present utility.

How did the human brain develop to such levels? Some factors are hard to ignore. Certainly Homo sapiens (and erectus and neanderthalensis) started out with broad evolutionary gifts. We must be wary of the tendency to think humans are only a mentally, not physically, gifted species, that we are slower than cheetahs, can't fly like birds, not as strong gram-for-gram as ants, etc. Such claims are true but misleading. We are robust generalists, large mammals capable, even without tools, of becoming jacks of all terrestrial, and aquatic, trades (unlike the great apes, we can swim). As has been remarked, no other animal can run a mile, swim a river, then climb a tree. Opposable thumbs make human hands perhaps the biosphere's best set of tools. Combine that with an omnivorous gastrointestinal tract and teeth good for both grinding and tearing, feet good both for running and quiet stalking, daylight color vision by eyes on a head held aloft by an upright stance freeing hands for manipulation and signaling, a larynx capable of subtle vocalizations and mimicry among our neo-primate, highly sociable natures.

Homo sapiens are not weak animals with a big brain; we are big, strong, multiply-talented social animals with a big brain. Not only our minds, but our bodies, were born to innovate.

But there must always be an environmental cause. What led to these steps in evolution? One factor seems to be climate change and its progressive exploitation by Homo sapiens. Cold and dryness seems to have promoted hominin evolution. There is evidence that the pulses of the Pleistocene ice ages did not merely produce high-elevation snow and ice, but drying across large areas of Eurasia and even north and east Africa. That means forest tended to become savannah, savannah steppe, and steppe desert. Essentially our precursors left forest for grasslands, that dangerous paradise of seeds and meat. Foraging for young plants, scavenging prey, and hunting for herbivores opens up an omnivores' heaven. But for this very reason grasslands are the scene of a relentless predator-prey game, all those carnivores seeking all those happy herbivores feeding on all those seeds and plants. They favor animals who see well, cover ground quickly, have fine hands and feet built for stalking. Crucial is the ability to survive predation at night without trees, which other primates cannot do (Geist 1978). It is possible that our gracile form was better suited to grassland hunting than the Neanderthals (Finlayson 2009). At any rate, our primate visual obsession with clues of social status was re-tuned to the objective signs of a vast taxonomy of plant and animal life in the steppe as cooperative members of a linguistic community. Grasslands were arguably the school for the human mind (Shepard 1973, 1998).

Another question is, having achieved robust body and brain size, how did we *keep* them? Species can develop a robust form while exploring novel environments, but diminish it in the face of the problem of efficient resource use among competing conspecifics in a stable environmental niche. Homo sapiens evidently *maintained a dispersal phenotype under maintenance conditions* (see Chapter 6). Geist suggests this required high-quality feeding for pregnant women and high-quality infant care, where multiple-births are rare, nursing (especially at night) delays the next pregnancy (lowering the number of infants per caregiver) along with strong male-female bonding hence male provisioning of the mother-infant pair, within highly cooperative small bands—all this accompanied by a method of storing novel information regarding infant care, tool-making, foraging, hunting, and social cooperation across generations, that is, *culture*. We had to remain exploratory social learners, even when staying put.

II. The Intentional Triangle

We can now see even more reason to doubt that any one thing made us human. Even if we were someday able to describe all the novel anatomical, behavioral, and mental traits of Australopithecus, early Homo, Homo erectus, Neanderthals, and early Homo sapiens we would be left with many distinctive traits, each with its own line of historical development. So what are we to look for? The hope is to identify a subset of our unique core trait(s) that likely played a role in generating or enabling many other core unique features, in effect, a nodal intersection in the developmental lines of multiple unique features.

There are some famous candidates for our core distinctive trait(s). We are *tool-users and makers*. This was once considered unique, but we now known that a variety of mammals and birds use rudimentary tools, even carrying them to the site of their use for the purpose. Our *intellectual abilities* are an obvious candidate; we have a capacity for discursive learning or "reasoning" that seems unique. We seem to have a distinctive kind of *agency*, being more open-ended or "free," and subject to moral evaluation. This may be related to some special *self-consciousness* (although here careful distinctions would have to be drawn from analogous nonhuman behavior). Certainly human *language* is far more complex than any nonhuman system of communication. We speak sentences composed of meaningful units (words) governed by syntactic rules of assembly, hence can form an unlimited number of sentences (what is called *recursivity*). How much of, and in which respects, human language could be exhibited by nonhumans is unclear.[10] Last, our use of signs extends to *culture*, which we create and transmit across time (although some claim evidence of generational transmission of local cultural learning among nonhumans, e.g., a troop of macaques in Koshima, Japan).[11]

There is an additional capacity that may be presupposed by several of these traits, and one highly discussed in the recent literature, *mind-reading*. This is the ability not merely to recognize and predict a conspecific's behavior—which many species must do—but to attribute mental states to the other. This can often be observationally and experimentally discriminated from mere behavioral prediction. A crucial part of our human equipment, it can be dissected into the attribution of perceptions, attitudes, goals, and false beliefs, which are acquired by human children at different ages. Mind-reading is explained by two rival theoretical perspectives, "theory of mind" and "simulation theory," the former claiming attributions presuppose a folk psychological theory or set of concepts about the mind of the other inferred from behavior, the later claiming the attribution is not primarily conceptual or inferential but an internal simulation on the part of the attributor called "empathy." The issue was made comparative by Premack's and Woodruff's 1978 essay, "Does the chimpanzee have theory of mind?."

Many mammals remember objects in spatial locations, follow and anticipate object permanence, even when unobserved, categorize objects on the basis of perceptual similarities, and use insight in problem-solving. They cognize or "know," as we will discuss in Chapter 10 (Griffin 2001, Kornblith 2002). Many recognize fellow troop members, kin, and dominance relations to self; predict behavior based on emotional signals; use social communication; cooperate in problem-solving; and learn by observing others (Tomasello 1999). Nonhuman primates are additionally able to recognize "tertiary" relations—that two other primates are related, e.g., submissive to dominant—as well as analogous relations—x and y are related like b and c (Tomasello 1997).

However, based on a mammoth survey of primate testing, Michael Tomasello claims nonhuman primates do not recognize the *intermediaries* of intentional or causal relations, they do not understand that the relation of antecedent and consequent, or stimulus and response, may entail an intermediate causative force or agent intention, hence the result might be preceded by a variety of antecedents (Tomasello 1997). They do not ascribe intentional states, or at least complex intentional states, like belief states, to others. He notes the host of simple, nonlinguistic communicative acts that apes cannot perform: "In . . . natural habitats nonhuman primates: do not point or gesture to outside objects for others; do not try to bring others to locations so that they can observe things there; do not actively offer objects to other individuals by holding them out; do not intentionally teach other individuals new behaviors" (Tomasello 1999, p. 21).[12] And there appears to be an affective dimension to all this. Jane Goodall herself, after many years of experience with chimpanzees in the wild, wrote, "I cannot conceive of chimpanzees developing emotions, one for the other, comparable in any way to the tenderness, the protectiveness, tolerance, and spiritual exhilaration that are the hallmarks of human love . . . [this] may represent the deepest part of the gulf between them and us" (van Lawick-Goodall 1971, p. 194). Might there be a kind of social relating, with both cognitive and affective features, that forms a key difference between ourselves and our closest living relatives?

A century ago G. H. Mead argued against the current of most of modern philosophy that communication is logically and temporally prior to mind. A human mind is a set of activities which emerge through social interaction, rather than the other way round. His most famous innovation was "significant gesture." Gestures are the communicative behaviors that animals produce to enhance the process of "mutual adjustment." One organism responds to another's act, or to an environmental condition, by a signaling device that changes the situation communicatively: A, rather than biting B, barks or growls; rather than running away, B shows a submissive posture. The gesture is the initial phase of an act (e.g., growling before biting) which functions to call out a response on the part of another organism (Mead 1962, p. 44). Many animals, including humans, communicate by using gestures.

But, Mead argued, humans alone also engage in *significant* gesture, producing a gesture that has become or functions as a sign. For Mead a significant gesture is still the initial phase of a developing act, but now it means or stands for the later phase. This grants the gesture objective status for others and the self or gesturer as a part of the environment, whereas for the nonhuman animal the gesture is "invisible to the gesturer, as it were, because it is merely part of his or her physiological or neurological state" (Baeten 1999). Mead explains the way the sign means with the example of coming upon a bear's fresh "footprints" in the woods. A dog sniffing the print feels fear, in effect is afraid *of* the footprint. A human takes the print as a sign which calls up a host of aspects of the bear and the potential situation which play a further role in the human's response (Mead 1962, pp. 120–21).

Mead's essential contribution is his account of what is required for significant gesture to arise. Gesture as sign is only possible if A responds to its *own* gesture *from the perspective of B*. It does so "implicitly"—we might say, "out of gear"—rather than explicitly. This also means A must regard herself *as an object* from the viewpoint, or attitude, of B.

> Gestures become significant symbols when they implicitly arouse in an individual making them the same response which they explicitly arouse, or are supposed to arouse, in other individuals, the individuals to whom they are addressed; and in all conversations of gestures within the social process . . . the individual's consciousness of the content and flow of meaning involved depends upon his thus taking the attitude of the other toward his own gestures. (Mead 1962, p. 47)

Mead went on to analyze human play and games as the venues in which we are trained to occupy and shift among the roles and standpoints of others, thereby viewing the self and its gestures from the perspectives of multiple individuals. This eventually allows us to act in terms of a "generalized other," the representative of any interlocutor as such in our own thought. All acts embody perspectives, and all perspectives, or potential forms of action, select out features of the natural world as an organism's environment. The status of environmental features for perspectival acts is objective but relative to perspective; e.g., acorns are objectively food, but only because there exist some animals who eat them. Mind *is* the process of significant gestures, and the self is the organization of a human organism's set of attitudes toward environment, and toward itself from environment, as expressed in significant gesture.

While I argue Mead is basically right, qualifications of his view are necessary. Mead regards the following list of attributes as unique to humans: a) significant gesture; b) mind; c) recognition of the "objective" status of a gesture (of another or self); d) ability to take the perspective of the other on the self; e) ability to cognize meanings themselves or possess or have "ideas;" f) awareness or even possession of a self; and g) the ability to produce and consume cultural objects or activities. This list is the right place to look, but there are at least two problems with it. The first we already know: there is no reason to restrict mind or consciousness to humans. The second is whether significant gestures are uniquely human. Determining the difference between a nonhuman social organism emitting and responding to a communicative gesture, and its treating that gesture *as a sign* in Mead's sense, is not easy to discern. However, today it certainly appears that a few primates (e.g., at least Washoe, Koko, and Kanzi, cf. note 10) have acquired significant gesture through human training. If this is true, then Mead's significant gestures are by themselves insufficient to distinguish the human and their relation to the other entries on the above list is more complex than he allowed. We will return to this.

Strikingly, recent research on early childhood bears comparison to Mead. Separately, Tomasello and Peter Hobson, drawing on their own studies and those of others, point out that in the first few months of life the human infant is frequently drawn into dyadic, affect-laden, facially mimicking "protoconversations" with adults. Hobson makes the point that this begins with the biological-emotional tie to the caregiver. Emotion and cognition develop together in the "musical" rhythm of the protoconversation, where the infant finds that another's attitude and act answer its own. Nonhuman primates do not engage in this intense face-to-face relating; among primates it is new with us (or earlier hominins). Indeed, failure of the protoconversation due to the nonresponsiveness or "still face" of the caregiver, has been shown to generate problems for the infant. Through a series of studies over the last four decades, Tronick and others have analyzed early infant-caregiver interactions in minute detail, involving both normal and depressed and pathogenic caretakers (see e.g., Tronick 2007). Failure of caregiver responsiveness is immediately noticed by very young infants, and they respond with changes in behavior. There seems to be a genetically inherited pattern of communicative turn-taking in infant behavior. In his "Mutual Regulation Model," Tronick argues the physical and emotional homeostasis of the infant can only be managed dyadically, that is, socially, leading to what he calls "dyadic states of consciousness." Just as we know that humans are uniquely "neotenic," born in an unusually dependent state that lasts for years, we seem genetically constructed to require social interaction even to maintain our *internal* state within tolerable parameters, at least during infancy.

Then at nine to twelve months infants begin to interact triadically in events of *joint attention* or joint collaboration regarding some object.[13] The adult introduces a toy. The child looks where the adult looks; acts on objects the way the adult does and toward the same goal; engages in dialogical turn-taking in collaboration or vocalization; uses "declarative" (rather than imperative) deictic gestures like pointing and holding objects up to be seen or handed off; clearly attributes to the other an intended goal along with differentiation of means from goal, hence recognizing alternative choices of means as intermediaries; and engages in "role reversal" imitation, using symbolic objects toward the adult as they have been used by the adult toward the child.

Crucially, the infant is able to take the perspective of the other, including her attitude. For example, when the caregiver introduces an initially distressing object—say, a monkey toy—the child first looks worried, then looks at mother. Mother smiles, amusedly handling the toy. Then the child smiles and handles the toy. The object has become acceptable by virtue of the child's taking up the mother's attitude toward the object. Hobson calls this early identification the "Copernican Revolution" of human mentality (Hobson 2002, p. 73). The object comes to be recognized as a single entity that has multiple meanings for self and for the caregiver. Some objects of joint attention become, Tomasello claims, "communicative devices" loaded with "communicative intentions." Joint attention is the beginning of *shared intentionality*, which Tomasello conceives through the empathy model. He suggests that the human child is hardwired to be motivated to engage in these activities; joint collaboration is certainly affect-laden and pleasurable. While nonhuman primates are capable of attributing perception or knowledge and goals to another (cf. note 12), *shared intentionality appears uniquely human*.

We should mention in passing that recent neuroscience has shown just how "hardwired" empathetic mind-reading can be. While we undoubtedly make inferences using concepts or "theories" of the other's mental contents, the fortunes of the empathy hypothesis have been recently bolstered by the discovery of *mirror neurons*.

When observing an act or a facial expression, neurons in the areas of the observer's brain that would fire if the observer enacted the expression herself, become excited. When I hear or see evidence of a mental state in another, my brain reenacts that state in an out-of-gear sense. So there is a scientifically recognized neurology of an internal modeling of the observed dispositions of others. Now, joint attention and empathy do not account for everything; surely cognitive development, social learning, and inferencing contribute to the growing child's ability to explain behavior. But it seems that joint attention and our endowment of mirror neurons are central to what will eventually become a normal mind-reading human individual. It seems that, as Karsten Stueber remarks, "Nature does not have the problem of other minds" (Stueber 2006, p. 142).

Returning to joint attention, we can depict it using a diagram of Hobson's, reproduced below with my modifications (Figure 9.1). As Hobson pointed out, in the infant-other-object relation the infant and other each attend to an object (A, F), to each other (B, C), and the child attends to the other's attitude toward the object (D). (Hobson's original diagram consists of the lines that in my diagram are labeled A, D, F, with one double-headed arrow in place of my B and C. His arrows are unlettered.) The infant also attends to *how* the other attends to itself (E), and the other attends to how the child attends to itself (H) and to the object (G). (The other's

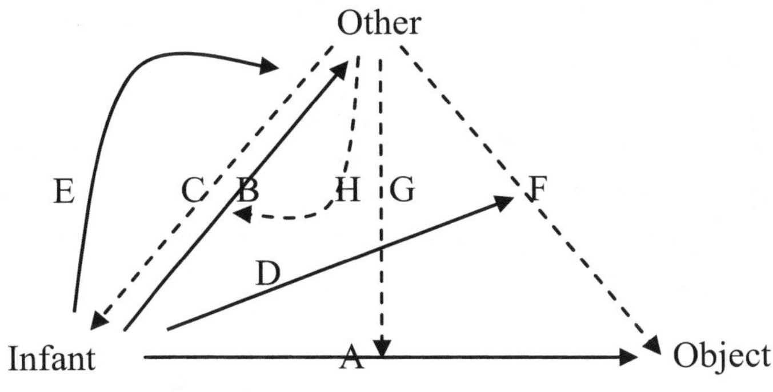

Figure 9.1. The Intentional Triangle (Modified from © Peter Hobson *The Cradle of Thought: Exploring the Origins of Thinking*, Pan Macmillan 2004 page 107, fig. 2)

attentions are dotted arrows, the child's solid.) The result is the child adopts the other's attitude (F) toward the object. This three-cornered relationship may be called *the intentional triangle.*

Let us analyze these relationships a bit more. To attain or incorporate the perspective of the other, the child does not merely see the object as it would from the other's position in space. Using the empathy model of mind-reading, it must be able to imitate or generate and hold the other's attitude within its own experience while not ceasing to have its own experience. The child must hold in one act of awareness, or switch quickly between, attention to: the object (A); the other (B); the other's attitude toward the object (F); and presumably the other's attitude toward itself (C). Taking this complex from the child's point of view, given an environment charged by the other's cognitive-emotional enabling presence (C, G, H), A plus B seems to be the basis or initiation of the experience, but then the child's D, perhaps with the help of E, allows the child *to incorporate F either into A or along with A* as an alternative to it. Hence, given the other's enablement, (A,B) + (D,E) → (F + A).

Later, at 18 to 24 months, symbolic play develops at the same time as early language. The child can pretend, say, a spoon is a car. This means the child can hold both meanings together, and begin to name them. Words serve as signs—communicative devices—for perspectivally specific selections of objects in the world. Calling a hat a "hat" is to take one of the attitudes toward it which pick it out; pretending it is a "house" is to take another. Pretending cannot happen unless the child can distinguish its perceptions from the play-meaning; if not, the spoon would really be a car, which is a mistake or a delusion. Also the meaning of the word used has to be "reversible," the same for the child and the other. Hobson considers this the beginning of thinking, that is, the ability to call up the sense of something *not present*. It is also the beginning of the development of a self for Hobson, since self arises as an entertainer of perspectives. "A person needs a self in order to think," he suggests (Hobson 2002, p. 206). Eventually, at four years—as most agree—the child is able to ascribe mistaken beliefs to the other's mind, as well as make distinctions between appearance and reality. The later, Hobson suggests, is a generalization of the perspective of others, the perspective anyone would take if they were in the right position to judge, the perspective that trumps others (Hobson 2002, p.146). The child's cognitive and sign-manipulative ability, coupled with the fuller learning of language, eventually makes it part of that uniquely human product, what Tomasello calls "cumulative cultural evolution," systematic teaching of the young and hence accumulated learning over multiple generations.[14]

Reviewing development up to 24 months, we can distinguish four apparent, mutually implicated achievements (not in temporal order). First is *recognition of the other's attitude*, or attribution of an attitude to the other, presumably through an empathetic internal modeling of the other's perspective, holding it along with its own. For the child at any moment already has an attitude of its own, so to recognize the other's as the other's it must model that attitude while holding it distinct from its own attitude. Second is the *incorporation of the other's apparent attitude toward something into self*, or "taking on" the attitude of the other on some external object or on the environment.[15] Third, this can hold as well for an act of the child, when the child takes the other's attitude toward its own act, hence on *a gesture taken to mean the same for other and self at the same time*. These are Mead's significant gestures, loaded with Tomasello's "communicative intentions," in short, something functioning as a sign. Fourth, all this implies or enables an object, the other, and the self to be *recognized as single entities capable of multiple jointly recognized meanings*. The spoon can be a spoon in one perspective, a car in another, manipulated in terms of its meanings, re-created or awarded novel roles by an agent and/or other, while not losing its common objective status as a spoon.[16] Likewise self and other are capable of multiple jointly recognized roles and perspectives—e.g., playing mom or dad or doctor—while yet remaining the same object.

Mead's notion that human mentality emerges from the ability to take up the standpoint of the other coheres with this account.[17] There seems to be a distinctive human form of mentation which does not merely involve or require communication in the coordination of activity, but is itself communicative. The human brain, coupled with caretaker interaction and culturally inherited language, has managed to *socialize animal intentionality*. Not only does the individual depend on the group for survival and satisfaction, as in many other species; not only does the individual spend much of its time concerned with social others and their relations, as in a few others. Our very thought process and self are social. For *the others are in my head*, in my internal sphere as well as in my public practices; my mind represents them, I think from their perspectives, I occupy their roles, and I have a new possession, a self, which emerges out of its relations to them.

The intentional triangle touches on all the candidate core human capacities mentioned earlier. First, to attend to one's body, behavior, and attitude from the perspective of another, to merge this with one's own experience of one's body, feelings, images, and thoughts, and to distinguish multiple roles for or perspectives on one's body and behavior, is to be conscious of a self. Eventually this leads to Damasio's autobiographical consciousness, which requires that the self own its narrative history. This self owns its past and its anticipated future, throughout its various roles and perspectives, enabling us to experience ourselves as temporally spread in a way different from other creatures.

Second, this throws light on one of the most difficult topics. When coupled with language, self-consciousness endows us with the novel ability to treat the events of our own mental processing *as objects* and manipulate them, stopping a flow of mentation long enough to select a thought, feeling, or image as a distinctive object. We can label them with signs, usually linguistic, and occasionally direct the process. As noted in the preceding chapter, some use "thought" for any propositional mental content, or "thinking" for any mental state whatsoever. But, as suggested, using "mental processing" for the relating of mental states and contents in general, whether feelings, images, or ideas, allows us to reserve "thinking" and "thought" for something narrower.

Hobson describes a bout of his own thinking: while looking out the window he remembers images and facts and feelings, anticipating future possible happenings, considers an event from one perspective or in one connection, then another, then another, all while perceiving the scene in his study and out the window. This is a kind of thinking. He is experiencing a series of intentional contents, many non-analogical, some imagistic, all perspectival renditions *of* and some *about* feature(s) of things (including himself) that occasion further judgment. He describes this as an internal conversation among various shifting perspectives on a variety of memories, perceptions, and anticipations. In this Hobson follows Mead, who saw thinking as an internal or "reflexive" conversation dependent on the experience of public communication. For Mead thought is conducted as a conversation among perspectives (Mead 1962, pp. 141–48, 191–2). Reflexive conversation internalizes the roles and perspectives of others, and with it the varied aspects of objects and states of affairs, that are objects of attention. Warnock suggests that "the possibility of taking up different perspectives is essential . . . to having a thought about something" (Warnock 1976, p. 171). For Tomasello, all manner of our experiences and their manifold features as actual or potential objects of joint attention, which is to say as actual or potential sites for and subjects of communication, can serve as the point from which other features of reality can be judged.

This is no place to engage the troublesome question of whether nonhumans think, or have ideas and concepts. Some nonhumans certainly perceive, learn, remember, emote, have core consciousness of feelings and images, cognize bounded equivalence classes, or categories, of objects and conspecifics, and they must mentally process, relate, and integrate, this welter of feelings, perceptions, and memories, including social messages, even imagine.[18] Some take this to imply they must have concepts, ideas, thoughts, and be thinking. But this is problematic, for if "having a concept" means merely behaving in a regular way toward selected objects—e.g., a type of food—it would apply *very* widely across species, to unencephalized invertebrates, perhaps even protists or plants. How then is human "thinking" and "having" ideas or concepts distinct? The preceding discussion suggests two things. First, failure to identify with others and their roles may mean nonhumans do not have a multiplicity of perspectives from which to draw an internal conversation. Second, what we mean by "having" an idea is not mere mental processing, nor rule-governed behavioral responses, nor the sheer presence of mental experience, but the *discrimination* and *preservation* of a mental content. That requires treating self as object, treating parts or instances of self as objects, and the use of signs to manage them.

Third, one particularly impressive feature of the thinking of a socialized, autobiographically conscious sign-user is a tremendously enhanced power of *imagination*. Imagination is the ability to entertain counter-factual intentional content. As noted, some nonhumans must have some power of imagination. But the human can manipulate the flowing experience it shares with other organisms possessing core consciousness, not only attending to but holding in mind and reflexively communicating, thinking about, selected features by adopting alternate views of social others. We are thus able to perform complex internal trial-and-error processing or what Campbell calls "vicarious natural selection," in which our imagined courses of action, in Popper's words, can "die in our stead." We are able to vary the features of a remembered or perceived situation, removing components from their context, constructing imaginary scenes for them. Our experience is thereby spread out in time, beyond the present, by the ability to hold the past with its unexplored, and to imagine a future with its yet to be explored, possibilities.

Fourth, it is perhaps least necessary to make an argument that taking the perspective of the other is necessary to our unique *cognitive abilities*. What the current context adds is how social these abilities are. We are able to regard things from multiple perspectives and to dialogue about them as they robustly are, independent of any one perspective. A community of linguistically-enabled joint-attending inquirers capable of multiple perspectives is the greatest mechanism for learning on Earth. Presumably this is as well crucial to our *tool-making*, not only because such has probably always been a cooperative, communicative endeavor, but because, as Dewey argued, objective knowledge of natural processes and instrumentality are linked (Dewey 1958). We must analyze how processes take place, and how parts combine to make wholes, in order to work on them, decompose them, and recombine materials into technological enhancements of our native manual and sensory abilities, and reciprocally, the analytic mimicking of natural processes is a crucial part of natural knowledge.

Fifth, humans seem to have a unique kind of *agency*. As we saw, teleological behavior, action guided by ends-in-view, is not restricted to human beings; the fear or desire and the image of the object can cause behavior in the nonhuman animal. But there is a different, more complex level of human action. Whatever else the self is, in the context of social communication it understands itself as an object and, as Damasio says, narrates to itself that identity over time as possessor of its experience and responsible for its acts. As self-conscious we carry a cumulative history from the past into the next act, all as ours, and anticipate a welter of different, fragmentary possible futures. To ascribe an *act* to such an agent is to say it manifests the disposition and choice of the actor, as something done and not merely undergone, hence as the agent's responsibility (Oakeshott 1991). We cannot blame or praise the conduct of an organism that is not an autobiographical agent. The concept of freedom, analyzed by philosophers, political activists, and cognitive scientists, is too complex to be discussed here. But we can at least say that part of the folk notion of freedom in human society is simply the condition of being a responsible agent making choices. We can have the latitude for such choice only if we can imagine alternative possible doings, multiple perspectives, and multiple roles.

Last, such agents are capable of acting on the basis of *reasons*, prospectively public statements of their own belief-and-desire states, and of the conformity of those states to publically understood rules of possible actions. Reasons and reasoning are developed through the identification with others, and the social need communicatively to justify action before the perspective of the "generalized other." "Dual-explanandum" and "compatibilist" theorists argue that while a bodily movement can be explained causally in terms of somatic and neural, putatively "physical" causes, an action or behavior the bodily movement constitutes, requires reasons to be explained. When, to answer your question as to where I want you to go, I point, the extension of my arm gets a neuro-somatic explanation, but my "pointing" is a socially recognized act motivated by my reason that I want you to fetch something. At that focal level, the act is a meaningful social process, phenomenally explained by my beliefs and desires and yours, and perhaps functionally explained by the act's role within a higher level social system (e.g., if you are my employee). The reasons, or beliefs and desires, can be nonefficient causes of the act.[19] The reductive biological explanation of the movement is necessary but insufficient to what my pointing is and why I do it.

All these abilities must require a distinctive neurology, and without the right genetic endowment they could not arise, nor lead to much. This endowment gives humans the possibility of acquiring the understanding of others, language and cultural signs, and the human sense of an objective world for responsible agency, through taking the perspective of the other in its various forms. But the actualization is social. Humans have found a way to be more social than our most social fellow mammals and thereby more distinctive individuals than, for example, eusocial insects. We are a species of heightened sociality *and* heightened individuality. This space of experience and action can be depicted by an expanded version of the intentional triangle which shows the relation now of self-consciousness, other-consciousness, and world, with an internal vertex (or the peak of a pyramid coming out of the page) showing how some of the objects/events of the world, produced or handled by the child, become sign-objects/acts (Figure 9.2). This is the intentional triangle in which we live as humans.

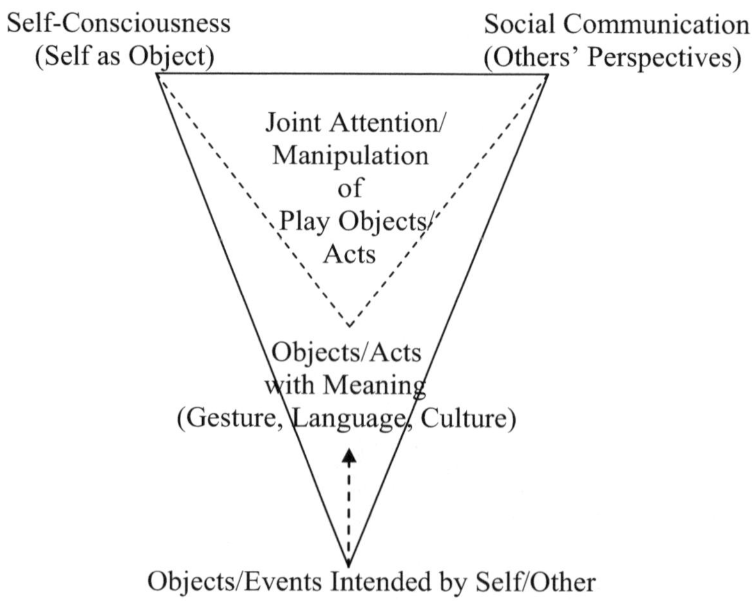

Figure 9.2. The Expanded Intentional Triangle

III. Meanings

All this implies humans have something special to do with *meanings*. There is much philosophical disagreement over the "meaning of meaning" in the past century of philosophy of language. We cannot review that literature here. But for naturalistic purposes it is important to avoid the common views of meaning as either solely mental or linguistic (which are very different things). We need to formulate a minimal, background conception of meaning. For this, we will again turn to Buchler, for just as he constructed a pluralistic metaphysics, he also formulated a uniquely pluralistic concept of human judgment.

By "judgment" Buchler meant not propositions or beliefs or applications of rules to particulars, but any human utterance, production, or act of any kind, any adoption of a stance with respect to things. Following a tradition going back to Aristotle, he distinguishes at least three kinds of judgment: doings or "active judgments," makings or "exhibitive judgments," and sayings or "assertive judgments." These have different norms of validity, respectively, goodness, aesthetic compulsion, and truth (Buchler 1951, pp. 51–57; Buchler 1955). Running for a bus, arranging furniture, and making a statement are judgments. The three judgments are *functions*; a diagram in a technical manual is a picture, but functions to say or assert; saying "I do" in a wedding serves as an act; a poetic saying serves as an aesthetic exhibition. Judgments always actualize or enact a perspective, a selective ordering of traits of some set of complexes. Buchler's aim in this theory was to forge an account where knowing, validity, meaning, and rationality would not be restricted to assertive or propositional thoughts and sayings, but would apply indifferently across all human utterance and production. We also remember Buchler's account of possibility from Chapters 1 and 4. Possibilities are as real, as experience-able, as capable of making a difference to other complexes as are actualities. Buchler used his account to define similarity, universals, and kinds: a *similarity* between complexes is the possibility of sameness in some respect (e.g., the redness of the red apple and the red wall); a recurring pattern of shared possible traits is a *kind* (the set of apples of any color); a general or *universal* is that complex in respect to which other complexes exhibit a similarity (redness). If a complex's possibilities in an order exhibit invariance, either continuous or periodic, that invariance is a *law*.

My suggestion is that a meaning is *a coherent set of perspectivally selected possibilities of a complex or set of complexes in some order.*[20] The meanings of a complex *are* the selected sets of possibilities, perspectivally determined. Like all possibilities they are not spacetime-occupying or energy-possessing. A meaning is "coherent" or "rule-governed," that is, incompatible possibilities cannot be part of *one* meaning. One may say they are a type of *structure*, just as shape, rhythm, or periodicity are real natural complexes, but not systems; the spherically shaped planet is an actuality, the spherical shape itself is a possibility studied by mathematics. Possibilities and meanings are traits of natural complexes, hence themselves natural complexes as real as any other. They exist, we can know, use, and manipulate them, they are publicly recognizable, and they can make a difference in our behavior. But meanings are relative to human activity: they did not exist before human minds (or some other autobiographically conscious minds), although the possibilities *referenced by* the meanings often have existed. There is nothing odd about this; food and behavior did not exist before organisms, verbs and religions did not exist before humans. Last, as possibilities meanings cannot be *efficiently* causal by themselves—otherwise there would be no un-actualized possibilities. Meanings are causal only in formal and final senses; they make a causal difference only when they play a role in the teleological causation of a sufficiently sophisticated mind.

My claim is not that we can mean only possibilities; we certainly can mean actualities, actual entities, processes, relations, etc. But words refer to actualities *through* universals, hence possibilities. The particular wall I see is actually white, but both *wall* and *whiteness* are possibilities. "Cahoone the philosopher" refers to me. The name "Cahoone" may be understood as a rigid designator (Kripke 1980) which does not refer to possibilities, but "the" is an article whose meaning is a set of possibilities, and "philosopher" is a kind, again a set of possibilities.

Other animals with complex brains must deal with possibilities in some sense. The remembered, now possible event is crucial to recognize—e.g., wariness in entering an area where a predator was seen the day before, the growl that indicates a possible attack, etc. Natural kinds are one set of possibilities that matter particularly—the deer must beware not only of the particular coyote seen yesterday, but this darker-colored, smaller version today, while not fearing a raccoon. But even in these cases, the animal seems to be generalizing a retained response, the basis of which is an instinct or fixed action pattern—i.e., fear of the tokens of "coyote," now generalized. The list of possibilities that matter, hence are responded to, is therefore very limited; what might be food, what might be a coming storm, what might be a predator, what might be a mate. One is tempted to say their limited number is small enough as to be handled without a complex representational device, e.g., mental representation of sets of possibilities *as* meanings.

So while there are doubtless nonhumans who must respond to some of the meanings of events, to respond to the events as having meaning, we alone have the means and capacity to represent and attend to *the meanings themselves*, hence to consider and manipulate possible meanings. Our uniquely componential, syntactically-constructed and recursive languages present the clearest example. Every word, sentence, gesture, and speech act we utter is shot through with possibilities. Each refers partly to things and events that are *not* present. It is my claim that on Earth humans alone are able to cognize and manipulate meanings like objects. Meaning manipulation allows us to do something remarkable, to objectify and communicate about complexes that are non-actual, and thus able to communicate about the actual through, or in the context of, the non-actual. Selected equivalence classes of possibilities of complexes become objects for minds and sometimes causally efficacious in human judgment.

Meanings arise from acts of social communication in an environment; hence semantic externalism is justified here. For while my *having* of ideas is not public, the meaning *thereby had is*. That is the lesson of Dewey, Wittgenstein and the last half-century of philosophy of language. Meanings are objective, externalist, intersubjectively rule-governed complexes, and can play a causal role when employed in that context. They arise in human social communication and creation and consequently in the autobiographically conscious mind. Following Peirce as well as Mead, it is the sociality of meaning that makes it capable of granting us heightened objectivity. Here I am making no general claims about the validity of our knowledge. And certainly any smart animal growing up alone in the wild will have its feelings and images and habits disciplined by environment. But in complex human ideation, it is not so much the environment as the community and its perspective on the environment, the "generalized other," which provides the individual mind with the constraints of objectivity (Mead 1962, pp. 154ff).

The developmentally primary and most ubiquitous technology of meaning is significant gesture or signs, hence language. As Mead put it, "The [human] mind holds on to these different possibilities of response . . . and it is [this] ability to hold them there that constitutes his mind" (Mead 1962, p. 135). Most fundamentally these are gestures or spoken (or manually signed) words and sentences, later written words and sentences. Peirce famously defined a sign as *something* which *stands for something* to *someone*, the sign's *ground*, *object*, and *interpretant*, respectively. The interpretant is itself another sign. Peirce distinguished three types of interpretants: the *emotional*, or a feeling; the *energetic*, an action or process; and the *logical*, or the idea the sign stands for in the interpreter's mind (Peirce 1931, 5.467–.91).

Many thinkers, even after a hundred years of semiotics and philosophy of language, continue to use "sign" and "symbol" in disparate ways. We cannot explore this adequately here, but we can say the main disagreements lie in distinguishing human linguistic signs from, on the one hand, "natural" signs, like a cloud or track in snow, and on the other hand cultural phenomena, like a work of art. We needn't deny that some phenomena in nature are "proto-semiotic," sign-like, or can function like a sign, e.g., the genetic "code." But a sign in the proper sense is a narrower phenomenon. As significant gesture, a sign is a vehicle of communication and recognized as such. It is *about* something, or stands for something, thus it communicates a message from someone to someone (although the someone may be anonymous, indefinite, or potential). Expansions of the notion of sign are derivative of this primary meaning, which is communicative, representational, agent-emitted and agent-interpreted. DNA, clouds, tracks, art works, and musical phrases are signs in a derivative sense.

Communicative signs represent in the sense of saying or asserting. To the extent that something functions as a sign, it functions as a saying or assertive judgment, and we are concerned firstly with its logical interpretant. Assertions *say* their meanings. I do not mean only declarations are signs; drawing a sketch of an accident solely for the purpose of showing what happened is saying. But significant gestures, words, and sentences represent or say their meanings, rather than show or enact their meanings. Language, among all modes of culture, thus has a special relation to meanings. On the one hand, significant gesture arises early in each human life and funds language as its most impressive early development, just as language appears to have arisen phylogenetically before artifactual culture. On the other hand, words label meanings; they stand for them. The art work does not "stand for" its meaning, and its capability of multiple interpretation contributes to its cultural value. (This is not to gainsay metaphor, which is exhibitive, and suffuses linguistic practice.) If thinking is reflexive communication, than perhaps the best rudimentary way to characterize thoughts is as neurally represented meanings.

Now we can return to the question of nonhuman significant gesture. Contrary to Mead it appears that significant gesture can to a limited extent be acquired by some nonhuman great apes (e.g., Washoe, Kanzi, Koko. cf note 10). Some nonhumans do have the neuro-mental-behavioral capacity to acquire use of some significant gesture. So this is evidently possible without being, or "becoming," human, which is to say, behaviorally indistinguishable from, e.g., a five-year-old speech-impaired human who has been taught sign language. What implications does this have for the use of Mead's account as an hypothesis for a core part of human distinctiveness? First, we must remember that, since Washoe et al. acquired this ability through human training, it remains the case that significant gesture seems only to exist in the context of *human society*. It would not be odd, for Mead or us, to regard significant gesture as primarily—in logical and temporal terms—a social rather than individual possession. Second, the linguistic behavior of Washoe et al. does imply that a distinction must be drawn between the acquisition of some significant gesture, which means a relatively small number of signs and the ability to take the perspective of others on *those* gestures, and an ongoing, systematic, open-ended use of significant gesture. This would mean achieving the perspective of others on some acts is not tantamount to living in and experiencing the world through the intentional triangle. It may be that significant gesture should be seen as a means or vehicle of communication that may help to stimulate the intentional triangle, and in which it later plays a crucial role, but evidently can exist without it. That is, the learning of a relatively limited set of significant gestures, gestures treated and experienced as signs, including linguistic signs, is possible without the systematic use of such signs and ongoing occupation of the perspective of others and generalized other that emerges in human childhood. Once "in" the triangle some usages of gesture become uniquely human. What seems special to humans is the ability to *function in* the triangle as a kind of platform for experience and behavior.

To summarize, complexes, including natural complexes, have traits, among which are possibilities. There are some invariances among sets of possibilities of individual complexes and sets of complexes. When social communication picks out and manipulates such coherent sets of possibilities, they are meanings. The meanings *are* the selected possibilities. The bearer of a meaning in communication is a sign; signs are vehicles of social communication, used to pick out and manipulate meanings. It is equally true to say that the meanings are emergent upon social communication as upon sign use. But broader even than signs or language is culture. In cultural making, meanings come literally to structure the human environment.

IV. Culture

We recognize, handle, and create meanings in a variety of ways. As noted we use significant gestures which label or stand for or represent meanings. We also create art works, fashion a home, chose a course of action, endorse a policy, or perform a ritual. These also mean, but they are somewhere between, on the one hand, significant gestures, which are designed as communicative vehicles, and, on the other, natural events or objects which have meaning for us, like clouds or tracks. Exhibitive and active judgments, hence art and action, produce entities or events or circumstances that mean, and the fact they mean is essential to what they are (unlike clouds or tracks). They are created at least partly because of what they mean. Yet rather than *standing for* a meaning as a vehicle only of assertive communication, they *embody* meaning. They mean by becoming the subject of further human judgments, which interpret their meaning (Buchler 1955, pp. 153–70). They communicate, but do not transmit. The painting of the wheat stalk, and viewing of that painting; the planting, reaping, harrowing of the wheat for food; and the description and classification of wheat in language—these all display, or actualize, or assert the possibilities of the wheat. Not all signs are cultural, and not all cultural objects or events are signs. But like signs, everything cultural *means*.

Cultural products and their meanings constitute an order of nature emergent upon human social communication. This is the final layer of complexity brought by humans, one that locates individual and group behavior in an artificial environment emergent from the joint manipulation of meanings, studied by linguistics, cultural anthropology, literary and art history, architecture, philology, religious studies, cultural studies, philosophy, and history itself. Humans are the cultural animal. In saying this I follow the great twentieth-century philosopher of culture, Ernst Cassirer (1962, 1965), as well as the "cultural realism" of Joseph Margolis (2001). Here I will argue that the cultural character of human judgment and (in the following chapter) knowledge is compatible with their inescapably natural character.

As noted above, there are other species that manifest unique group practices, versus species-wide behavior, passed across generations via learning rather than genetic inheritance, like the Koshima macaques. This is sometimes called "culture." But as I have argued elsewhere, we must distinguish culture from society (Cahoone 2005). If not, we would have to say any social change *is* a cultural change and vice versa, and deny that one can change independently of the other or affect it—not to mention, there could be no such thing as a *multi*-cultural society. Societies can be defined as relatively stable groups of unrelated conspecifics for whom membership makes a difference to individual behavior. Society is in effect a horizon of interaction and interdependence, and requires rules of behavioral intelligibility and propriety. Cultures are horizons of meanings and ends, more specifically, the collection of practices, narratives, and artifacts that constitute the mediate significance of acts. Human societies have, but are not, cultures. The two overlap and interact, but are not the same. A macaque troop is a society.

When it acquires potato washing its *society* has indeed changed and it has passed on social learning. Human societies change in this way as well. But human societies also have cultures that are yet qualitatively distinct.

For culture is *making things that mean*. Here a distinction has to be made. Some of a human society's culture informs the grammar of propriety and intelligibility required for at least more complex social coordination of actions. We socially coordinate differently than other animals: we use signs and their meanings, especially language. This is one level or phase of culture. But there is more to culture than this. There is compelling evidence for making a distinction between the kind of meanings and sign-use involved in everyday communication, which must hold for every and all human social groups more or less throughout their existence, and a more restricted and intense use of symbolism. Evolutionary anthropologists cite the distinctiveness of behavior in ritual, mythical narrative, art and creation of icons (Chase 1999, Watts 1999). Whereas hunting-gathering pre-literate peoples converse socially much the way we do, during ritual the use of language, dance, and ornamentation, and sign-use becomes narrow, loud, intense, highly structured, and repetitive. We find evidence of such intensive cultural practices first in the Upper Paleolithic era, along with greatly enhanced variety and specificity of tool construction, in the Great Leap Forward (Diamond 1999, pp. 39-41; Dunbar, Knight, and Power, 1999). So whereas the *mundane* level of cultural activity, manifest in everyday practices, narratives, and artifacts, surely is intertwined with social intelligibility, propriety and behavioral coordination, there is as well a more *symbolic* layer of cultural rituals, icons, and myths serving as "thick" social ends which become generic motivators.

Culture is equally comprised of all three forms of judgment—doings, makings, and sayings—which is to say practices, artifacts, and narratives. But in the case of human culture all these are *made*. They are stable, enduring products, not merely evanescent actions or sayings. I am not saying the function of cultural products is solely exhibitive; some products function to exhibit (e.g., art), others say or represent (e.g., narrative) and others act (e.g., social practices). Nor is my point that making has general priority over the other modes of judgment. It is that when we refer to "culture," we are referring to the cumulative, constructive role of judgments in a history, hence focusing on their role as makings which have meaning and intrinsic value as ends for those responding to them.

Cultural objects, whether artifacts, narratives, or practices, exist and endure as meaning-endowed complexes emergent upon material individuals or human activities or both. This is my way of expressing Margolis' cultural realism, which holds that cultural phenomena are real, intentional or meaningful entities, processes, structures, or properties, "emergent" upon—and in his terms, "embodied" in—material or biological phenomena (Margolis 2001, pp. 134–41). In addition to their material properties as constructions or performances, they have meaning-properties independent of their enjoyment or creation, like many other sets of possibilities. But these can only be actualized and play their causal role through the experience and judgment of humans (or a similarly endowed species). The interpretation of these meanings displays social variance and relativity; they are not subjective or idiosyncratic, but emergent upon the relations of social groups of human judges to their referents.

On Earth we alone live by fabricating in speech, in act, with our hands, always with our minds. This ability to create culture is an enormous alteration of evolution (Lorenz 1974). All life evolves by environmental selection of genetically based or influenced phenotypic features. But humans alone are also systematically Lamarckian, inheriting acquired or learned characteristics via stored public signs, artifacts, and practices. This does not mean biology has been transcended. The cultural emerges from the mental and biological and remains dependent on them. It means that we are the first species (unless this was also true of Neanderthals and/or Homo erectus) systematically to augment the contribution to our evolution by natural selection. The context of human existence then becomes heavily artifactual; language, myth, art, historiography, and science, as Cassirer put it, becoming our newly added "worlds" or environments, emergent upon bio-material-physical environments (Cassirer 1962, 1965).

In this sense culture is the business of *Homo faber*. We create meanings-objects, tools, ornamentations, ledgers, legal rules, clothing, time and space measurements, art, dance, myth, and ritual. We take bits of our experience and treat them as things separate from their context of arising. Our objectivity is manifested both in heightened intending of object-invariances and in creating objects. A visitor from another planet would notice humans incessantly arranging, forming, and producing, whether sewing and building, talking and singing, dancing and planting. We make tales and verbal-propositional representations, we structure behavior into the form of practices, we form tools and artifacts. This creativity is, at least on Earth, the property of humans alone. Not that any of this is un-natural. Culture is *the way* modern (post-50,000 ya) Homo sapiens live in nature. Just as nonhuman nature contains structures, processes, possibilities, pasts, purposes, behavior, learning, experiences, and communication, in the human case it also contains significant gestures, signs, languages, meanings, ideas, art, rituals, narratives, and religions. The current state of Earthly nature cannot be understood with recognizing that fact.

With culture comes one last thing: *history*. The retention of cultural complexes over time is history. Social groups have a history because they have culture, hence what Tomasello calls "cumulative cultural evolution." For with all of this making goes the necessity of transmitting this increasingly complex competence and knowledge across generations, and with it the storytelling and eventual historiography of the social group. No other species has a history in this sense, that is, a historiography. Species develop, and undergo changes, but do not generationally transmit a record of accumulated novelty of behavior and artifact. Chimpanzees did not have a medieval period, from which chimpanzee-modernity could emerge. From social communication to fabrication of an artificial culture, we recognize, manipulate, and make meanings. We do not leave nature but supplement it by drawing out unrealized possibilities. That is what a naturally cultural animal does. So one of things *Homo faber* makes is *Homo historicus*. The creative animal is the one with a history. Part of our history is the changing nature and extent of knowledge. To that piece of our history we now must turn.

10

The Evolution of Knowledge

The present chapter does not describe a stratum of nature. It addresses one kind of animal capacity whose distinctive human expression is particularly important for our project, namely knowledge. While the jobs of epistemology and metaphysics are different, they entail each other.[1] The naturalism described in the preceding five chapters implies that human knowledge emerges and functions within nature. True to that perspective, below I will accept a naturalistic or evolutionary epistemology. But doing so raises two philosophical questions. First, it clearly seems that human knowledge is a cultural phenomenon, as implied by the preceding chapter, and not solely a natural phenomenon. How can these claims be compatible? Second, one may wonder about the justification of the kinds of scientific knowledge that this work has deployed. It may seem that placing knowledge in a biological and/or a cultural context might trouble a *realist* view of the very knowledge I have used in this study. Epistemic realism is the view that true judgments are true with respect to what they judge independent of the judgment. If human knowledge is biologically and/or culturally determined, can we still believe it is true of its objects as they obtain independent of human judgment? Especially since, as we saw, if knowledge is cultural, we must admit it has changed historically. How can the knowledge this book depends on survive such an admission?

What follows cannot be a fundamental or general epistemology. My job is more modest. I will suggest that a coherent, minimally realist epistemology is available to the present naturalism. Given the multiple, layered orders in which knowing functions, I will argue that, rightly conceived, a naturalistic, evolutionary epistemology (Section I) and the historico-cultural location and funding of knowledge (Sections II to III) are compatible with each other and with a realist understanding of knowledge, including scientific knowledge (Section IV). Knowing too, like mind and culture and meanings, is part of nature.

I. Natural Epistemology

A preliminary problem in characterizing knowledge is deciding whether "knowledge" will refer solely to something human. Philosophers typically focus their investigation on "propositional" knowledge ("knowing that"), while admitting as well the existence of practical knowing ("knowing how") and recognition or familiarity (e.g., "Betsy knows Paris"). But it can be argued that animals of many species "know" (Kornblith 2002). Knowledge is one of the things some organisms must have to survive; contemporary epistemology is seriously derelict in failing to incorporate animal knowing.[2]

Cognitive ethologists think of knowledge not in propositional terms but as information acquisition and storage exhibited in behavior change. While appealing in its avoidance of anthropocentrism, such widening of the scope of "knowledge" raises a problem. All organisms behave and, along with cameras and compact discs, store information. To accept that knowledge is behavior change due to information acquisition threatens to apply knowledge to *all* living things and perhaps non-living systems as well. All life is irritable and responds to stimuli, hence all life is information-gathering and responsive, without mind, brain, or even nerve cells. But are we to say the amoeba lurching from a noxious stimulus has "knowledge"? To avoid this, while nevertheless accepting a conception of knowledge that will apply to both humans and encephalized nonlinguistic animals, I suggest the following baseline parameters.

Most generally, knowledge is a capacity acquired by living organisms (Buchler 1955, p. 33). So only living things know. It is possessed by the whole agent (perhaps my immune system "learns," but it doesn't "know"). It is something the organism acquires, not genetic or instinctive. Knowledge has to be the product of learning, and must be stored. Further, I suggest knowledge is related to cognition and perception, hence requires an animal endowment; it is not mere sensitivity, which is evident in all living things. Such knowledge should be thought of as a capacity for something, for experience, action, and/or production; gaining knowledge is a gain in that capacity. Last, knowledge must be valid, whether true or adaptive or functional. These are, to be sure, very different normative terms, but for now vagueness must suffice. If the expression of the capacity an organism has gained is mistaken, maladaptive, hallucinatory, or false, it is not a gain of knowledge. I propose then as the broad sense of knowledge, applying to any animals who know, that knowledge is *learned, stored, adaptive dispositions* for a variety of manifestations (action, experience, communication, etc.) that directly or indirectly *enhance adaptive fit or the successful functioning* of the organism in its environment. We can, and do, apply such a formulation to human knowing, e.g., when we speak of "know-how." But humans also have a form of knowledge that is distinctive, in which we are particularly interested: assertive or propositional knowledge that is supposed to be *true about* its objects.

The idea that human cognition takes place in a biological context is not unusual. Any naturalist must have such a notion. And if humans evolved, our cognitive capacity presumably did as well. Darwin made suggestions in this direction. So did Peirce, in his 1878 essay "The Order of Nature," and James, in the final chapter of his *Principles of Psychology* (Peirce 1992e, James 1950).[3] But while there were contributions *avant la letter* (for the history, see Campbell 1988a), the major progenitors of evolutionary epistemology were four thinkers of the second half of the twentieth century, the philosophers Karl Popper and W. V. O. Quine, the ethologist Konrad Lorenz, and the psychologist Donald Campbell. Others have followed them (e.g., Kornblith 1985, Shimony and Nails 1987, Radnitzky and Bartley 1993).

Popper postulated that, to use his book title, "Conjectures and Refutations" or trial-and-error learning, the production of guesses plus openness to disconfirmation, is *the* most crucial method of knowledge, from advanced animal learning to human knowing and science. He wrote,

> Assume that we have deliberately made it our task to live in this unknown world of ours; to adjust ourselves to it as well as we can; to take advantage of the opportunities

> we can find in it; and to explain it, *if* possible. . . . *If we have made this our task, there is no more rational procedure than the method of trial and error—of conjecture and refutation*. . . . (Popper 1963, p. 51, author's emphasis)

Quine's approach to what he called "naturalistic epistemology" famously suggested that human knowing is a natural, biological, phenomenon, and ought to be studied as such. This entails rejecting foundationalism in epistemology. As Quine wrote,

> I shall not be impressed by protests that I am using inductive generalizations . . . and thus reasoning in a circle. The reason . . . is that my position is a naturalistic one; I see philosophy not as *a priori* propaedeutic or groundwork for science, but as continuous with science. I see philosophy and science as in the same boat—a boat which, to revert to Neurath's figure as I so often do, we can rebuild only at sea while staying afloat in it. There is no external vantage point, no first philosophy. (Quine 1968b, pp. 126–27)

Lorenz made central contributions to evolutionary epistemology. His remarkable 1941 essay, "Kant's Doctrine of the *A Priori* in the Light of Contemporary Biology," argued that the twentieth-century neo-Darwinian synthesis of natural selection and genetics provides an answer to the philosophical problem of knowledge along the lines of a biological Kantianism. Knowledge is, Kant claimed, an active process of construction. That led Kant to posit an unbridgeable gap between appearances and unknowable things-in-themselves. But Darwin provides a bridge: the subject's construction of appearances has *itself been selected* by things in themselves. Cognition is a natural capacity naturally selected, an adaptive mechanism produced by evolution. Things in themselves have a causal relation to our cognition. Lorenz later produced *Behind the Mirror*, a systematic account of cognition's evolution from amoeba to humanity (Lorenz 1973).

But it was Campbell, inventor of the term "evolutionary epistemology," and to whom Lorenz and Popper both admiringly refer, who provided the most telling suggestions as to the evolutionary methods and mechanisms of human cognition (Campbell 1988b). He began with his notion of *blind variation and selective retention* or BVSR (Campbell 1960). He insisted that for trial-and-error learning to occur, at some level there must be a blind (or random) generation of roughly independent alternatives, among which some mechanism, operating with stable criteria, selects and retains one option. This is arguably the method of evolution itself, in which the blind variations are mutations, the retention is inheritance of genes by the next generation, and the selection is performed by the environment on the phenotypic properties of the organism. In this context knowledge must be defined broadly: "any process providing a stored program for organismic adaptation in external environments is included as a knowledge process, and any gain in the adequacy of such a program is regarded as a gain in knowledge" (Campbell 1960, p. 38).

The BVSR model was early criticized by Herbert Simon and his coworkers. They invoked as the "British Museum Algorithm," the famous thought experiment of Émile Borel and then Eddington about the probability of a troop of nonhuman primates typing at random to produce the books in the British Museum (Newell, Shaw, and Simon 1958).[4] Simon and his colleagues argued that such "brute force" searching is impractical in biological time frames, that heuristics or hypothetical strategies must guide the search. From today's perspective this is a "is the glass half-empty or half-full?" debate. Campbell's key point was that *at some level and within some temporal unit* a random exploration or generation of trials must take place. But that could mean a truncated search among a set of options established by a prior selection or heuristic or both. Campbell accepts Simon's heuristics as long as they themselves had to be hit upon by *some* prior trial-and-error process (Campbell 1960, p. 393). Wimsatt's recent conceptions of trial-and-error *and* heuristics, applied to nonhuman and human learning, rests on an integration of the insights of Campbell and Simon (Wimsatt 2007).

The most primitive or fundamental form of BVSR is blind locomotor search. The organism moves in space until it bumps into favorable or unfavorable objects or chemical or temperature gradients. But the major role in this process at more complex levels is played by "vicarious selectors," internal mechanisms which substitute for or replace external events of moving-and-bumping-into, allowing selections short of injury or death. Heuristics are vicarious selectors. Whereas the paramecium's blind search could as easily make *it* food, distance receptors allow a higher organism to scan until it finds a representation of beneficial conditions, and only then moves to achieve them. Vision plays an especially useful role as substitute for locomotor search, even if it is always abstractive and presents "fringe imperfections" requiring tactile checking.[5] The point is, distality matters, and so does objectivity understood as invariance across organismic states and orientations. As Campbell wrote, "Continuing the evolutionary paradigm, we can note that the higher the level of development the higher the degree of distality achieved . . . and the greater the degree to which external events and objects are known in a manner independent of the point of view of the observer" (Campbell 1985, p. 54). This independence is relative; there is no utter independence of all observer perspectives. The environment "edits" the organism's search process. Among higher animals we also find instinct and habit. The two interact in phylogeny: habit became a "selective template around which the instinctive components could be assembled. . . . This can be conceived as an evolution of increasingly specific selection criteria, which at each level select or terminate visual search and trial-and-error learning" (Campbell 1988a, p. 407). Later, memory and thought become the stage in which "thought trials" are the vicarious selection mechanism. The great advance of mind is to present an internal representation of the environment which permits BVSR to operate articulately and vicariously. The more active this process—the more Kantian rather than Lockean—the better. Intelligence makes it possible, in Popper's words, for our hypotheses to "die in our stead."

Campbell points out the oddly useful case of "monitor-modulating" systems which generate perceptual judgment that conflicts with the data *in service of objectivity*, a non-phenomenological but ontologically useful biasing mechanism. When an unchanging object is seen against differently shaded backgrounds, we perceptually judge that there is a change in the object's shade when there is none. This shows an ontological bias in our visual recognition system toward positing stable, enduring objects rather than discrete images. To localize holistic alterations, the brain artificially constructs a likely ontology. As Campbell writes, "These features make our experienced image of the world more vividly real and complete, but do so by an artificial reconstruction of that world" (Campbell 1987, p. 183). Campbell's point is that this is a useful bias. Another familiar case is our perceptual judgment of cinema, where the discontinuity of individual frames is read out by our visual system, and as a result we "misperceive" the photographic presentation but thereby have *a more objective perception of the filmed action*, which, after all, was continuous (at least within the bounds of a directorial shot). Knowledge needs to make yes/no decisions sometimes when sensation finds a "fuzzy" situation, and often those decisions achieve a higher level of objectivity.

Campbell did not shirk from the normative-epistemological implications of naturalistic epistemology. He accepts that all knowing is pattern-matching, and selection of the internal, vicarious pattern must be made relative to alternatives at some level of precision, often through "triangulation." (As Popper pointed out, *every* measurement of a continuous quantity is falsified if the level of accuracy is raised indefinitely.) The fundamental epistemic idea, Campbell claims, is "fit," drawn from evolution. The cognitive apparatus must be both plastic *and* rigid if fit is to be achieved and maintained; plastic enough to modify itself with respect to its objects, rigid enough to maintain a structure that will be stably causal within the organism, and perhaps reproducible or communicable. What "fits" has a representational but pragmatic cast; in the human case these are "rule-systems" or "action recipes." What is selected is a "physically implementable action-rule structure" (Campbell 1987, p. 176). The rules take the form, "If x is observed, then do q; if y, then r," creating an "if-then" patterning of the environment. This is an activist or pragmatic, Kantian or Peircean kind of knowing: "A highly detailed passive reflection of the environment is, by itself, of no use to the organism, and approximations to it appear very late, if at all, in the evolutionary branching" (Campbell 1987, p. 176).

There are also structural or internal selectors, for example, coherence, hence nature's editing is often, Campbell admits, "indirect" (Campbell 1987, p. 170). As a result, "we should expect a gap between scientific beliefs and the physical world comparable to that which we find between animal form and ecological niche" (Campbell 1987, p. 172). Furthermore, the historical evolution of knowledge, as of anything else, works with the resources available.

Nothing is designed all at once for perfect fit; what adapts at any moment is the cumulative product of a particular history of past adaptations. Recognition of these internal factors does *not* lead to doubts about the validity of knowledge, however. Campbell quotes a marvelous passage from Lorenz arguing that we cannot doubt the limited, perspectival, and fallible but real and objective validity of the judgments of the naturally selected cognitive apparatus. The organism "constructs" its hypothesis, meaning its exploratory heuristics, but under the pressure, and through the selection, of an un-constructed world.

> This central nervous apparatus does not prescribe the laws of nature any more than the hoof of the horse prescribes the form of the ground. Just as the hoof of the horse, this central nervous apparatus stumbles over unforeseen changes in its task. But just as the hoof of the horse is adapted to the ground of the steppe which it copes with, so our central nervous system apparatus for organizing the image of the world is adapted to the real world with which man has to cope. Just like any organ, this apparatus has attained its expedient species-preserving form through this coping of real with the real during a species history many eons long. (Lorenz 1941, pp. 186–7)

What then distinctively characterizes *human* knowledge from an evolutionary perspective? Many things, but we may select a few. First, it would seem that we have especially high capacities for distal perception (vision and hearing) and for tactile confirmation of boundary coincidence (especially dexterous hands and feet), hence a great capacity for objectivity. Second, as seen in the preceding chapter, we are *self-conscious, social sign-users*, able to take the perspective of others on the self and its components, to represent features of experience as capable of description under multiple meanings and to manipulate these with others as a social practice. Language gives us the ability to label, hence think about, particular features of experience, to analyze and distinguish not only perceptual components but memories and motivations. This increases our ability vicariously to select actions, to compare anticipations of the consequences of actions with perceptions. And obviously, we can now *talk about* all these, making possible articulated group inquiry comparing our truth claims with those of others. Linguistically specified group learning among a community of inquirers is arguably the greatest mechanism for increasing knowledge on Earth. Finally, the language in which we perform such inquiry, along with all our constructed practices, narratives, and artifacts, is cultural. Human knowledge is a cultural phenomenon.

II. Cultural Knowing

Now, we cannot say, nor need we, that *all* human knowledge is cultural. I know when my cheek is struck by a fist, and this knowing may have vanishingly little to do with culture. But all linguistic expressions of knowledge, all propositional knowledge and know-how, whether in sayings or practices or makings, are cultural, and the passing of knowledge from generation to generation is through culture. Thus virtually all human knowledge we can identify as such, in its development if not acquisition, propositional or not, operates with and through cultural meanings and media characteristic of a particular historical community (Cahoone 2005). As noted, Homo sapiens are naturally cultural, the Earth's preeminently Lamarckian animals.

This raises a famous problem. Embedding human cognition and reason in culture, hence *cultures*, is just what cultural relativism, perhaps the most widespread form of anti-realism, does. It might appear that what is knowledge or even what is *true* is relative to local cultural inheritance. I have argued to the contrary that is no good reason to believe that there is a distinctive philosophical problem of "cultural relativism." By that I mean the location of human knowing in cultures introduces no special realism-threatening epistemic problem beyond the problem of a jointly biological, psychological, and social creature trying to know. The philosophical "problem of knowledge" holds true with or without culture. A fuller discussion can be found elsewhere; here I will make the basic point that the cultural embeddedness of knowledge does not imply relativism (Cahoone 2005, ch. 7).

Cultures are not consistent belief systems with "foundations," or "final vocabularies" (Rorty 1989). A "belief system" could only mean a theory, that is, a controlled, limited product of method, which forges a consistent structure of beliefs. Only communities of methodic inquirers—and even then, only in their professional activities—have consistent belief sets with foundations or a hierarchic structure controlled by key beliefs. Human cognitive networks are by no means so consistent or systematic. Knowing operates by applying different congeries of cognitions to different events; to paraphrase Peirce, our belief or knowledge seeks to cover the world or data in the way that clumps of kelp seek to "cover" a region of ocean surface. There are plural, discrete networks of interwoven, mutually supporting strands, some of which then merge together to cover more surface area. None of the clumps "ground" the others and there are freestanding clumps, although the overall tendency is toward expansion and integration.

Cultures are pluralistic historical products and exhibit similarities as well as differences. Diversity *within* cultures and diversity *between* them are not of a totally different order. All cultures rest on deeper similarities of the human process, in biological need, social existence, and environmental dealings. This does not mean we cannot speak intelligibly of identifiable, distinct cultures. The claim instead is that we cannot say that cultural difference is of an order beyond other sources of cognitive differences, that somehow overwhelms the possibility of intercultural intelligibility. Cultures change over time, hence must develop their own internal resources for understanding cognitive change and diversity. All but the most historically isolated cultures have had to evolve means for interacting with visitors and other cultures. In each case they have traditionally evolved "thin" (Walzer 1994) and "vague" (Nussbaum 1990) languages or versions of themselves that can bridge "thick" cultural differences. Cultures are inter-translatable: members of culture X can ask themselves regarding the initially mysterious speech or act Y of a member of culture Z, "in what game of theirs is Y a reasonable move?" or "what would I be doing/saying in doing/saying what they are doing/saying?" (Simpson 2001).[6] The point is that, with rare exceptions cultures develop the internal resources for translation, mutual negotiation, and transformation. Culture provides a telic or "thick" social frame for human activities, but the frame is plural, complex, interactive, changing, open, and adverbial. In Campbell's terms, there is plasticity as well as rigidity.

Thus we cannot imagine cultures as cognitive enclosures. Culture is the *way humans know*, not a filter that blocks cognition's access to objectivity. Awkwardly put, cultures are not the kinds of things, relativity to which creates a barrier to realist knowledge. This denial of the anti-realist implications of culture does not of course mean that what culture adds to human behavior and cognition is easy to understand. On the contrary, the level of complexity of this uniquely human layer of knowing and acting is great. Here we face a causal thicket in which the location of cultural objects, e.g., artifacts, narratives, and icons, in a pluralistic nature of multiple orders makes the decision as to the direction of causal arrows impossible to determine systematically (Wimsatt 2007). The socio-cultural causal thicket is in some respects far more difficult to understand than the bio-psychological thickets of minded organisms and the ecological thickets of inter-species relations in the biosphere, for in the human socio-cultural thicket we have biological, psychological, social, and cultural processes in interaction.

III. A History of Reason

Just as many creatures have knowledge, many are genetically programmed with particular methods of investigation or means of acquiring new knowledge. That part of our human investigative repertoire that is unique to us is cultural. What we call *reason* is part of that cultural transmission of knowledge-acquisition. I have argued that *reason is emergent upon culture* (Cahoone 2005). We are commonly in the position of having to decide among competing judgments. Reason is a kind of judgment by which humans investigate and adjudicate other social judgments, whether those judgments are active, exhibitive, or assertive. Human societies must adjudicate claims. When we do so methodically or systematically, according to rule-governed practices we develop, that is reasoning. Reason is *methodical meta-judgment*.

We know from historiography that knowing has changed over the course of the human sojourn, and with it the method of adjudicating among competing judgments. It is an historical fact that such methods have changed over the past ten thousand years. Reason has a history. An obvious example, and one specifically relevant to the present study, is that modern science is an historically distinctive method of knowledge. As noted at the outset, I cannot here give a complete epistemic justification of science, but I am at least responsible to characterize the sense in which modern science rightly claims an epistemic advantage. Doing so requires presenting a truncated history of reason, that is, of the distinctive forms of reason employed in human history. We can suggest such distinctions based on the historical scheme of anthropologist and philosopher Ernest Gellner (Gellner 1988, Cahoone 2005).

Gellner divides the human sojourn into three eras, the segmentary or hunter-gatherer, the agrarian-literate, and the industrial or modern.[7] Segmentary societies are primarily small and local, even if they involve migration in pursuit of wild herds. Their religions are typically polytheistic and animistic, and not soteriological; they don't need "saving," because religion is so intermixed with society that a social member can hardly fail religiously. They arguably have an ecological metaphysics of power or value circulating among natural forms (Cahoone 2006).[8] They are highly egalitarian, compared with later agrarian civilizations, even in the sense of rough equality of what Ivan Illich called "vernacular gender" (Illich 1990). This was the exclusive condition of all human beings for the vast majority of Homo sapiens' existence until the invention of agriculture, and for some to our day.

Gellner claims that segmentary culture is characterized by the undifferentiated normative governance of cognitive activities; each act is beholden to a multiplicity of value-constraints that actors do not differentiate. Using Buchler's theory of judgment from the preceding chapter, this means exhibitive, active, and propositional judgment, and their norms, the beautiful, the good, and the true, are undifferentiated. As Gellner remarks of the vast majority of human existence, "Language is not merely rooted in ritual; it is a ritual. . . . Most uses of speech are closer in principle to the raising of one's hat in greeting than to the mailing of an informative report" (Gellner 1988, p. 51). The villager may approve the statement "It is raining" because the village shaman predicted it would rain, even though it is *not* raining. The point is not that she lacks hearing or rationality, or even fears the shaman, but that propositions, including "It is raining," serve not one illocutionary function, but several; they are as much reaffirmation of a social tie, or ritual performance, as truth-functional report. This means 'truth' and 'goodness' have not been fully differentiated as norms. I call this "omnivalence," where human judgment indifferently serves multiple rather than segregated norms of judgment.

This does not mean segmentary peoples lack knowledge. They had vast empirical knowledge of plant and animal species and their seasonal locations and activities; they were the true Baconian empiricists and Linnean taxonomists. While omnivalent or multifunctional judges, still in their mundane practical activities they were excellent empiricists. We must imagine that their cognitive and cultural norms developed to be consistent with uncountable generations of observation in cyclically stable habitats (plus the lack of written records meant that ancient changes were poorly remembered). So their ritualistic propriety rarely had to be trumped by their empirical knowledge. The shaman did not often claim it was raining when the sun shined.

Gellner calls the type of reason characteristic of segmentary societies *Durkheimian*. Durkheimian reason is the consideration of plural saliences through analogy to narrative exemplifications, deployed with minimal differentiation among the values or norms of judgment. The reasoning used to decide amongst judgments is pluralistic, entertaining multiple considerations, without our hard and fast distinctions of irrelevance (e.g., our belief that argument from authority is a "logical fallacy"). In segmentary life *society is everything*. This does not mean nature is nothing, for we are speaking of a form of society that exists in relative harmony with its natural environment with a minimum of novel artifice, and with divine characters multiplying along the border between the two. Wisdom in this context is the ability to interpret narrative tradition, to reason from a variety of validities expressed in practices, usually via analogy to paradigm cases, in order to adjudicate current disagreements or problems.

Agriculture and writing ushered in agro-literate civilizations. Both technologies led to social hierarchy, for food now depended on land and grain stores that would have to be defended by a martial elite—who thus became the owners of land—and reading became a coveted skill of a class of scribes, usually religious, sometimes vying through their "cultural" authority with the coercive or political authority of the swordsmen. Thus was society typically divided into three classes or estates, pithily expressed by Gellner as those who work, those who pray, and those who fight. To that description of the agro-pastoral-literate epoch we can add a conception which Gellner himself notes but does not fully exploit. German philosopher Karl Jaspers coined the term "Axial Period" for the remarkable global explosion of philosophical-religious genius that occurred within 300 years before and after the midpoint (or "axis") of the first millennium BCE, including: the Hebrew Prophets Isaiah and Jeremiah, the Persian Zoroaster, the authors of the Hindu Upanishads, the Buddha, Confucius, Lao-Tzu, and the Hellenic Greek philosophers (Jaspers 1953). The Axial Age brought, in southwest Asia, the flowering of monotheism of a personal God, and in south-central Asia an impersonal immantentism, but in each case a transcendent conception of Divinity along with a need for salvation, calling the individual conscience to confess its fealty to the Ideal. Once the religious task has changed from performance of ritual to belief in doctrine and obedience to authority, religion becomes something to which social members can be re-called by prophets and periodic revivals. Doctrine and narrative are now codified in classic texts.

One way of putting this is that the Axial Age developed the momentous perspective of *logic*. By this I mean something wider than, but including, the treatises on normative thinking produced by Aristotle and ancient Hindu and Buddhist philosophers. I mean that the standards for human belief, formerly narrative, now are based in *Jasperian reason*, the capacity to judge judgments via explicitly formulable rules supposed to be society-and culture-transcendent. The model of validity is Platonic. The observed particular and the practical decision are valid because they embody or participate in a transcendent model. Reasoning must relate worldly events and possibilities to rules independent of the processes in question. Normative models, now textual, can only be known by those with special knowledge, i.e., literacy. One might say, in a somewhat Derridean vein, that Axial or Jasperian rationality is an exploitation of the implicit possibilities of writing. They carry high culture, which urbanized agrarian societies distinguish from low or folk culture. The hierarchy of reason nicely matches the hierarchy of caste and power. For the first time, society and culture can be criticized as failing to meet transcendent standards. Validation is participation in, emulation, embodiment, or instantiation of what is beyond sensory experience and social convention. Thus society is no longer everything; it is now merely *almost* everything, but validated by and controlling itself in reference to Something greater.

The third era of human history started a mere three centuries ago in central and western Europe and North America, if one tracks the change in cognition (the scientific revolution), two centuries ago in politics (republicanism and liberalism) and economics (market economies), perhaps only one in art (if our focus is modernist urban culture). There are many accounts of modernity, but it is at least inconceivable without a new way of directing economic activity, a novel form of knowledge of nature, itself eventually applied to centralized or cosmopolitan organization of practical affairs, and a new post-aristocratic politics of equality.

Whatever account of modernity one favors, we may follow Gellner in labeling its new form of reason *Weberian*, after Max Weber. Weberian rationality is the instrumental rationality by which practices and claims achieve justification in the context of goals and explicit premises. Hence efficiency in achievement or logicality of procedure rationalizes any particular act or claim. Adjudication is highly differentiated, "modular" and "uni-functional." Truth, goodness, beauty, salvation, and process-norms like efficiency and rationality are utterly separable. The key is the differentiation and dependence of the validity of judgments on contextual orders, of which instrumental dependence on practical goals is one type. Incommensurability of norms permits commensurable judgments *within* the discourse of each norm. In the case of natural science, the method becomes the search for hypothesized invariances sufficiently precise to yield, at the appropriate scale, predictions subject to publicly accessible observational disconfirmation. The task of understanding the world is distinguished from social fealty, status, moral duty, aesthetic satisfaction, and salvation. No one can claim cognitive legitimacy or truth for a result *because* it would be good to believe it, will make society operate better, or is more beautiful. Truth is specialized, traditional hierarchy undermined, the distinction between high and low culture dissolved. This is the differentiation of value spheres cited by Weber (Weber 1972), the loss of effective encompassing metanarrative for Lyotard (Lyotard 1979), and the fragmentation of modern society as "a whole that is less than the sum of its parts" in Luhmann (Luhmann 1982, p. 238). Human life and culture remain omnivalent in the sense of multifunctional, but now many of their activities are differentiated by mode. We retain the earlier forms of reason but the Weberian remains the leading edge of modern cognition. This context for understanding modern reason will help us, not to justify realism, but to articulate its requirements, argue for their plausibility, and show how the scientific knowledge used by this study is coherent with it.

IV. Realism for Naturally Cultural Knowers

Epistemic realism holds that true judgment is true with respect to what it judges, hence objects independent of the judgment are part of what makes the judgment true. Realism minimally requires five conditions that are, by themselves, rather reasonable. First, that true assertive judgments be valid with respect to what they judge (Cahoone 2005). Truth is an object-relational property ("object" here used in the broadest sense). I suggest we cannot make sense of a truth that doesn't hold in this way, for it is implicit in the notion of a judgment being "true of" something. Second, the relevant character of what is judged, about which the judgment is claimed to be true and of which contrary judgments cannot be true, must obtain independently of and make the judgment true. The judg*ing* cannot make it true—this is one of the features of the unique kind of judgmental validity that holds for assertions we call "truth" (unlike exhibitive or active validity, such as saying "I do" in wedding ceremony, which does render itself valid). Third, as "a" what, the object cannot be truly judged by contradictory judgments in the same sense; within assertive judgment or inquiry p and $-p$ cannot be said of the same thing at the same time in the same respect, as Aristotle put it. That is born out in our behavior, for wherever contrary assertive judgments are asserted of one object at the same time in the same sense that is regarded as a problem we try to resolve. Fourth, this implies a linear relation of the object-as-judged across differing or changed but inter-translatable semiotic nets, hence the rejection of incommensurability. This rejection is justified every day by bilingual individuals, whether fluent in French and Swahili or Newtonian and Einsteinian. Incommensurability is an artifact of contingently chosen incompatible languages, to be resolved through translation via a more comprehensive or *relatively* neutral language, neutral with respect to the languages in question (not neutral in general or *per se*, there being no such thing). Fifth, this implies the unity of inquiry, a unity not of *science* as promoted by the logical positivists, which required the reduction of natural knowledge to physics, but of *sciences*.

Now, for more than a half century philosophers and social scientists have been harping on the biasing of our judgments by society, culture, and history. Language games (Wittgenstein), theory-embeddedness of observations (Hanson, W. Sellars, Goodman), background theory choice (Carnap, Quine), non-objective factors in scientific progress (Kuhn), political history (Foucault), thick semiotic complexity (Derrida), final vocabularies (Rorty), and the onto-theology of the West (Heidegger), all render our knowing suspect. We do not have "immaculate" perception "immediately" of "presence" or objectivity, and we lack a view from "nowhere," since our minds or practices "construct" the world we encounter. As I have argued elsewhere, these criticisms are significant, but are commonly conflated and overblown (Cahoone 2002). Rightly understood, they do not derail a chastened, minimal realism. Why not?

First, our knowing is *fallible and only approximately true*. Following Peirce, certainty is out, not only for its impossibility but its undesirability. Our knowing of the world is always incomplete and improvable and cannot be conceived otherwise. It is also never arbitrarily precise, but only valid over a range of determinations. Approximation and vagueness are not disqualifying; they are inevitable. We never know everything about anything. But this is fine. Realism is entirely compatible with fallibilism.

Second, our knowing is *funded, perspectival, and mediated, but because adverbial, still direct*. Our cognition is always funded by past learning, perspective, language, and culture, and its relation to the object is never immediate, as Peirce rightly saw in 1867, long before Heidegger, Wittgenstein, Quine, or Derrida (Peirce 1992a). Immediate would mean, following Buchler, un-analyzable. But mediated cognition can still be *direct*, as "direct realism" claimed (Sellars 1922, Putnam 1994). We perceive and know the independently existing object, not merely a representation. Buchler writes of the "procept," the object of experience: "A procept is the *existence itself* . . . *in so far* as it is relevant to an individual as individual. . . . The complexes of nature are not 'presented' to experience. They occur, and when their occurrence involves an individual they *constitute* experience" (Buchler 1955, pp. 122–23, author's emphasis).[9]

Mediacy and directness are compatible because our cognition is *adverbial*. Gadamer's point remains: the complex cultural and practical modes of human historical cognition are the *ways* we know reality (Gadamer 1960). That no more means such cognition "obstructs" information than that the cephalopod's tentacle "blocks" its experience of a crustacean. Some means are better than others, and one means may serve to correct another. It is true, as Quine and Putnam argued about models on the basis of the Lowenheim-Skolem theorem, that there is no *one* true description of any thing or all things as a whole (Putnam 1981, pp. 32–38, 217–8). Inquiry seeks and validates *a* true account, not the *one and only one* true account.[10]

Third, knowing is *active but not constructive*. Activity and passivity, like stasis or stability and change, are relative terms. Our cognition of the world is certainly active, but active in certain respects and not others. As Lorenz put it, the horse's hoof does not "construct" the steppe. Autopoiesis cannot mean a construction that allows no constraining influence of environment. The use of "construction" in epistemology is an aesthetic metaphor gleaned from German Idealism and Nietzsche. In the human domain, construction occurs in exhibitive judgment, in engineering and art. We do not construct experience: we shape, refract, and select. We not only *do*, but also *undergo*, assimilate, respond, and adapt. Cognition *is* constructed—it does not *do* the constructing—by the organism in order to be sensitive to forces and factors it cannot control. Our response is edited by the environment. A truly "constructive" cognition could only lead to death.

Last, philosophy from Hegel to Quine and Derrida arguably makes the knower and the known inseparable. That is, a critical examination of the sources of the experienced and known world cannot reach "things in themselves." As James famously put it, "The trail of the human serpent is over all." A very sophisticated version of this view comes from contemporary American philosopher

Joseph Margolis (Margolis 1995).[11] Margolis refers to the embeddedness of judgment in the various anthropic cognitive strata, the mutual implication of language and world, as *symbiosis* (Margolis 1995). The traditional epistemological attempt to discover objectivity with no taint of interpretation or bias or cognitive selectivity, like the attempt to determine the full repertoire of human cognitive capacity without reference to its objects, is impossible. We never face an "un-languaged world" of reference uncolored by our knowing or an "un-worlded language" we can catalogue independent of what exists. In a sense, this recognition lies in Hegel, Dewey, James, and Merleau-Ponty, for whom the traits of the knower and the traits of the known are emergent from the primordial interaction of subject and object.

But there are problems with this view. The first is straightforwardly naturalistic (even though Margolis accepts naturalism and an evolutionary justification of human knowing). Unless natural science is grotesquely wrong, experience, perception, knowing, culture, and history are latecomers to a universe which existed long before discursive life. From the perspective which puts the symbiosis first, *that* judgment (the preceding sentence) is a construction or production, a product of the interaction. From a naturalist and realist perspective, experiencing and knowing obtain in a context they do not construct. Human judgment is perfectly capable of judging that something obtains completely independent of human judgment. Out of our symbiotic experience with the world, human cognition is able to posit some features of the environment that must be independent of that symbiosis, and which are relatively invariant, distinguished from environmental features that are less invariant. This is what we call objectivity. Surely we cannot judge that something obtains that is *unrelated* to our judgment, since the judgment is itself a relation of some kind. What is known is knowable, whatever is posited or discriminated has some relation to the positing or discrimination or the agent thereof. And surely, the interaction of subject-object is the context in which the judgment emerged, and on which its evidence depends. Nevertheless, part of our cognitive business is the discernment of relatively more objective, hence independent and invariant, environmental features. Humans alone can make our symbiotic interaction with the world an object of second-order reflection, then posit those features of the object independent of us as responsible for its causal contribution to the interaction. The achievement of this capacity is what evolutionary epistemology seeks to explain, by regarding the interaction of subject and object, and our ability to discriminate independent-invariant features causing the symbiosis, as itself a product of natural evolution (and, Margolis and I would both add, cultural evolution). The subject-object interaction is itself a natural product. The symbiosis is epistemically, not metaphysically, prior.

Second, and implicit in the foregoing, the degree of entanglement of the object in its interaction with the human judge varies (Cahoone 2005). Symbiosis is *graded.* That the adverbial means of judgment and what is judged are ultimately inseparable does not mean they are not *incrementally* separable. For example, it is not true that given disconfirmatory evidence any component of our worldview is equally liable for rejection or replacement. Neither scientist nor cabbie behaves that way. We never reach the asymptotes of un-languaged world or un-worlded language, but we daily strip bits of world or language from any given confrontation with experience. That is one of the ways we learn. We vary perspective, mode of access, observers, hence "thin" the embeddeness of our judgment in particular humanly bequeathed backgrounds, thereby discovering relatively greater invariance or objectivity across them. We are always triangulating on the basis of characterizing an object from more than one perspective—indeed, some nonhumans do this, at least through varying sensory location—a metaphor that works as well for reflexive as for social communication.

Our networks of judgments and modes of access overlap most completely around the most robust, least subject to control, hence most thin and universal features of the world as we know it. These are the minimally interpreted, the least dependent on a particular modes of access, the least embedded. The thicker and more particular, the more the judged or known object is colored by interconnections with diverse other judgments; while the thinner and more universal, the more the judgment abstracts from the particularities of other judgments and objects. Now we can see that "thinning" means, first of all, institutional abandonment of multifunctional or omnivalent judgment, and then, within an assertive or propositional judgment that has been differentiated from other modes of judgment, the adoption of cognitive methods and heuristics which narrow their judgment of the object toward sensitivity to those least controllable (most susceptible to disconfirmation) and most universal (most invariant among perspectives).

That is what natural science and mathematics do. Science is, as Cassirer argued, the most transparent of our symbolic forms (Cassirer 1965). This does not mean the empirical sciences constitute the most elevated or complete model of human knowing. There are forms of cognition that simply cannot be accommodated in a scientific model, hence science cannot be the method of choice for every object of cognition. Nor is science *utterly* thin or presuppositionless or neutral. But it is uniquely sensitive to certain kinds of facts, namely the causal interactions of natural complexes with shared human experience. Regarding those objects it *can* address fruitfully, it is the thinnest, the least embedded in culture, the least symbiotic, hence most capable of cross-cultural travel. It is also the least satisfying, the least able to serve cultural needs of determining the ends and meaning of human existence. This was Max Weber's point, that we gain and lose from modern rationality. We have developed a powerful method of inquiry extraordinarily sensitive to disconfirmation by what causally interacts with us as distinctively biological, mental, and cultural-historical agents, capable of continuous improvement, at the cost of omnivalent unification of our cognitive interests and methods. Modern natural science, upon which the current study depends, is that form of historical knowledge which grasps the most invariant possibilities of types of natural complexes by a method that generates hypotheses for public disconfirmation, thereby characterizing its objects in a way that is more independent with respect to bio-, psycho-, social- and cultural differences than any other form of knowing.

But it can't do everything. It is not the only kind of knowledge. There are legitimate cognitive aims it cannot fulfill. It is at a loss when the concern is omnivalent, when we seek to cross the bounds of purely assertive and practical or aesthetic query. And it falters when we enter causal thickets, namely, the bio-psychological thicket of a human individual's behavior, the socio-ecological thicket of the interaction among biospheric species and their societies, including our own; and the socio-cultural thicket of human group behavior and expression (Wimsatt 2007). There is no mysticism here, just massive complexity. There are many disciplinary orders that can be fruitfully brought to bear on the phenomenon in question, all of which have something legitimate to say. Science is privileged, but like all privileges, this one is limited.

Thus, as we have seen in the preceding five chapters, nature is not solely physical, even though everything in it must depend on the physical. Nature includes not only physical entities but complex material organizations, mathematical lawfulness, possibilities, life, need, behavior, intelligence, purpose, societies, minds, meanings, signs, and knowledge. Human knowing, a subset of animal knowing, is markedly distinct both because of the unique sensory, motor, and neurological funding of Homo sapiens, and because our species developed a uniquely social, Lamarckian, cultural form of production, which opened up a novel way of knowing that achieves both a more complex objectivity and is historically cumulative. This too is nature.

Part III

Naturalistic Speculations

11

A Ground of Nature

In Chapters 1 through 10 I have shown that naturalism, rightly conceived, can handle features of reality we robustly know, from the physical to the cultural, and as such is our most probable approximately true description of reality. This chapter and the next are a speculative coda to this study; they make inferences regarding important metaphysical topics on the basis of naturalism that go beyond the likely bounds of empirical inquiry. In the current chapter I explore two interesting facts about nature recognized by physical cosmologists. First, our universe appears to be past-finite. It *began*. This is the theory of the Hot Big Bang we saw in Chapter 5. Second, our universe's development has been highly improbable. Our universe is overwhelmingly inorganic, driving toward higher entropy, based in quantum indeterminacy, and filled with chance cataclysms; but it still exhibits far more organized complexity than we have a right to expect given the laws of physics. One of the factors responsible for this structure is that a number of fundamental physical constants, which are not determined by law, all have very special "fine-tuned" values. I will argue that contemporary cosmology's explanations of these two facts are either implausible or less plausible than positing a Ground of Nature, or God, as their explanation. I am claiming that the old arguments from cause and design, properly modernized, together form a robust argument for God. While the account is my own, others have tried to formulate a modern version of arguments for God on this basis (see for example Clayton 1997; Craig and Smith 2003; Russell, Murphy, and Isham 1993).

Some will object that the universe's origin and improbability are simply inexplicable facts neither requiring nor open to rational explanation. This is a plausible concern. But notice that current physical cosmology does *not* refuse to explore these matters. Explaining them has been a primary task of cosmologists in recent decades. They do try to explain, or explain away—in the sense of rendering moot—the origin and improbability of our past-finite universe. I agree with them that the physical facts deserve explanation, but disagree that their physical hypotheses can supply the explanation.[1]

It will also be objected that to speak of the Ground is to step outside naturalism. But that depends on the character of the posited Ground. For example, the Ground may be part *of* nature, that part which causes the rest. Some may say that within naturalism one can only make causal inferences *similar* to those made in natural inquiry. This I accept. As a part of nature, I am making inferences as to the cause of the initial state and certain features of the physical order, on which the other natural orders are dependent. The inferences are motivated by the evidence that nature is past-finite and has some features that, with great probability, could not have arisen randomly. Such inferences can only be contingent and probabilistic. There are cosmological theories that would, if true, invalidate my arguments or render them ineffectual. I will argue they are implausible or unlikely. Nor do my arguments imply that the evolution of the universe as we know it was deterministic, necessary, or inevitable. Last, in this chapter I will leave the nature of the Ground largely indeterminate, avoiding traditional attributions like omnipotence or omniscience. The only features I can assert are those required for the causation of natural orders. My point is that it is respectable for a naturalism to posit a Ground of Nature. What *kind* of Ground is an even more speculative topic that awaits our final chapter.

One last caveat is unavoidable in discussions of recent physical cosmology: very little is confirmed or universally accepted. New models and approaches are being created with great rapidity, some questioning the standard Big Bang model itself. I will of necessity avoid discussing what theory will eventually come to encompass both QFT and GTR, string theory or quantum gravity. Some of the theories below come in alternative models and versions I cannot present here. We will have to argue on a level that does not hope to adjudicate such matters.

In the first section I will give my argument for a Ground as origin of the universe, and consider cosmological alternatives, particularly the claim that the universe could have come from nothing. In the second I will give my argument for the Ground's setting of the fine-tuned constants, and consider other alternatives, particularly the notion of many universes, and the model of a recycling universe. But because the evidence and theories that address these questions sometimes overlap, I will not always be able to keep the two issues segregated.

I. Explaining a Past-Finite Universe

In the Western tradition nature as a whole has been commonly thought to be static or non-evolutionary, even if engaged in some periodic or cyclic motion. It either existed eternally, as Aristotle held, or was created, for the Judeo-Christian-Islamic tradition, but in either case was not evolving. Only in 1929 did Hubble discover that this appears wrong, because the universe is in the process of expansion, meaning that galaxies are moving away from each other, implying all was once very concentrated. The 1965 discovery by Penzias and Wilson of the background microwave or heat radiation that fills the universe in every direction made that conclusion very likely. The claim that the universe is past-finite is also buttressed by a thermodynamic argument. According to the Second Law, closed systems move toward, not away from, a condition of highest randomness or entropy. Like any closed system, this is presumably true of the universe; whether it will reach it is uncertain. If the universe were eternal in the past it would have had infinite time to reach equilibrium, hence it would have reached it already. But it has not.

Our current "standard cosmological model," called Λ-CDM (Λ or lambda stands for a cosmological constant, CDM for cold, dark matter), assumes "the cosmological principle," that the regions of space we can observe are no different from the regions we cannot observe, hence that the universe is homogeneous and isotropic. Our observable universe appears to be a Friedman-Lemaitre-Robertson-Walker expanding universe, with very flat curvature overall, that began in a Hot Big Bang, expanding from a condition of very small volume at very high temperature about 13.7 billion years ago. We must add to this two recent discoveries: most of the matter in the universe is cold (moving slowly) and dark (unobservable); and expansion is now accelerating, presumably due to dark (unobservable) energy, in effect, a cosmological constant (Λ) that imbues empty space itself with energy and hence "repulsive gravity" (as we will see).

But there are complications here, and while very widely accepted the Big Bang model is not immune to criticism. As noted in Chapter 5, Penrose and Hawking showed that, given plausible assumptions, the Big Bang must have begun in a singularity with infinite temperature and energy density, hence infinite curvature. This is itself a troublesome instant to include in any account that must use GTR. The first 10^{-43} seconds of the universe is the Planck era, of which, absent an adequate theory of quantum gravity, we know virtually nothing. At 10^{-37} seconds it is now widely believed (except for proponents of a cyclic universe) that the universe underwent a massive, abrupt inflation, from 10^{-60} to10^{20} m^3 in size. As inflation ended at around 10^{-35} seconds, continued cooling *may* have led to the symmetry breaking of the strong from electroweak forces—if they were originally unified according to grand unified theory, for which there is little contemporary support. We have more reason to believe that electromagnetism broke from the weak force at 10^{-12} seconds, giving us our universe's four forces, and a continued cooling and Hubble expansion. There are today serious questions about all but the last of these phases. In what follows I will take the reliable core of the Big Bang account only to mean that our observable universe began a finite time ago in a condition of very high temperature and density—whether there was a singularity at infinite temperature and density or not. As Vilenkin himself writes, "one thing is clear: the universe as we know it could not have existed forever" (Vilenkin 2006, p. 210, n. 4).

Having stated my version of the philosophical argument for the existence of a Ground of Nature elsewhere, I will merely summarize it here (Cahoone 2009). We have good reason to accept the view (from the ancient philosopher Parmenides) that Nothing can neither *be* nor *refer*. "Nothing," in the Greek sense of *ouk on*, or sheer absence (as opposed to *me on*, indeterminate or potential being, or "no-thing") can have no reference in any possible world (Tillich 1951, p. 188). It also has no experiential content, and, most relevant for the current work, no meaning in physics. Absence of matter is not Nothing in a universe that is mostly dark energy. The quantum field vacuum is a state of non-zero energy from which particles continually emerge and into which they are annihilated. Space is three aspects of a four-aspect system called spacetime with its own metric structure that causally interacts with mass-energy. So, even what we imagine to be Nothing—a black empty space—is *not* Nothing in the sense of utter absence.

Now, if we cannot refer to Nothing, or, if "Nothing" does not refer, then we must follow the Lucretian principle, "*ex nihilo nihil [fit]*," "from nothing, nothing comes" (Lucretius 1921, book one). If Nothing can never be affirmed, we cannot say something came from it. Nothing cannot explain anything.[2] If so, and if our universe is past-finite, what sense could it make to say *the universe* came from Nothing?[3] If that is implausible, there must be, or have been, a Ground of Nature, which is to say with Aquinas, an *uncaused cause* of the universe (*Summa Theologiae*, Part I, Ques. 2, Art. 3). Unlike Aquinas, I argue not that an infinite regress of past causes is irrational, only that it is apparently false. *Caused* causes of the universe would be other universes that are either eternal or past-finite. If they began, the origin question returns, unless one resorts to a series of polytheistic demiurges, which doubtless have their own evidentiary difficulties. So if the universe began and did not come from Nothing, there must be something which could exist without it, something without a beginning, which caused it. *Either* the largest or oldest physical ensemble containing our universe as a part or phase is eternal, *or* there is an uncaused first cause. In short: Aristotle or Aquinas.

We will momentarily review the potential physical objections to my argument. The most sensible *philosophical* objection would be something like this: "We all must accept that, as Wittgenstein said, 'Explanations have to end somewhere.' You yourself say they end with the Ground. But that is an explanation too far. Rational inquiry must stop with the Big Bang or the singularity that may have been its source. If we probe further, we end up with a mysterious Ground insusceptible to physical, hence rational, inquiry." Here looms a potentially unending debate about the aims of inquiry and the burden of proof. My only defense is that the censorship of that extra step can make no obvious claim to rational superiority.

The question is whether it is legitimate to inquire into the cause of our past-finite universe's initial state. If it is legitimate to inquire into all states of the reality we inhabit, including the first, then it seems natural and rational, for a philosopher at least, to inquire one more step, and somewhat unnatural and irrational to post an intellectual "No Trespassing" sign. In taking that farther step, we are confronted by the mysterious nature of the Ground, and the techniques of physical inquiry must falter. But that is the inevitable price of admission, not a sufficient reason to refuse the step, if we accept that the boundaries of inquiry exist to guide the discovery of truth rather than insulate extant views from challenge. We must choose our poison: either a likely although unverifiable model of the initial physical state that leaves it unexplained—which may be proper for physics, but less so for metaphysics—or a further explanation that steps into territory inaccessible to physical inquiry. I find the latter more rational; I would at least argue that it is not *less* so.

A Universe from Nothing

The first physical alternative to this argument was proposed in 1973 by Edward Tryon (Tryon 1973). The quantum vacuum is the scene of continuous emergence and disappearance of particles. Tryon suggested that the universe may have begun as such a fluctuation. This might seem impossible because the mass-energy of fluctuation-emergent particles is inversely related to their longevity, so for any such universe to avoid violating conservation of energy it would have to be extremely short-lived. However, Tryon noted that gravitational fields have *negative* potential energy which in a closed universe precisely matches, hence cancels, their mass-energy.[4] In that sense a closed universe has zero energy. It is then conceivable that vacuum fluctuations could produce a long-lived universe. As Guth summarized the idea, "*everything* can be created from nothing" (Guth 1998, p. 15).

Others pressed Tryon's idea further. New attempts in quantum gravity, coupled with inflation, led to two proposals for how the universe could emerge from "literally nothing." In 1982 Vilenkin analogized the creation of the universe to quantum tunneling (Vilenkin 1982, 1984). In quantum mechanics there is a nonzero chance for a particle to arrive at a state for which it did not have the requisite energy. If we graph the potential energy required to achieve that state, the quantum particle "tunnels through" the potential barrier (in Figure 11.1, from A to B) instead of bouncing off (like C). If *a* is the universe's radius, the square of the wavefunction Ψ(a) then gives the probability for a universe of radius *a* to appear on the positive side of the barrier, in existence. The portion of the process where $a < a_0$ is regarded as "Nothing."

Then in 1983 J. B. Hartle and Stephen Hawking published their "no boundary" proposal (Hartle and Hawking 1983). They summed all possible relevant spacetime paths by which a universe could evolve to find the path with the highest wave amplitude, hence highest probability to occur. That path is represented by a three-dimensional sphere in which time has become a spatial axis with imaginary values; we may imagine it as a sphere whose North Pole (P), where $t = 0$, is the initial state of the Universe or Big Bang, and whose South Pole (Q) is the "Big Crunch" at the end of the universe, if there is one (see Figure 11.2).[5] We now have a representation of the universe without a singularity or boundary, the first and last states of the universe being just two points on the sphere.[6] As the authors famously wrote, "the boundary conditions of the Universe are that it has no boundary" (Hartle and Hawking 1983, pp. 2961 and 2965). As Hawking later summarized, if so,

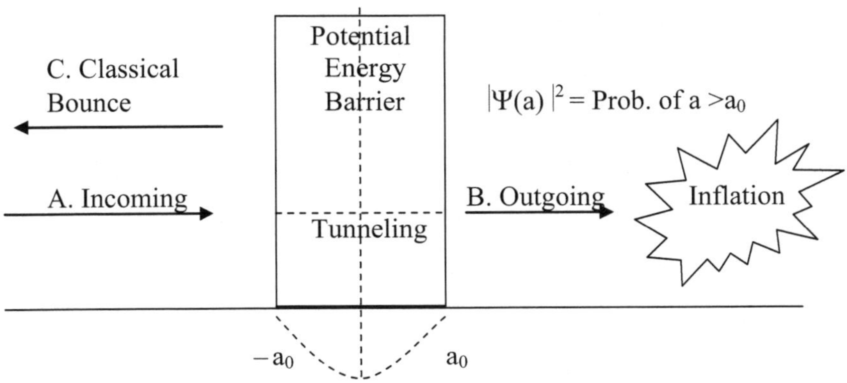

Figure 11.1. Vilenkin's Tunneling Hypothesis

"the universe would be self-contained and not affected by anything outside itself. It would neither be created nor destroyed. It would just Be" (Hawking 1988, p.13).[7]

There are some physical questions raised by these proposals. The wave function of the universe, it seems, must state, not the likelihood of our universe's existence compared to Nothing, but its likelihood compared to all other possible universes, whose sum (with ours) must be one. Treating time as an imaginary value must either mean that time is running backward or deny the physical reality of time and hence the Second Law (Cahoone 2009, pp. 786–7). But philosophically, the most glaring question regarding these theories remains: is their "nothing" really Nothing? The philosopher of physics Tian-yu Cao makes the point that quantum events "require a background space whose specific properties make fluctuations possible . . ." (Cao 2004, pp. 192–3). Quantum field theory presupposes a four-dimensional Minkowski spacetime in troubling contradiction to general relativity. As John Polkinghorne notes, to speak of quantum tunneling requires the quantum vacuum (Polkinghorne 1989, p. 59ff). Indeed, the language of Vilenkin and Hartle-Hawking proponents suggests a rather substantive "Nothing." Physicist Frank

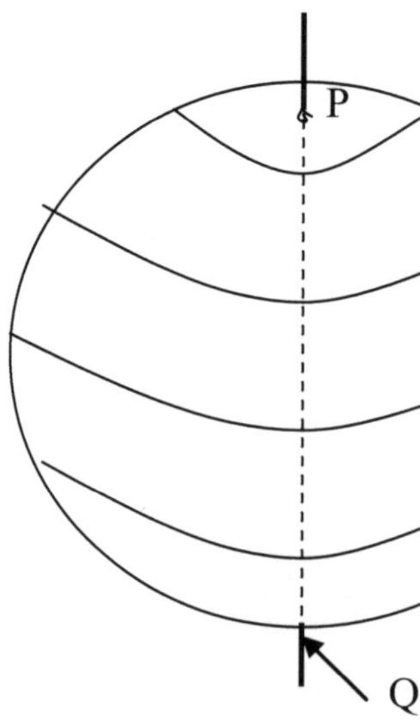

Figure 11.2. Hartle-Hawking "No Boundary" Proposal

Wilczek writes, "Our answer to Leibniz's great question 'Why is there something rather than nothing?' then becomes " 'Nothing' is unstable' " (Wilczek and Divine 1989, p. 275). Quantum Geometer David Atkatz remarks that the universe "nucleat[ed] from the *eternally existing nothing* . . ." (Aktatz 1994, p. 625, my emphasis). Victor Stenger supports Vilenkin and Hartle-Hawking, arguing that it is "natural" for nature to come to exist: "The transition nothing-to-something is a natural one" (Stenger 2006).

Here the philosopher must object. Nothing is *not* simple or complex, stable or unstable, temporal or eternal, natural or unnatural. *It just is not.* Either we are speaking of a state with properties and a causal role, hence speaking of something, or we are not speaking about anything.[8] It seems inevitable that, as Cao goes on to say, the theories in question must posit a "quantum nothing," an extant quantum vacuum or some other physical state that precedes $t = 0$, which is *not* Nothing in the sense of utter absence. If *that* state is past-eternal, we would have returned to the Aristotelian option.

Lastly, what about physical law? Quentin Smith, a supporter of Vilenkin and Hartle-Hawking, admits that their cosmologies require the "Platonic-realist theory" that the laws of nature exist independently of the universe (Smith 1998, p. 82). To claim there is a transition from Nothing to Something requires that the transition be rule-governed. So the laws must hold when the universe is *not yet*. Vilenkin admits "The Laws of physics must have existed even though there was no Universe" (Vilenkin 2006, p. 181).

This gives the game away entirely. The laws of physics are not Nothing, particularly when the whole thrust of the new cosmology is to build structure into the laws so as to leave nothing to initial conditions. As Guth writes, "If someday this program can be completed, it would mean that . . . the laws of physics would imply the existence of the universe" (Guth 1998, p. 276). Its existence would be *necessary*. But how could changeless eternal laws, essentially mathematical structures, *cause, initiate* and/or *fund* the mass-energy of a past-finite universe which they *pre*-exist?[9] If the laws not only structure but initiate and fund the universe, and exist when it does not, they begin to sound suspiciously like a Creator.

It is naturally the goal of physicists to explain the existence of our universe without reference to something outside the bounds of physical inquiry. In effect, their goal is to *avoid* an argument for a Ground. Whatever the virtues or vices of the new cosmological proposals, it is not true to say that they distinguish themselves from theistic arguments by following that traditional goal. Such was already available, if one is willing to stop one's explanatory regress with the initial state of the physical universe, which is to say, to *explain physical states in terms of other physical states.* Recent cosmologists aim to do more; they want to explain the universe without reference to contingent initial conditions, partly because those conditions are likely unknowable, but also in hopes of determining them by law, rendering them necessary rather than contingent. This task has led them to try, not merely to explain physical facts in terms of other physical facts, but *to explain the existence of physical facts per se.* They are literally trying to answer the metaphysical question, "Why is there something rather than nothing?" This is a novel ideal of physical explanation. Guth recognizes that "the attempts to describe the materialization of the universe from nothing . . . represent an exciting enlargement of the boundaries of science" (Guth 1988, p. 276). In suggesting that, absent his no boundary proposal, physics cannot show why this particular universe was selected, Hawking writes, "Was it all just a lucky chance? That would seem a counsel of despair, a negation of all our hopes of understanding the underlying order of the universe" (Hawking 1998, p. 133).[10]

Such "despair" results from holding two contradictory hopes: to explain everything about the largest and/or oldest physical ensemble, including its fundamental laws and constants, leaving nothing to unexplained contingency (e.g., initial conditions), *and* to keep physics closed, with no reference to an "outside." *One cannot do both.* Physics can rightly eschew reference to an agency independent of the universe, but in doing so it must leave some physical conditions unexplained. If our local, past-finite universe is the only universe, physics is left with no explanation for its existence, or for the nature of its initial state and governing laws; there will always be parameters and laws regarding which we can say *what* they are but not *why* they are what they are. Even an encompassing past-eternal multiverse which, like Aristotle's universe, avoids any need for a cause, still must have some unexplained constraints.[11] The new cosmologists have sought more, to explain the physical *per se.* But without reference to an outside that is impossible. One cannot explain how X comes into being without reference to not-X. An existing universe can govern itself, but it cannot govern its own creation.

II. Explaining an Improbable Universe

Constants are the fixed numerical values of certain quantities that play a role in laws or rules of interaction, which have to be put into such rules by hand as a result of observation instead of being fixed by law. Fundamental constants would be those playing a role in fundamental laws. Some of these would be *c* (the speed of light), *h* (Planck's constant for the unit of quantization), *G* (Newton's gravitational constant), the strengths of the other basic forces and the masses and charges of fundamental particles.

Now, any measurement of any property or value can yield any number, depending on the unit used. The distance from home plate to first base can be any positive real number we like, depending on the unit: 90 feet, 30 yards, 1080 inches, etc. What remains constant, and what really matters, is the relationship among values, in this case, lengths. Whatever units we use the distance from home plate to first base in baseball is about one and a half times the distance from home plate to the pitcher's rubber, about 3/10s the length of an American football field, etc.

Many have noted that our universe is improbable, given the laws of physics, and this is at least partly because many fundamental physical constants are "fine-tuned." This means that the values of those quantities occupy very narrow ranges in relation to each other (e.g., mass of the electron compared to mass of the proton, electromagnetic force strength compared to gravity). If the "fundamental" values in question did not occupy these narrow ranges, the universe would not be just a little bigger or smaller or older or hotter, but would lack stars or main sequence stars, heavy elements, planets, life, or all the above. Another way to put this is that a reasonable set of our actual fundamental constants occupy an extremely small region of the phase space of any universe with the same laws, forces, and elementary components as ours but whose values could randomly differ. The improbable rightly stimulates scientific investigation. Many scientists regard this investigation as central, e.g., Davies (1982), McCrea and Rees (1983), Barrow and Tipler (1986), Smolin (1997), and Barrow again (2002).

The Improbabilities

One set of improbabilities concerns the large scale structure of the universe and its expansion. The universe is remarkably homogeneous and isotropic, that is, any given large region is much the same as another, and the universe (e.g., the background microwave radiation) looks the same in all directions from Earth. This is unexpected because the rate of expansion since the Big Bang has put sections of the universe further apart than light would have had a chance to travel between them in the available time—hence no mutual causal influence has been possible between those sections since very early in cosmic history. For example, already at the Planck time no volume greater than 10^{-35}m could affect or be affected by another. The current universe is composed of at least 10^{80} such "particle horizons" which have never been in contact since that time. So the universe could not have developed its homogeneity and isotropy over time through interaction; they must be the causal result of the extremely uniform state close to the beginning. To reach current isotropy, the "anisotropy" at the Planck time, or directional differences in expansion rate between any points on the expanding sphere of the universe, had to have been less than 1 part per10^{40}, hence be exact to 40 places to the right of decimal point. The sameness had to be *astronomically* exact.

We also know from general relativity that mass and pressure will change the curvature of space. The background curvature of our universe appears to be very close to zero, its average density indeterminately close to its critical value for Euclidean flatness, Ω (omega). To reach the currently observed flatness, those values would have to have been less than $1/10^{60}$ different from their critical values at the Planck time. A change of 10^{-60} would have led to a very different universe. But its density was not *perfectly* uniform either. For such uniformity would have made it impossible for gravity to create stars and galaxies. P. C. W. Davies writes,

> It is hard to resist the impression of something . . . capable of transcending spacetime and the confinements of relativistic causality—possessing an overview of the entire

> cosmos at the instant of its creation, and manipulating all the causally disconnected parts to go bang with almost exactly the same vigour at the same time, and yet not too exactly coordinated as to preclude the small scale, slight irregularities that eventually formed the galaxies, and us. (Davies 1982, p. 95)

Particularly crucial is stellar formation. Hawking and Penrose argued that the formation of black holes is overwhelmingly more likely than main sequence stars, like our Sun. For normal stellar formation, the ratio of Ω to G could have been no more than one part in 10^{60} different from its current value at the Planck time. Any larger difference and all the matter of the universe would have accumulated in black holes.

Life can only exist near stars that are neither Blue Giants—hot radiating stars—nor Red Dwarfs—cool convective stars. The fact that there are main sequence stars, like ours, Brendan Carter pointed out in 1974, is due to the remarkable relationship between the strengths of the gravitational and electromagnetic forces. The relation between the gravitational fine structure constant α_G, which is $Gm_p2/\hbar c = 5.9 \times 10^{-39}$ (m_p is the proton mass), and the electromagnetic fine structure constant, which is $\alpha = e^2/4\pi\varepsilon\hbar c = 1/137$ (where ε is the electromagnetic force constant), is $\alpha_G \approx \alpha^{12}\,(m_e/m_p)^4$ (where m_e is the electron mass). This expresses the somewhat mysterious fact that gravity is far, far weaker than the other three forces. Carter determined that the window of variation in the relation of the two that allows for the formation of main sequence stars like our Sun is 5.9×10^{-39} to 2.2×10^{-39}. That is a remarkably small figure, the difference being $1/10^{39}$ of the higher value. Slightly weaker α_G and all stars would have been Red Dwarfs, slightly stronger α_G and they would have been Blue Giants (Barrow and Tipler 1986, p. 336).

Nuclear chemistry plays a particularly important role in constructing a universe in which life is possible. Hoyle hypothesized, and was later confirmed by experiment, that there must be a resonance level of Carbon-12 near 7.7MeV (MeV is a million electron volts, KeV a thousand). Resonance is the tendency of a system to amplify its oscillations about certain discrete frequencies, so that the system yields more energy output per energy input. He hypothesized this because the stellar nuclear reactions involving hydrogen, beryllium, helium, and carbon—namely, $3He^4 \rightarrow C^{12} + 2\gamma$ (γ are gamma particles), $He^4 + (99 \pm 6)KeV \rightarrow Be^8$ and $Be^8 + He^4 \rightarrow C^{12} + 2\gamma$—could not account for the amount of carbon in the universe. So, he reasoned, there must be a fortuitous resonance level accounting for greater than expected carbon production. The resonance level proved to be 7.656 ± 0.008MeV, *barely above* a resonance level of the combined beryllium and helium at 7.3667MeV. It also turns out that O^{16}, which would be produced by adding He^4 to C^{12}, has a resonance level at 7.1187MeV, which is just below the resonance of C^{12} plus He^4 at 7.1616MeV. That means if the O^{16} resonance level were slightly higher, C^{12} would have combined with He^4. There is thus an extremely narrow window within which the chemistry of hydrogen, beryllium, helium, oxygen, and carbon must fall, without which there would be virtually no carbon in the universe.

Some physicists have tried to quantify the collective improbability of the universe's constants. Penrose constructs a diagram that depicts the phase space of all possible universe evolutions, $\mathbf{P}_u$ (Penrose 2004, p. 730, fig. 27.21). Boxes inside the diagram represent particular initial conditions or parameters, the volume of a box giving its probability. (The phase space of a typical gas would show that the box occupied by equilibrium is almost equal to the entire phase space.) Our universe and its history is represented by a squiggly line with its $t = 0$ tail end in a tiny, lowest entropy box, which then oscillates through larger boxes, leaving its "now" arrowhead in a moderately sized box. Penrose insists on figuring in gravity, as many accounts of the entropy of the universe do not. The early universe was in a material, electromagnetic, and thermal equilibrium, but *not* a gravitational one, so its total entropy was in fact low (see Chapter 5). Now, substantial black holes might have an entropy per baryon of 10^{21}. Our universe has 10^{80} baryons, or total entropy 10^{101}. Using the Bekenstein-Hawking entropy formula for black holes, if all were caught up in the highest gravitational entropy state, that would give the universe an entropy of 10^{123}. So in the Penrose diagram, the volume of P_u, the phase space of possible universes, would be 10 to the 10^{123} – 10 *raised to the power of* 1 followed by 123 zeros. As Penrose puts it, if we imagine the Creator as picking out just the right possible universe in the phase space by sticking a pin in its box—like pin the tail on the donkey—then, "The Creator's pin has to find a tiny box, just 1 part in 10 to the

10^{123} of the entire phase-space volume, in order to create a universe with as special a big Bang as that we actually find."

Lee Smolin performs a related calculation. Starting with *G*, *c*, and *h*, he asks himself what the values of the proton, neutron, electron, and neutrino masses (represented by the subscripts *p*, *n*, *e*, ν, respectively), the Planck and cosmological constant masses, and the range and strength of the four forces, must be to make a universe in which stars can live for more than a billion years. To do this, the m_p/m_{planck} ratio must be 10^{-19}; the probability of that occurring is one in 10^{19}. Complex stable nuclei require that protons and neutrons have comparable mass, roughly a thousand times the also comparable electron and neutrino masses. The odds for the three ratios m_p/m_n, m_p or m_n/m_e, and m_e/m_ν each being within tolerance levels is 10^{-22} cubed, or 10^{-66}. The odds for m_p to be near its value is 10^{-9}. The probability for the universe to exist long enough for stars to evolve the cosmological constant or vacuum tension expressed in mass must be less than 10^{-60}. The probability of the weak and electromagnetic forces achieving their actual ratio to the strong force (taking the strongest in each case as the standard) is each one in 100. The probability of the ranges of the strong and weak forces to that of electromagnetism (the widest) is each one in 10^{40}. Combining all these probabilities, the likelihood for four force strengths, four force ranges, and four particle masses to be within the tolerances to create a universe of billion-year-living stars is unimaginably small: one in 10^{229} (Smolin 1997, pp. 401–02).

Let me be clear. The fact about the universe that needs explanation is not order or harmony or beauty or humanity. The fact to be explained is the universe's improbable complexity, of which life is one example. That complexity in turn hangs, not on any one fundamental physical constant occupying a small range of values, but on the coincident tuning of *many* constants to the range required for the complexity of the universe we observe. The occupation of a range of values by *a* constant is an event or fact. What is surprising is that all these facts occupy narrow ranges that *have something in common*: their necessity for the observed complexity to evolve. This design argument differs from older versions in that it is not a choice between design and randomness.[12] Current physics must assume that at least some laws of physics and some constants are unchanging during the entire history of our universe, since only that presupposition allows us to make guesses about the early universe.[13] The alternative to accepting a designing Ground is *not* believing the constants evolved through random interactions, because to say so is to claim the constants evolved. Note finally that my argument is probabilistic: that the constants occupy their common range is improbable, not impossible, hence that a Ground of Nature fixed the constants is merely probably true.

Thus conceived, the design argument is a special kind of argument from cause: it is based in the claim that the coincidence of constants requires explanation, that it is likely that something caused it. It is a contingent fact for which science has not yet found an explanation internal to our universe. As noted, current cosmology regards the coincidence of constants as an improbability that calls for explanation; attempting to explain it is *not* a mistake. As Smolin suggests,

> If we are to genuinely understand our universe, these relations . . . must be understood as being something other than coincidence. We must understand how it came to be that the parameters that govern the elementary particles and their interactions are tuned and balanced in a such a way that a universe of such variety and complexity arises. (Smolin 1997, p. 55)

One possibility raised by some physicists to weaken the improbability of our universe is that the "dials" may not turn separately, the constants may not be independent. For physical reasons unknown to us, the fact that one constant occupies its narrow range may be dependent on the fact that another does. If so, that would lower the number of possible combinations and hence the improbability of the observed values. This is indeed possible. The question is, just *how much less improbable* could such considerations conceivably make the universe? Suppose correlations among the dials reduce Smolin's 10^{229} by 10^{100}, twenty orders of magnitude higher than the number of baryons in the universe, to 10^{129}, close to Penrose's 10^{123}. Would that do it? How about another 10^{100}, for a total improbability reduction of 10^{200}? That would make the likelihood of the observed constants one in 10^{29}. Would the design argument lose its probabilistic rationality if the improbability of our universe evolving were a mere one in 100,000,000,000,000,000,000,000,000,000?

The Anthropic Principle

While a number of cosmologies serve as alternatives to my argument, a special notion employed by several of them needs explaining first. The original function of the anthropic principle was to recognize a selection effect or bias in our interpretation of evidence. Trying to explain some striking observations, particularly of the size of the universe, in 1961 Dicke pointed out such observations could only occur after stellar formation, explosion, and seeding of the universe with carbon had led to the evolution of intelligent observers. In short, the only universe that *could* be observed is one which had been expanding for at least 10 billion years (Barrow and Tipler 1986, p. 246).

Unfortunately, in recent decades the anthropic principle has often been stated in confusing ways, e.g., "the universe must have traits *x*, *y*, or *z*, *because of* the presence of intelligent life." The anthropic principle directly concerns, not the universe, but our observations of it. The Dicke example does not explain why the universe is old, but why all our observations should be expected to find it old. To say the universe "must" be old "because of" intelligent life cannot mean that the presence of intelligent life causes the universe to be old, only that intelligent life justifies our inference that the universe is old.

The anthropic principle is relevant to explaining the improbability of our universe only if there are or were other universes with other constants (with which Dicke was not concerned), but which could not be observed because *their* constants preclude intelligent life. The force of the principle is that our observational sample (our universe at this time) may be unrepresentative of the population from which it is selected (all universes). That requires a population larger than the sample, hence more universes than just ours. My argument for a Ground does *not* use the anthropic principle, not because there is anything wrong with it (used conservatively), but because explaining the universe's improbability through a Ground has no need for it. It is the alternative ways of explaining the coincidences, below, which hypothesize many universes and explain the observation of our local universe by the anthropic principle.

Inflation, Brief and Eternal

Inflation was first proposed by Alexei Starobinsky in 1979, then independently by Alan Guth, and later revised by Andrei Linde, Alex Vilenkin, and others. Inflation is the claim that the early universe's expansion underwent a brief but enormous acceleration very soon after the Big Bang, around 10^{-35} seconds, followed by regular Hubble expansion. The theory holds that the vacuum space of the early universe was filled with very dense energy fields, which Guth ascribed to Higgs particles, others to scalar fields today simply called the "inflaton." Several models have been proposed in this evolving notion based on QFT. There is a quantum field vacuum with non-zero energy. Theories of cosmological symmetry breaking—holding that the electromagnetic and weak force were once unified, or that all three nongravitational forces were unified—must posit a far more energy-dense quantum vacuum. Call the lowest energy vacuum of our universe the "true" vacuum. Inflation theorists posit that there existed in the very early universe, at the end of the Planck era, a very dense "false" vacuum—a scalar, or nondirectional energy field—"false" in the sense it is far denser than the true vacuum. Such a vacuum is equivalent to ascribing energy density to spacetime itself, akin to Einstein's idea of a cosmological constant.

The remarkable fact is that a cosmological constant would produce *repulsive gravity*. Repulsive gravity rests on the idea that gravity can create negative pressure (or tension). Gravity (or positive gravity) bends spacetime inward, but there are conditions that result in a bending outward, or expansion, of spacetime. In GTR the mass-energy tensor is dependent on pressure. Normally, if we have a cylinder of gas molecules, and stretch it to greater volume, we expect the density of the contents to decrease. But if there is a cosmological constant, a mass-energy density of vacuum space itself, then that is *not* the case; mass-energy would depend only on the constant, not change in volume (Guth 1998, pp. 172–3). The work done in stretching the container should be equal to $-p(dV)$, negative pressure times change in volume. But in the case of space itself, since energy density doesn't go down, energy conservation requires there must be an additional $p(dV)$ worth of force pressing *outward* to make up for $-p(dV)$. Negative pressure produces expansion if the outward force is greater than the inward tending force (positive gravity). So negative pressure yields expansion or outward curvature.

Inflation claims that this false vacuum expands massively as repulsive gravity defeats positive or attractive gravity. As it expands it decays toward a true vacuum by releasing true vacuum bubbles. Because the scalar field's mass density is fixed, it does not lose energy density as it expands; otherwise inflation would shut down immediately. This massive expansion of mass-energy does not violate energy conservation because gravitational potential energy is negative, and balances the gain in mass-energy (cf. supra note 3). At the same time, quantum fluctuations during decay constitute inhomogeneities in the distribution of mass-energy. Upon reaching the true vacuum, energy is released as masses of particles.

Inflation is widely accepted today, largely because that it is the only way we know how, consistent with other theories, to explain features of the universe that otherwise could only be assigned to inexplicable initial conditions of the universe, for example, the universe's current size while allowing the pre-inflationary universe to expand slowly enough that spatial regions could be causally connected by information traveling at no more than the speed of light. Properly adjusted, this explains homogeneity, isotropy or uniformity in all directions and nearly flat curvature; with the addition of quantum fluctuations it may also explain the inhomogenities of our universe. The hope is to make initial conditions of the universe irrelevant to the post-inflationary universe.

But Guth, Linde, and Vilenkin have gone further to hypothesize a multiverse of *eternal* or *chaotic* inflation. In such views inflation begins in many places and does not end, producing quasi-independent "island" universes in a vastly larger multiverse. The argument is that if the rate of the false vacuum's expansion is greater than its rate of decay, more new areas of false vacuum keep popping up. This implies there will always be more regions inflating than decaying, hence a continual and endless increase in the number of universes. The improbability of our universe and its constants can then be explained as a selection effect by the anthropic principle.

We are observing one of the few habitable universes. In fact, the view of those endorsing eternal inflation is that there are an *infinite number* of universes. Every possible constant-defined universe, meaning a universe not violating the laws of physics, occurs an infinite number of times. Each second, the number of pocket universes in any area of false vacuum is multiplied by 10^{37} (Guth 2007, p. 11–12). Indeed, Guth writes, "In an eternally inflating universe, anything that can happen will happen; in fact, it will happen an infinite number of times."

Such a hypothesis would indeed explain the improbabilities of our universe. But at a price: *it is the most extravagant physical speculation imaginable*! There is no evidence that the multiverse hypothesis is wrong, but also none that it is right. Guth remarks that the objection that universes cannot interact, and so observational evidence for other universes may be impossible, does not make the hypothesis unscientific, since science often entertains and even accepts hypotheses without observable evidence. It is true that scientists often entertain hypotheses where no evidence is available, if the hypothesis is the best available explanation of observations, if it coheres well with well established theories, and if there is the chance of more robust confirmation in the future, e.g., predictions that someday *might* be tested. The hypothesis of an infinite number of unobservable universes presses this license rather far. It swamps the speculative nature of theism, which posits one Ground to explain one observed universe. The justification for adopting eternal inflation is that it alone can avoid ascribing some of the improbable features of our universe to initial conditions or their cause (if they have one). The enterprise of positing an open-ended mechanism for creating our observed universe, which can be altered at will in order to make the condition of the universe before inflation irrelevant, brings to mind the story of the fellow looking around on the pavement beneath a streetlight. Asked what he is doing, he replies that he lost his keys nearby in the grass. Why then is he looking on the pavement? Because the light is better here, he answers.

Additionally, inflation's relevance to the *origin* of the universe is dubious. Vilenkin, Guth, and Borde have argued that in any inflationary universe where the average Hubble expansion rate is greater than zero, past-tending world-lines slow due to relativistic effects, yielding a finite past (Borde et al. 2003). Vilenkin writes, "past-eternal inflation without a beginning is impossible" (Vilenkin 2006, p. 175). Today, Guth, Linde, and Vilenkin agree that the multiverse is past-finite. If so, the multiverse is the most extravagant conceivable answer to the fine-tuning problem, and no answer to the origin problem.

A Recycled Universe

One alternative that might address both the origin of the universe and its improbability is the hypothesis of a past-eternal "bouncing" or expanding-then-contracting universe. It was proposed in 1922 by Alexander Friedmann, briefly embraced by Einstein and further developed by Richard Tolman in the early 1930s. Tolman showed that a closed universe could continually cycle without violating the Second Law *if* across cycles the universe continually increased its size, rate of expansion, and the length of cycle. Entropy would then increase without limit. But in 1957 Herman Zanstra realized that this meant earlier cycles must be shorter, and so a cyclic universe could not be past-eternal.[14] The model has been recently revived by Paul Steinhardt and Neil Turok (2007). It is meant to be an alternative to the inflationary multiverse. Steinhardt and Turok insist that nothing in the inflationary theories, brief or eternal, addresses the most basic problem in cosmology: the explanation of the origin of our observable universe. Such views leave a Big Bang, with its singularity, hence a beginning of all time, in place without explanation. They also believe use of the anthropic principle is unscientific.

Their alternative is ultimately dependent on the form of string theory called M-Theory (M stands for "membrane"). It posits two nine-dimensional membranes or "branes," each with our three usual spatial dimensions plus six additional spatial dimensions curled up in unobservable small knots within them, bounding the tenth dimensional space. Our observable universe is *one* of the two membranes. The branes vary in size, smoothness, and proximity. An attractive "springlike" force operates between them in the compacted 10th dimension.

What Steinhardt and Turok seek is a mechanism that would smooth and flatten the universe *before* our Big Bang, and entail no singularities. They identify the potential energy of the attractive force between the branes as the source of the dark or vacuum energy which is currently accelerating our observable universe. Their view is that the branes collide and separate cyclically, each collision being a collapse or "Big Crunch" and each separation a new Big Bang. Each Bang is followed, on our membrane, by a radiation-dominated phase, a matter-dominated phase, and finally a dark-energy-dominated phase, as it stretches, widens, and varies in smoothness. As the branes approach, some of their potential energy is converted into kinetic energy; at the collision, some of the kinetic energy is converted to hot matter and radiation, hence the Big Bang. This energy density is much larger than the remaining potential energy/dark energy. Only after billions of years would the energy density fall low enough that the potential energy would dominate again, leading to dark energy acceleration, which would smooth and flatten the universe. Because dark energy decays in their model, our membrane reaches a limit of smoothness and size, then begins to contract. The contraction takes place in the compacted tenth dimension; in our usual three dimensions the membranes stay stretched. The membranes approach each other leading to the next collision. Each cycle lasts about 10^{12} years. This cycling allows a process of smoothing and flattening our universe to a near vacuum, albeit with the necessary inhomogeneities for structured matter to form.

What makes the whole scheme thermodynamically possible is that there is no limit to negative gravitational potential energy, from which every bounce or collision draws kinetic energy, so more and more space and material can be created, hence the size of the space of our membrane increases with each cycle. While entropy is always increasing, entropy per unit of spatial volume remains constant. Steinhardt and Turok believe that some other traits of the universe also evolve across cycles: "some properties thought to be constants, like the masses of elementary particles, the strengths of the various forces, and the cosmological constant could actually vary over very long periods" (Steinhardt and Turok 2007, p. 166). They suggest the cosmological constant was once very high, hence in the early cycles stars could not form at all. But after some number of cycles at a given level, the vacuum energy quantum-tunnels to a lower energy step. It is in the nature of the dark energy decay that more cycles take place the lower the dark energy density, so the lower steps have the greatest number of cycles. They figure we are now on the lowest "step" of that density, and that 10 to the 10^{100} cycles can occur on this step (ten to the 10^{100th} power) (Steinhardt and Turok 2007, p. 249). Such an account replaces inflation entirely, avoids singularities, gives dark energy, the most important cosmological discovery of the past two decades, a central place in its mechanism, and avoids the need for the anthropic principle to explain the observable universe's improbability. But how well does it handle the origin and improbability of our universe?

First, there is the question of the past-eternity of the universe. If there was a beginning to the cyclic universe, then the old problem applies; for that reason, eternity has historically been part of the intellectual appeal of the cyclic model. But is the process of Steinhardt-Turok cycling past-infinite? The authors are not clear. On the number of past cycles, they write, "Perhaps the number is infinite. Or maybe there was some 'beginning' . . . after which the universe was driven toward regular cycling behavior . . ." (Steinhardt and Turok 2007, p. 244). While they occasionally say time is infinite, they do not assert as one of the implications of their model that there *must* be a beginning.[15] It would seem the membranes and their spacetime relations must be past-eternal, there being no account in the theory for their evolution.

Second is the problem of the fine-tuned constants. On the one hand, Steinhardt and Turok note that over very long periods of many cycles constants will evolve, although they present an account only of the cosmological constant. Our membrane will stretch over time, entropy will increase, and black holes (which will be preserved through bounces) will accumulate. On the other hand, they repeatedly say that universe-cycles will be remarkably similar, "nearly identical." Throughout the universe's history—or at least on our low step of dark energy—each cycle creates a universe with the necessary fine-tuning of constants to make an observable universe like our own. They write, ". . . the cyclic model predicts that everywhere in space has a distribution of galaxies and stars similar to what is seen from the Earth. . . . Virtually every patch [of space] produces galaxies, stars, planets, and life, over and over and over again" (Steinhardt and Turok 2007, pp. 241–2). They suggest that the current structure of our universe, its forces and character of elementary particles, arose over the course of many cycles as the "most efficient" way for passing though the Crunch/Bang, somewhat echoing Smolin's proposal of "natural selection" of universes (Smolin 1997). They regard this likelihood of the production of habitable universes a virtue of their theory: "All other things being equal, a theory that predicts that life can exist almost everywhere is overwhelmingly preferred by Bayesian analysis (or common sense) over a theory [the inflationary multiverse] that predicts it can exist almost nowhere" (Steinhardt and Turok 2006, p. 12).

Now, either all or most of the universe cycles have our constants, or they do not. If they do not, then we are thrown back onto the anthropic principle as an explanation of our observation of our improbable cycle. But if they do, the question of fine-tuning is merely transferred to *the entire chain* of universes. Why are the membranes and their relations structured so as to produce again and again universes with favorable constants? The larger ensemble of cycles has been made into what Robin Collins calls a "many-universes generator" that itself must have the structure necessary not only to produce cycles but cycles statistically *likely* to be habitable (Collins 2002). The improbability issue is then simply transferred to it.

Last, we must note that the "endless universe" remains, if less speculative than the multiverse, still impressively so. An unobservable mathematics of eleven dimensions and two cyclically colliding membrane spaces generates a potentially endless number of universe phases a trillion years old each. Even our current low-cosmological constant era would contain 10 to the 10^{100} cycles of universe evolution.

Conclusion

How can we explain the origin of our past-finite universe *and* the improbable setting of so many of its constants? The origin could be answered by asserting emergence out of Nothing, or the eternity of some larger ensemble containing or older process initiating our universe, or by an uncaused Ground of Nature. The improbability could be answered by either an anthropic selection from an enormous, randomly generated population of universes or past phases of our universe, or a Ground of Nature that fixed our constants. I have argued that emergence from Nothing is implausible in principle, and that the anthropic selection of our improbable universe out of infinite universes is the most grandiose hypothesis imaginable. Current theory is *not* compatible with a past-eternal multiverse. Regarding the cyclic hypothesis, current versions neither claim nor rule out past-eternity. An eternally cycling hypothesis *might* conceivably eliminate the origin question, but it either leaves the improbability question unexplained or returns us to an anthropic selection of our cycle from an infinity of cycles.

It seems to me, from the point of view of metaphysics, the most plausible hypothesis is that the initial state of our observable, past-finite, improbable universe was caused and had its peculiar constants fixed by something uncaused, a Ground of Nature. The disadvantage of this hypothesis from a scientific point of view, that it posits a Ground that is at least partly independent of the physical world *per se*, is not so great in light of the fact that its competitors posit an infinite number of universes or universe cycles that are in principle unobservable.[16] The Ground would have to: be uncaused; fund and initiate the universe; and fix at least some crucial features, like the laws of physics and the fine-tuned constants (we leave open whether this "fixing" is an act or a consequence of its nature). Nothing else about the Ground, or the process of Creation, need be assumed for the arguments in this chapter to work. Some may object that in principle the existence of such a Ground is itself inexplicable, so its hypothesis is not rationally superior to positing an infinity of universes. It is true that, as noted above, any line of explanations must end somewhere, and, at least in regard to explaining our universe's existence and improbability, must end in something that is inexplicable in terms of something outside itself. I believe positing one unobservable Ground of Nature is rationally more plausible as an explanation for what we observe than positing an infinity of unobservable universes. And I see no reason why the probabilistic hypothesis of such a Ground is incompatible with a physics that expects only to explain all interactions among physical realities, nor with a theology that respects our most sophisticated knowledge of nature.

12

Natural Religion

So you want to live "according to nature?" Oh, you noble Stoics, what a fraud is in this phrase! Imagine something like nature, profligate without measure, indifferent without measure, without purpose and regard, without mercy and justice, fertile and barren and uncertain at the same time, think of indifference as a power. . . .

—Friedrich Nietzsche, *Beyond Good and Evil*

Even more purposeless, more void of meaning, is the world which science presents for our belief. . . . That man is the product of causes which had no prevision of the end they were achieving; that his origin, his growth, his fears, his loves and his beliefs, are but the outcome of accidental collocations of atoms . . . that all the labors of the ages, all the devotion, all the inspiration, all the noonday brightness of human genius, are destined to extinction in the vast death of the solar system, and the whole temper of Man's achievement must inevitably be buried beneath the debris of a universe in ruins—these things, if not quite beyond dispute, are yet so nearly certain that no philosophy which rejects them can hope to stand. Only within the scaffolding of these truths, only on the firm foundation of unyielding despair, can the soul's habitation henceforth be safely built.

—Bertrand Russell, *A Free Man's Worship*

The more the universe seems comprehensible, the more it also seems pointless.

—Steven Weinberg, *The First Three Minutes*

Throughout the history of Western thought, the question of whether we are strangers in a strange land, alien from the natural world, has hinged in part on the question of the place of value in nature. Religious traditions have plied each side of the issue with different ends in mind. For the dualist tradition, we are continuous, not with the world, but with the Source of the world. A modern nature reductively understood as chance macroscopic outcomes of purposeless efficient causal interactions of elementary particles makes our alienation within nature more difficult to avoid. It has been commonplace to claim that the greatest achievement of modern science was to banish Aristotle's final causes from nature. The quotations above, one from an existentialist, one from a scientifically-inclined analytic philosopher, and one from a physicist, lead to the same destination: the universe is devoid of purpose or direction, values and purposes are ours alone, and when they seem to arise from nature they are our projections (Cahoone 1999). Bedfellows as strange as twentieth-century existentialism and positivism agreed on this.[1]

The European eighteenth century had an intellectual tradition called "natural religion" or sometimes "natural theology," most famously associated with the Englishman William Paley. It attempted to formulate religious ideas from reason, experience, and nature, compatible with the science of the day, rather than from revelation, faith, or private experience. Recent natural theology is heavily based in evolutionary or process philosophies, above all Whitehead. At any rate, the tradition of attempting to forge a notion of religion based on inferences from nature continues to this day, contributed to by a number of philosophers, scientists, and theologians, for example, John Cobb, Philip Clayton, Donald Crosby, Ursula Goodenough, Erazim Kohák, Robert Neville, John Polkinghorne, Robert John Russell, Jerome Stone, and Wesley Wildman.[2]

As of this writing we in the United States seem to be in the midst of a spat over the relation of science and religion in intellectual circles, marked by rhetorical excess on both sides. Less than a century ago, few thought religious faith or speculation incompatible with the highest level of scientific work, as evident in the writings of Darwin, Einstein, and Schrödinger. If one prefers the view of someone closer to our time, actually a player in the early stages of the polarized atmosphere in which we now live, we can listen to Karl Popper from a 1977 paper. His reflections, which include a remark drawn from Darwin's view, deserve to be compared with the quotations above.

> I am on the side of science and of rationality, but I am against those exaggerated claims for science that have sometimes been, rightly, denounced as "scientism" . . . I am against intellectual arrogance, and especially against the misconceived claim that we have the truth in our pockets. . . . the success of Darwinian natural selection in showing that the *purpose or end* which an organ like the eye seems to serve may be only apparent has been misinterpreted as the nihilist doctrine that all purpose is only apparent purpose. . . . Although Darwin . . . [showed] that what appeared to Paley as purposeful design could well be explained as the result of chance and of natural selection, Darwin was most modest and undogmatic in his claims . . . [writing] "I cannot think that the world, as we see it, is the result of chance; and yet I cannot look at each separate thing as the result of Design. . . . With respect to Design, I feel more inclined to show a white flag than to fire . . . [a] shot." The question may not be within the reach of science. And yet I do think that science has taught us a lot about the evolving universe that bears in an interesting way on Paley's and Darwin's problem . . . I think that science suggests to us . . . a picture of a universe that is inventive or even creative; of a universe in which *new things* emerge, on *new levels*. (Popper 1987, pp. 142–43, author's emphasis)[3]

In what follows, no systematic natural theology can be attempted, nor will the reader find ruminations at the level of sophistication of a Darwin, Einstein, Schrödinger, or Popper. All I mean to suggest is how the question of values in nature leads to the deeper question of whether nature exhibits direction or purpose, and how this might cohere with a minimal set of claims about the Ground and its relation to nature. After a discussion of the "fact-value problem," Section II suggests characterizations of the Ground and Creation. Sections III and IV propose that nature has, as a matter of contingent fact, a direction, and take from this a (not *the*) principle of value which has at least a modest connection with the human prospect. Throughout, my proposals will be highly speculative. Whatever their worth, I hope to throw doubt on the belief that the quotations presented at the outset of this chapter are inevitable for naturalism.

I. Facts and Values in Nature

Along with the dualistic metaphysical view described above goes a special problem, the "fact-value dichotomy." The most famous component of this view is sometimes called "Hume's law," that one cannot infer an "ought" statement from an "is" statement.[4] Second is G. E. Moore's claim that good is indefinable because it is a simple, primitive, and non-natural concept, hence it cannot be pleasure as hedonists, Epicureans, and Utilitarians suggested. A third component of this tangle is the fallaciousness of "appeals to nature"—that something is natural does not by itself make it good. And we might add Max Weber's famous argument for the separation of value and fact that scientific knowledge does not *by itself* validate the political or moral views of scientists who hold them (Weber 1972). The cumulative effect of these claims has been to make talk of values in nature, if not impossible, at least dubious. Nature, it appears, is value-neutral; values are solely human.

In the last thirty years revisionists have argued that the dichotomy is overblown and fraught with exceptions. There are in fact some "is" claims from which we *can* infer "ought" claims. There are natural language predicates which arguably span fact and value; for example, murder and cruelty are in principle morally wrong because the very terms refer to something illicit; the decision is whether an act *is* murder instead of simply killing, or harmful instead of cruel. Factual claims about function, how something works, imply "ought" claims without any trouble. As Alasdair MacIntyre liked to say, the knife is *supposed to be* sharp. Lastly, as Hilary Putnam, himself inspired by pragmatism, points out, our very investigations of reality, hence our statements of fact, are suffused with "epistemic values," not only truth, but consistency, simplicity, reasonableness or rationality, and even aesthetic satisfaction (Putnam 2004).

Now, this revisionist response to the alleged dichotomy is certainly correct. The putative dichotomy is riddled with important exceptions. Further, as we have seen, there certainly are values in nature, if for no other reason than that there are valuing organisms which value themselves, select actions for their value, and inherit a valuational system. But stating this fact begins to indicate that such considerations do not address the underlying problem which, one suspects, continues to fuel belief in the difficulty of ascribing value to natural fact. For it seems to me the yet unaddressed core of the fact-value problem is, as Max Weber knew, that it is part and parcel of modern rationality to validate conclusions in terms of distinct norms and premises internal to some activity, two of which are assertion of fact and assertion of value. As argued in Chapter 10, modern rationality is uni-functional in Gellner's sense. This does not establish an unbridgeable fact-value dichotomy, but a modern recognition that crossing between fact claims and value claims requires in most cases a distinctive justification, that the assertion of natural fact is by itself insufficient to justify statements of value (except in the special cases the revisionists rightly noted). After all, a central problem of meta-ethics is how *any* statement of good or right can be non-circularly justified at all. The real problem of fact versus value is that, even if there are values or goods in nature, that by itself that does not mean we *ought* to value those values, that they are intrinsic values we would be irrational not to honor. Can normative human values be justified by our knowledge of functions or values or value-loaded-facts in nature?[5]

It seems that an affirmative answer would require enunciating a context in which natural value is embedded and in which we are also located. This does not mean finding values where the natural system promotes our interests, which are real enough but instrumental rather than intrinsic, and so do not answer our question. Rather, we would have to be able to say that a natural system serves an end or embodies a value that *also* is, or should be, an intrinsic value for us. That would seem to require that it be a natural end or value that we ourselves serve or embody. But to ask this is to begin to raise a daunting question indeed: does nature in general have a point or direction or end which, as participatory components of nature, we ought to respect or embody?

That may seem an unnecessary or even unfair inflation of the issue, but it is not. This can be illustrated by a disagreement within contemporary environmental philosophy, between anthropocentrism and biocentrism. The latter argues we can justify intrinsic values in nonhuman living things. The former holds that there are values of nonhuman things which we ought to honor, but those values lie in the instrumental use of such things by and for us. Anthropocentrists then hope for an enlightened self-interest on our part, e.g., the recognition that we benefit from environmental diversity.

Here is a question for both camps: *is stellar nucleosynthesis good in itself?* Was it a good thing that stellar nucleosynthesis emerged in cosmic history? When it occurred, was that an event with a value *we* now ought to recognize? From the perspective of anthropocentrists that event was value-neutral and only became good once humans existed to endow it with value as a precursor to themselves. Biocentrists would have to say much the same thing, now shifted back four billion years: stellar nucleosynthesis was value-neutral until life appeared. But each side must, if willing to face the question further, conclude that if stellar nucleosynthesis started up and then stopped a bit later—a momentary hiccup in a few proto-stars—nothing of value would have been lost.

That seems inadequate. For one is tempted to say, if anything that has ever happened is or was good, nucleosynthesis is and was! Suppose that judgment is legitimate. What view of the observable orders of nature must *that* claim presuppose, if it is valid? To believe that nucleosynthesis is a good thing, in itself, would seem to require belief that *nature itself has a direction or purpose or end exhibited by nucleosynthesis, a direction or purpose or end that we participate in and ought to honor.* Answering the question of value in nature thus drives us toward the troublesome but inescapable question, in Weinberg's terms (above), of whether the universe has a "point." I will argue, on the basis of our *a posteriori* scientific observations, that the universe does, as a matter of contingent fact, exhibit a valued direction or point.

II. The Ground and Creation

When naturalism speaks of God, or a Ground of Nature, it does so *a posteriori*, as in the preceding chapter. Hume famously made two insightful criticisms of such arguments. The first is that causality *within* the world cannot be analogically extended to a cause of the world as a whole (Hume 1910, 1998). Related to this is the question of what kind of cause the Ground can be for a universe to which time, or at least the time we know, is *internal*. We cannot say that the Ground existed "before" the universe, or speak of creation taking place "at a time." Hume's lesson is that if one argues from the nature of the universe to God as its cause, one's conclusion must be faithful to the context of that inference. If we extend the intra-natural notion of causation *to* nature it must look something like causation of things *in* nature. Hume's second argument is that because the world is only partly, not perfectly, ordered, the design argument can prove only a limited, not perfect and omnipotent, god. I will take Hume's points to heart.

The Ground must be continuous with nature in some respects and discontinuous with it in others; continuous so it may physically cause the universe and be its source, but discontinuous so it may exist independent of nature and cause the universe's initial state. While we don't know how such a Ground creates or created nature, a cautious approach would suggest that to cause a physical world the Ground must *act physically*. At the same time, the Lucretian rule I used to argue for the Ground, that nothing can come from Nothing, suggests that there can be no creation *ex nihilo*.[6] Rather, it is suggested that universe must not only be initiated by the Ground, but must *come from* the Ground, its content must be funded by the Ground. Thus, if there is nothing coeval with, yet distinct from the Ground prior to the Creation, the Ground must create the universe out of itself.

That points, not inescapably but readily, to the doctrine of *panenthism*, which holds that the universe is in or of the Ground, even if the Ground is more than the Universe. This was arguably the view of Spinoza, as clarified later by Schelling, and recently advanced by Clayton (Clayton 1997). As will be noted below, to cause the universe's initial state, the Ground must (among other things) physically cause or fund it. The Ground must be at least partly physical. I am thereby joining Spinoza in claiming that at least one of the attributes of the Ground is physicality. Panentheism also helps with our notion of causality. The Ground obtains independently of the universe, but the universe is not independent of it. The causation of the universe is something that happens to, is done by, or proceeds in the Ground. It remains impossible to talk of the Ground "before" the creation because time is part of creation. But it is not necessary to do so.

My concept of the Ground might seem to raise problems for the kind of naturalism developed in this book, based as it is on Buchler's ordinal pluralism. Nature, as I have understood it, is natural complexes. The Ground must both be continuous with nature and discontinuous with nature, or at least with the orders of nature we know, like the physical, so that the Ground may cause physical nature while nevertheless existing independent of it. Those features of the Ground that are continuous with nature must be, in my language, natural complexes, but some features of the Ground that obtain independent of nature cannot be, if they obtain independent of nature. One may say that in so far as the Ground is independent of nature, it is capable of becoming natural complexes, but is not a natural complex. Or one could understand those features of the Ground that are discontinuous with the natural orders we know as yet another order of nature, from which the physical emerges. Neither of these options contradicts a naturalism that has avoided claims about the Whole or all complexes.

Now, *what are* the characters of the Ground which are continuous with the universe is of course difficult to say. Following our usual approach, we will proceed outward from the science. Students of quantum gravity assume that spacetime is an emergent phenomenon around the Planck time (Smolin 2001). We cannot understand efficient or material causality without energy. Perhaps then the physical aspect of the Ground is energy not located or extended in spacetime, but nevertheless obeying a law of conservation of energy internal to the Ground. This would mean the Ground was and is metaphysically continuous with the physical universe while not exhibiting spacetime or the determinations that generate the fields of quantum field theory and general relativity. It may be the Ground has some more direct internal connection to either, or both, the quantum vacuum and spacetime singularities. The option that springs to mind, then, is that the physical dimension of the Ground is vacuum physical energy. The suggestion is that, independent of Creation, the Ground exists in a non-spatio-temporal but energetic state of minimal determination akin to Anaximander's Boundless, Schelling's Absolute, or

Neville's Indeterminate Being-itself (Anaximander 1988, Schelling 1936, Neville 1968). The Ground is not constituted by the kind of determinations characteristic of natural complexes. But this cannot mean the Ground is *entirely* indeterminate or without traits, since that would make the Ground "Nothing" in the objectionable sense. It must be determinate in some sense(s).

In what ways the Ground is determinate independent of nature, my inquiry cannot say at all (in this sense, I follow Neville's 1968). But at least one negative remark seems appropriate, and this is our response to the second of Hume's famous criticisms above. Hume's premise, that the world is only partly ordered, today appears to have been right. Our universe includes objective chance, evident in quantum theories, and a tendency to disorder enshrined in the Second Law of thermodynamics. Our universe exhibits *both* disorder and order, as Peirce claimed. This means not that the teleological argument is invalid, but that it needs revision.

Bluntly put, there is no good reason to regard the Ground as infinite, omniscient and/or omnipotent. In recent theology, the most famous endorsement of divine finitude comes from Charles Hartshorne (Hartshorne 1984). He holds that God is the preeminent or greatest conceivable being, but not perfect, since that would preclude change. God is unsurpassable by any other being, but not by Itself. God is the most powerful being, but only as powerful as is compatible with the power granted the beings of Creation, hence not *all* powerful. God knows all past and present actualities, but only knows the future *as possibility*, which, for Hartshorne, is all the future is. God does not know the unknowable, namely, which possibilities will become actual.[7] In effect, God is the most perfect *becoming* rather than the most perfect being.

One reason such a view is theologically preferable is that the traits of omniscience and omnipotence are more trouble than they are worth. Even under moderate interpretations, an omniscient Ground would have to know all states of all elementary particles and all bodies composed of them, all their relations, properties, pasts, and all their possibilities at all times. An omnipotent Ground would have to be able to accomplish its purposes, whatever those are, *in a way different than it has*, e.g., with force equal to mass times velocity instead of acceleration, create life without carbon or water, minds without neurons, etc. We cannot make sense of such notions. Worst of all, the most daunting problems attendant on the monotheistic conception of God are the result of omniscience and omnipotence, it being difficult to reconcile the former with objective indeterminacy and human free will, and the latter with God's goodness. That is the problem of evil. With regard to any evil in the world, *if* God both knew it was going to take place *and* could have arranged things so as to accomplish all relevant Divine purposes without the evil in question, yet *chose* not to do so, the Ground could not be entirely good. Denying omniscience and omnipotence frees us to accept the reality of chance and risk, human free will, the reality of evil, and the goodness of the Ground. (And we do wish to ascribe sufficient goodness to the Ground as an inference from the goodness of the complex forms of being the Ground has initiated, to be seen below.)

My suggestion is that the Ground does not know and cannot control all the particular details of the vast interactions of the system it generated out of itself. Even if there is a Divine purpose, it has required a universe of objective chance, stochastic order, and the Second Law, in which only *some* of its features or events will achieve that purpose or its component steps, not all, just as the achievement of the plant's purpose requires that only a fraction of the plant's seeds or pollen will germinate and grow. If the Ground's knowledge and power are not infinite, they can still be unimaginably great—i.e., greater than the greatest knowledge or power imaginable, multiplied by the greatest positive integer imaginable. That seems great enough.

Now, for our past-finite universe to have arisen the Ground must have done three things: *initiated* or have been the efficient cause of the universe; *funded* the universe's initial state, as its physical cause; and *determined* both the fundamental laws of nature and the fundamental physical constants (one might regard this as providing the "formal" cause, although that is a complex issue). While some extant laws or interaction rules and constants may have evolved during the universe's history, there must be some laws or interaction rules and constants that held initially, and some of these must not change, e.g., energy conservation.[8] The laws (or some of them) and the constants (or some of them), along with the initiation and funding, must be acts or consequences or characteristics of, developments from, or emergent upon the Ground—which language is best is unclear. Regarding the process of creation itself, here we may take a suggestion from physics.

As we saw in Chapter 5, conservation laws were in the twentieth century discovered by the mathematician Emmy Noether to be symmetry transformations, that is, spacetime transformations under which the phenomena remain invariant. Some physical processes "break" symmetries (e.g., freezing water molecules line up in a definite framework). But some scientists stretched the notion even further: in the late nineteenth-century Pierre Curie could write, "*C'est la dissymetrie qui cree le phenomene*," "Dissymmetry creates the phenomenon" (Curie 1894, p. 393). A phenomenon can be distinguished from local background states by its lesser symmetry.

This way of thinking can be applied more generally to the components of creation. Nothing or Non-Being has been the historical antithesis of Being in Western philosophy. But I argued in the last chapter that "Nothing" or "Non-Being" has nothing to refer to. If the difference between something existing and not is *not* the difference between presence and utter absence, what is the difference? One plausible answer is to replace the conceptual opposition of Being and Nothingness with the opposition of Determinate Being and Indeterminate Being, the distinction being *relative* (Neville 1989). To exist as a natural system would mean to maintain a *state*, the state of a relatively determinate something, that contrasts with background conditions or environment. What we mean by nonexistence is the cessation of distinctiveness or identity, for a thing to become indistinguishable from its environment.

Borrowing from the physics of symmetry-breaking and Schelling's notion of the *Ungrund*, we can suggest that the Ground independent of creation is neither One nor Many, because both 'One' and 'Many' require numerical definiteness or determinateness which the Ground may not have. This need not be a static notion; the Ground independent of creation may be in a state of oscillation or cyclic activity as in Schelling's "rotary of forces" (in the Greek imagination, God endlessly contemplating God). It may be that that the Ground is characterized by vacuum energy, perhaps the quantum field vacuum, or whatever eventually becomes the predecessor of the quantum vacuum in the yet unachieved adequate theory of quantum gravity. Perhaps today's quantum vacuum is that part of the Ground which is nearest to its original state, enduring in a primordial, dynamic, but cyclic condition. The creation, that is, the Big Bang, would then be a stoppage in that cyclic activity, the Ground *breaking its own symmetry*, yielding a massive spike of energy to expand, creating spacetime as it went, to cool under the rule of time, or the Second Law. The universe is an outpouring into discriminable forms of the original energy that was and remains a property of God.

Whatever the plausibility of these speculations, at a minimum we may say that the Ground of Nature as understood herein should be conceived as just that: Nature's Ground, the uncaused complex or order (partly natural) from which the rest of nature emerges, responsible for nature's existence and law-governed but improbable development.

III. Does Nature Exhibit a Direction?

We can extend our claim that natural existence is an asymmetry between a system and background or environment, to say, if not about all systems, then about complex, atomic and supra-atomic systems, that the existence of any system is a state of dynamic tension. Time, which is to say irreversible processes following the Second Law, which is to say increasing entropy, is the necessary background for any system. Natural existence is a kind of struggle against time, which is to say against the irreversible march to equilibrium or the elimination of distinctive structure. Put in ordinal terms, things prevail in the natural order only on the basis of the tension between entropy and structure. As Rolston remarked, noted earlier, "In nature there is a negentropic constructiveness in dialectic with an entropic teardown, a model of working for which we hardly yet have an adequatie scientific, much less a valuational theory" (Rolston 1988, p. 199). The existence of discriminable beings, supra-atomic structured beings, is thus beset by *hazard*. Physical realities are always in the process of falling toward thermodynamic equilibrium. The very existence of physical systems is in that sense a fight against entropy, probability, and/or time.

However, as recent scientists of complexity have argued, this is only to look at the glass as half-empty. It is also half-full, for the global tendency to equilibrium is what allows and fuels the local build-up of highly ordered systems far-from-equilibrium. The high-entropy environment provides the source of the complex system's energy and the sink into which the complex system must pump its own entropy (Schneider and Sagan 2005). Just as in genetics, Rolston writes, randomness or mutation is a diversity generator, the tendency toward equilibrium *enables* structure and complexity (Rolston 1988, p. 207). The background rush toward thermodynamic equilibrium is the pool of resources, the recycling, of material and energy from defunct pockets of order, or lesser symmetry, from which anti-entropic processes build up new pockets of improbable order. Entropy becomes what Schelling called the "dark ground" of the Absolute, and the various forms of order built up by the four forces of nature become the "Existence" of the Absolute.

The Neo-Platonic notion of devolution or ingression followed by evolution is not a bad metaphorical description of the history of the universe from this point of view. The universe began as an outpouring "chaos" in the sense of intense heat, expansion, and early on, thermal equilibrium. This is the process of emanation or involution, which meant the dispersal of radiation and, with cooling, hydrogen and helium atoms. Emanation then led to evolution in the form of concentration due first to gravitation, then electromagnetic processes, the build-up of improbable, complex systems, above all, stars. We only must adjust the ancient notion by saying that the transition from devolution to evolution occurs not once but *many times in many places*. The evolution of stars leads, if they eventually explode, to a new devolution or spread of the atoms of heavy elements. As systems move toward equilibrium or devolve, and some achieve it locally, in many locales order has built up. Any complex system is in a self-maintenance struggle that it will ultimately loose, but some make a creative difference to the structure of its locale that will outlive them. So, to join Persian-Zoroastrian with Indo-Buddhist figures, there may be an evolving Ground, but within that an opposition and struggle of "cosmic principles" that are nevertheless symbiotic: structure depends on entropy and cannot exist without it.

All this is of course very speculative. But the following is less so: *it is a contingent, a posteriori truth that nature appears to have a direction*. As this book has sought to show, increasing complexity based on stratified dependence and emergence is a—I do not presume to say "the"—moral of the story of the universe. That is, a major feature of reality is that it has evolved hierarchically layered types of entities, structures, and processes of vastly differing scales, as the product of cosmic, chemical, biological, psycho-social, and cultural evolution. In the universe's history the complex has generally come later. This claim cannot be made of every step in development; the process has been fitful, with frequent reverses. Not everything, indeed, not most mass-energy in the universe, exhibits this direction. Nor do we know where the universe is going. But we have a reliable notion of where it has been. My claim is *not* that everything in the universe is characterized by increasing complexity—total entropy cannot decrease—nor that there are laws or forces aimed at complexity, nor any "vital force" of complexity-building. That nature has developed progressively more complex forms is presumably the contingent result of the favorable physical constants and laws, and the interplay of entropy and structure. But local complexity is where the contingencies have led.

This is our final revision to the traditional argument from design. As we saw, the universe is indeed only partly ordered or deterministic, and its creation does not require that God be omnipotent or omniscient. The argument itself is probabilistic; the Second Law makes the undesigned achievement of complex structure improbable, not impossible. Furthermore, the universe's achievement of complex structure, even given design, need only be probable. The degree of design my argument requires is only that the original ensemble of spacetime energy, or whatever turns out to be the ontology of a reliable theory of quantum gravity, was constrained by the Ground so that the likelihood of the eventual development of the complex forms of order we actually see would be *sufficiently higher than* their otherwise enormous unlikelihood—to what number between zero and one I do not propose to say. It need not claim that the evolution of stars, life, and mind was certain or inevitable. Splitting the difference between quantum uncertainty and Einstein's claim that "God does not play dice with the universe," Ralston puts it this way: "There is dice throwing, but the dice are loaded" (Ralston 1988, p. 186). There is no need to assume that the development of our universe was deterministic.

If the development of increasingly complex systems is a fact about the universe's evolution, then this provides a context for a claim about value. Any enduring complex phenomenon from a hydrogen atom to a DNA macromolecule to a cell to a human is a locus of dynamic self-maintenance amidst competing forces that would otherwise reduce it to greater symmetry and lesser structure. This suggests that in nature organizationally complex existence is always an improbable achievement. I shall introduce a value-claim that is difficult to avoid: achievements, *qua achievements*, are valuable in themselves. This does not mean anything achieved is good, or any achievement is as good as any other, only that the achievement itself a good. In our universe the default process is falling to equilibrium. But in it we find remarkable local teleomatic, teleonomic, and teleological processes which depend on each other and on the default process. These telic processes all exhibit value in the context of the default process, achieving the rare, the improbable, the structured.

We can now say that *complexity, qua complexity, is good in itself.*[9] For it is the system's complexity which must be an achievement. Kohák, the environmental phenomenologist, recaptured the Augustinian claim that *esse qua esse bonum est*, being as being is good (Kohák 1987). He meant not that everything that is, is good, nor that, regarding anything we can think of, it would be better for it to exist than not to exist, but that the existence of things is good. In effect, this is to say that something is better than nothing. But, as argued here, "non-being" is not a possibility. So my analogous but narrower claim is that complexity, as complexity, is good. And the more complex, the more valuable. Why? Because complexity is improbable, achieved, because it requires work. This is not to say *only* what is complex or achieved or improbable is good; I am not attempting a general theory of value. The claim is that whatever else is valuable, and however often this value is trumped by other values, the complex has value.

While my argument does imply that the more complexity a system has, the more good or value it has, this is only *ceteris paribus* true. There is certainly useless and redundant complexity. There are different kinds of complexity, and there is the effect of the complex system on the complexity of environing systems. The evolution and/or introduction of a more complex species is a good, but not where it becomes an invasive species that obliterates the complexity of an ecosystem. Bureaucratic and military organizations and autocracies may seem more complexly structured than liberal democracy, but while they have more of *one kind* of complexity, they have far less of other kinds.[10] Diversity is a kind of complexity, characteristic of societies and ecosystems, where diversity may also be a good. But that a primate is more complex than a worm does not mean it is more "diverse" than a worm; it means it has more and different kinds of parts, a more articulated set of anatomical structures, especially a much more powerful central nervous system, hence undergoes more processes than the worm (like sophisticated play) and produces far more behavioral variability (including learned behavior). That does not mean the worm, or the plant, or the protist, lack value as ends in themselves or are less important to ecosystems; it mean the primate manifests properties beyond that of the worm, is a more difficult achievement of nature, hence intrinsically more valuable.

We cannot treat all the implications of this perspective here, but one deserves special mention. Some concerned with nonhuman animals object to the "anthropocentric" implication that a human being, because more complex, is more valuable than a nonhuman animal. But it is difficult to know what the pragmatic meaning of this objection could be. Even the most prominent systems of animal rights and welfare of recent decades accept that a human individual is entitled to greater moral concern than a nonhuman; neither Thomas Regan nor Peter Singer would suggest that in the triage choice between saving my daughter Rose or two cats from a fire, the obligation is to save the cats (Regan 1985, Singer 2002). They assert only that nonhumans have rights or possess moral value that must figure in human moral judgment, hence a *non-necessary* human interest (e.g., a nice hat) cannot justify harm to a necessary nonhuman interest (e.g., a beaver's life). To accept a continuum of degrees of moral consideration among animal species—e.g., in a situation of danger, if you must choose, save the pig instead of the spider—opens the possibility that humans, who are also on that continuum, have unique traits requiring unique treatment. All I am suggesting is that these comparisons have something to do with degree of organizational complexity of anatomy, neurology, psychology, behavior, and cultural capacity. To say we are a project of nature is to include us in that continuum. As Rolston writes,

> [F]rom a longer-range objective perspective systemic nature is valuable intrinsically as a projective system, with humans only one sort of its projects. . . . *The inventiveness of systemic nature is the root of all value, and all nature's created products have value so far as they are inventive achievements* . . . this is nature's most striking feature, one that ultimately must be valued and of value. . . . (Rolston 1988, pp. 198–9, my emphasis)[11]

The valuation of complexity may seem philosophically limiting, as if I were to say entropy and chance were bad, and unity or order or integration is good. But to repeat, disorder, difference, microscopic chaos, and the Second Law are the basis of everything in nature. There is no unity without disunity, no order without disorder; I mean this as an empirical, not a logical or philosophical claim. In a more philosophical and less physical sense, creativity, novelty, and freedom require randomness, objective chance, disorder, and irrelevance. Entropy's piper must be paid. Equilibrium, or lack of structure, is the default condition of natural existence. Nor do I mean destruction, or what Nietzsche referred to as the Dionysian principle, the urge to reduce wholes to chaos or rubble, is evil. But it does imply that it would be bad or evil if it were *generally* to contravene the bending of the arc of action toward the creation and maintenance of structure. Destruction of complex order cannot go too far without destroying the valuation of reality, unless it is linked to creation—which is how nature seems to work. The breakdown of the snake into its molecules is what feeds the build-up of structure that is the hawk's chick.

I will go further: we generally *do* regard complexity as a good. Once again we can turn to an environmental example. We often disagree about whether saving a wild species from extinction is a lesser good than preserving local economic interests. Some would even like to see the elimination of certain species, like the wolf, in effect arguing for simplification of local habitat because of the countervailing value of their properties or domestic herds. But notice what is *not* ventured in this argument: nobody argues that extinction or simplifying ecosystems is *good in itself*. No one says fewer species is just better. No one rhapsodizes simplicity: "If only the forest contained only one plant species and one kind of animal! Would that we could experience such simplicity and uniformity!" We can broaden the question further. Are there any geocentrists who bemoan that tragic latecomer, living organisms, for having covered the formerly pristine Earth with slime and fouling its atmosphere with oxygen? Would a universe without solar systems and stars be superior to ours? Why not prefer a quark plasma, or the simplest equilibrium state possible for the universe? It would not be *logically* impossible to hold up such alternatives as aesthetically or even morally better than the universe we now have. But no one does. It is very hard to say "yes" to simplification as an intrinsic, versus instrumental, good. Why? Because more complex systems are more highly valued.

We cannot fail to recognize that the universe has been in the business of evolving more and more complex forms of order, whatever other businesses it has. This evolution is stochastic and entropic, meaning the creation of order is selective and statistical, not universal or deterministic, and always occurs in locales at the price of increasing entropy elsewhere. The complexity evolved by the universe is a vast archipelago floating in a sea of near-equilibrium states. That may well be the only way complexity can arise. Thus we may speak more generally: the achievement and maintenance of pockets of complexity is a naturally good tendency of nature and one which we, as its products and participants, ought rationally to recognize and value.

IV. Post-Script: The Human Role

In this final section speculation will turn into personal confession, displaying without much argument what the author cannot help but think. Readers annoyed by such talk may wish to close the book here. But the author cannot. Not just yet.

I find myself unable to believe that creation exists in order for human beings to face a moral test, or thereby glorify the Ground. Nor can I believe the ancient Greek idea, recaptured by the German Idealists, that the Ground's chief interest is self-contemplation, that creation is necessary for the Ground's coming to know itself. There may well be *some* bit of truth in those views, but it is difficult to imagine the creation of the universe to be in service to either. I do not know, nor know how one might come to know, the purpose of creation, if there is one. But without conjecturing that purpose or goal, I find myself unable to doubt that whatever purpose there might be, *increasing complexity of nature serves it*. If there is purpose, if "purpose" is a plausible term, it is likely that the forms of increasing complexity play some role. There is too much difficulty, improbability, and design required for them to be unrelated to such purpose.

This is not to say that human beings are the pinnacle of creation. We have almost no knowledge of the conditions and possibility of life on extra-solar planets in our galaxy, and none regarding planets outside our galaxy. On such topics we are like pre-Columbian cartographers mapping the world; we should remember that two decades ago the first non-solar "exoplanet" had yet to be identified! There may be many other living planets, and many with similarly, or more, intelligent species; we do not know. We may never know. Still, we are very complex creatures relative to everything else we know. We must presume that complexity of life on Earth, and particularly of Homo sapiens, is relevant to, hence is meant to serve, whatever purpose there is in creation. It is impossible for me to believe that, whatever that purpose may be, the highest evident points of increasing complexity, the most hard won by physical, material, biological, psychological and cultural evolution are not a part of that purpose. But here again we would be one example of one relatively significant participant, among possibly many others, not by any means *the* point of nature's development.

Regarding Divine action, or more precisely, Divine *special* action in the world, while it may be possible—we would have an awful time trying to prove its impossibility—I imagine it must be a rare exception. The initiation, funding, rules, and elaborate fine-tuning of the universe imply an ensemble which is meant to generate complexity without much *ad hoc* intervention. I do not mean the Ground is a watchmaker or machinist, or inactive, perfect and immune to change. I have endorsed panentheism; it may be that Ground is constantly "acting," in some sense, in the world, and it may well be the Ground changes over time. I only mean that it is hard to believe that there would be any point in designing a set of laws and constants whose patient application over more than 10 billion years brings forth stars, life, mind, culture, etc., if accomplishing ends by edict were a common option. Occasional intervention might be a necessary lawful requirement of nature, although it ought then leave its mysterious footprints in the sand of scientific investigation. We don't know. But regardless, my guess is that *whatever point there is to the general flow of things ought to be evident in the general flow of things*. As to whether some more esoteric investigation of that general flow, leading to otherwise unavailable information, is possible, that is a separate matter. But I do not believe the achievement of cosmic purpose, if such there is, could hang on such esoteric achievements, alone or primarily.

The claim of purpose is compatible with the existence of indeterminacy, objective chance, and real hazard in the universe. Life may have started on Earth multiple times, and died out all but the last. We know that Earth had and lost more than one atmosphere. The Moon and the current orientation of the Earth are likely the result of an apocalyptic collision. There have been numerous great extinction events, and massive climate changes that pushed life out of vast areas. There were several Homo species, at least two of whom were contemporary with Homo sapiens for quite a while, with one of whom we likely interbred. All of them are gone except for us. Perhaps it is true that the universe might *not* have worked.[12] If the universe serves a purpose of the Ground's, it probably does so stochastically, in some of its larger and/or key features, but not in every particular. This means there is no reason to imagine a divine "plan" for me or my life or for any particulars. The existence of objective chance precludes that possibility. Whatever facts or events play a significant role in the unfolding of the universe with respect to fulfilling its "purpose," if that is the right word, one suspects they must normally be what happens either to galaxies, solar systems, planets, and biospheres (if there is more than one), perhaps key species and populations, not individual organisms. And as to our future, while the fate of the universe is unclear, the fate of human life on Earth must be limited by the Sun's expenditure of its fuel, which in one half to one billion years will cause an increase in solar radiation that will make Earth uninhabitable. (That should not be too immediately alarming for a species that has only been around a quarter of a million years; there are an untold number of unforeseen hazards, and opportunities, that will arise before that happens.)

But absence of a plan for particulars does not mean particulars cannot or do not play a role in the process. It seems we can say, with perhaps logical but not really practical uncertainty, that one obligation at least is clear: *we* bear an obligation to achieve, conserve, and/or enhance good. To use a vindicatory argument, if there are obligations, that would seem to be one. It is vague, of course, and its application is practically unclear, since there are doubtless many kinds of good which conflict. Nevertheless, I have claimed, complexity is one such good. Hence, *whatever else is good, maintaining and/or enhancing complexity is good.*

Complexity happens naturally, independent of us and without our help. But there are types of local complexity whose existence involves human beings and human action on the surface of the Earth. The existence of human beings and their societies must be, judging from our planetary neighborhood, a significant contribution to local complexity. Ecologically the biodiversity of local ecosystems and, as we now see, widespread features of the biosphere itself can be altered in ways harmful to local and global complexity. Hence our restraint can preserve complexity, but even more, our cautious participation can restore and actually enhance naturally arising complexity.

We ourselves create complexity. After all, as noted, on Earth we humans are *the* creative species. This may be the one respect in which we human beings, compared to the rest of our biosphere, bear similarity to the Ground, hence are in the "image and likeness of God." This is not to be overdrawn; we are utterly unlike the Ground in presumably almost all respects. Our creativity is not the same as God's, nor are we as creative as God. But as far as we know, there is nothing else, save similarly intelligent species elsewhere, capable of *deciding to make* and what to make, deciding to create something novel, and what goes along with that, able to decide to conserve or enhance what has been made. If we find similar intelligent beings in the universe, presumably this will mean that they, like us, have this quality, or have it to a greater degree.

It would be "natural" to find a place for a commonly religious notion, faith, in this account. For Paul Tillich, faith is the "courage to be" (Tillich 1952). There are two broad types of courage in history, the courage to be as an individual, and the courage to be as a part or a participant. His revisionist theology conceives of God as the Power of Being. Such Power is that Object, participation in which does *not* destroy individuality, unlike devotion to social or economic causes, something less than a matter of "ultimate concern," hence idolatry for Tillich. So, once the Power of Being itself is taken to be the matter of ultimate concern, then it is possible for the best features of both forms of courage, as a part and as an individual, to be synthesized. Borrowed but different from Tillich, the faith I suggest lies in *courage to participate in the evolution of complexity which seems crucial to the purpose of nature, hence creation.* Our creative activity would seem at least plausibly to be a participation as individuals and communities in the preservation, creation, and enhancement of complexity. Faith is the courage to be a part of this natural evolution, which for humans means to be in the process of making in the broad sense, making by acting, saying, and forming. It is a faith in the significance, value, and meaning of the evolving reality arisen from the Ground, and in the meaningfulness of one's contribution to it.

There are myriad ways of making a contribution to this process of maintaining or enhancing complexity. Humans make culture, or rather, culture is the register of our artifactual creation. Marriages, language, art, domestication of plants and animals, writing, history: we build and construct clothing, houses, vehicles, tools. We are, as Geist puts it "condemned to art" (Geist 1994). We create human communities and societies, complex organizations, and new human beings, not only through sex, but by parenting and the communal activity of education. When not engaged in economic forms of production or raising children and tending to local community, we ornament, garden, narrate, tinker. To a limited extent, we make ourselves, that is, we alone are capable through learning across the span of a lifetime to skew or select among paths of self-development what will be a more complete or complex or valuable human career, which then forms the basis for an enhanced contribution to community and society. In all this we participate in the creation of complex systems on the Earth's surface which, even if they have parallels elsewhere, must be cosmically rare.

To do socially valuable work, as Emerson says, to make a patch of garden, to create children or educate the young, to form part of the glue of a human community, to make cultural artifacts and meanings, to advance human technology or science, to build a better mousetrap or to divert funds from mousetraps to parks or babies or mosques, all these are creative makings that matter.[13] There are in addition rare history-changing figures who create or maintain the large scale structures of human societies, new technologies or advance knowledge or practice, in effect creating those kinds of progress which enhance human benefit, and/or dialectically compensate traditional needs in the face of such progress. Then there are those who seek, and may succeed, in achieving a higher, socially-transcending insight into the world process itself. All this, I believe, counts. *Homo faber* manifests itself in many ways.

Now, one may feel this fragmentary picture to be a meager religious notion. Doubtless that is true. Nothing has been said here to justify personal existence beyond death, or the retention of the person by the universe or Ground in some form, including in the memory of God. Nothing has been said that implies God or Nature redeems human suffering and loss in any direct or personal way. The sociologist and religious thinker Peter Berger once offered a marvelous passage on the nature and sense of the religious.

> A child wakes up in the night . . . from a bad dream, and finds himself surrounded by darkness, alone, beset by nameless threats . . . in the terror of incipient chaos the child cries out for his mother. She will take the child and cradle him in the timeless gesture. . . . She will speak or sing to the child, and the content of this communication will invariably be the same—"Don't be afraid—everything is in order, everything is all right." . . . Yet this common scene raises a far from ordinary question, which immediately introduces a religious dimension: *Is the mother lying to the child*? (Berger 1970, pp. 54–55)

We might label this the "frightened-child-in-the-night" test for religious content, pragmatically understood. Now, I am not proposing a philosophical religion, a kind of postmodern Deism. But if the meager suggestion above has any meaning that can rightfully be called "religious," it ought to say something to this test, hence to that child. To participate in the world process of creative complexity as a human individual who contributes to family, neighborhood, society, culture, ecosystem, or biosphere is certainly to gain *meaning*, now in the existential or salvational sense. But nothing in it grants certainty of immortal reward or redemption or the erasure of tragedy. What I have described is patently not as reassuring to the child as saying, "Yes, everything is alright. In fact, God will not let anything bad happen to you, and neither you nor mommy and daddy nor anyone you love will ever *really* die." I am not objecting to such reassurances; even if they were to be false, lying to a frightened child is not the worst thing in the world. But what could we say to the child that would reflect the view developed above? It might go something like this:

> Yes, everything is alright *now*. Bad things can and will happen, though. So we will love you and take care of you until you are big and strong and smart. Many things are unknown to us in this vast, mysterious world, through which you will have the most exciting journey. But if, when you grow up, you love and are loved, and work and learn and create, you will have a *good* life. You will be part of something amazing, the continuous recreation of the world. We think God wants you to do that. And by doing it, you may learn more about this world and its purpose than even mommy or daddy. I know this is not easy to understand right now. But until you do, I'll be right here.

Notes

Chapter 1

1. Shimony credits Jerome Lettvin of MIT, along with Peirce and Aristotle, as inspiration for the phrase (Personal communication).

2. See Bernard D'Espagnat's otherwise magnificent *On Physics and Philosophy* (2006).

3. Donald Campbell wrote, "Rather than logically complete ostensive definitions being possible, there are instead extended, incomplete sets of ostensive instances, each instance of which equivocally leaves possible multiple interpretations, although the whole series edits out many wrong trial meanings" (Campbell 1988b, p. 415).

4. In logic a meta-language serves to define terms from a particular object-language, like "truth," that cannot be defined in that object-language without yielding contradictions.

5. As for "views from somewhere," in Michael Ruse's phrase, there are a lot of them. Buchler's view represents a metaphysics that would be invariant across virtually any translation of perspective or every "somewhere" (Ruse 1995, p. 154).

6. Buchler follows Peirce in rejecting nominalism, the claim that only particulars exist. For Buchler an absolute particular would have to be absolutely unique, hence disconnected, unrelated, and devoid of similarity to any other thing (Buchler 1990, p. 179).

7. For Buchler, not only can there be no round squares, but no Catholicism without sacraments, no chess without queens or kings, and no talking insects (Buchler 1990, pp. 134–42, 255–56). Others disagree on whether such constraints are compatible with Buchler's ordinalism. See Wallace 1990 and Weiss 1990.

8. Randall and Buchler both described their metaphysical schemes as "objective relativism."

9. While Buchler's metaphysics is not naturalistic in my sense, his theory of human judgment is. See Buchler 1951 and 1955. We will turn to this part of his work in Chapters 9 and 10.

10. Thus I am not asserting what Buchler called an "unrestricted" view of nature as "provision of complexes" (Buchler 1990b). Also, Buchler frequently referred to a small set of particular orders as examples, e.g., facts, fiction, processes, and histories, which functioned as a set of secondary categories in his system. I will in effect consider nature as such a secondary category.

11. Quine wrote in his essay "On What There is" that "Wyman's overpopulated universe is in many ways unlovely. It offends the aesthetic sense of us who have a taste for desert landscapes . . ." (Quine 1953, p. 4), and in the same spirit in *Word and Object* described as among the philosopher's tasks "clearing ontological slums . . ." (Quine 1960, p. 270).

12. It might appear that to speak of the "universe" violates Buchler's injunction against talk of an "order of all orders." But it need not. We can mean by the "universe" all natural complexes—in my sense of "natural"—that we can discriminate.

13. The last views belong to the otherwise naturalistic James and Dewey.

Chapter 2

1. Strictly speaking, Spinoza was a "panentheist." Pantheism is the equation of God and the world. "Panentheism" is the contemporary name for the view that the world is in and of God, but God is more than the world. Spinoza claimed God has infinite attributes, only two of which are nature as we observe it (matter and mind).

2. These are mathematically equivalent reformulations of Newtonian mechanics, developed by Joseph Louis Lagrange in 1788 and William Rowan Hamilton in 1833. "Phase space" is a mathematical representation in which every point on a plane or in a space represents a possible state of a system's variables.

3. These two categories eventually became three (Whitehead 1933, pp. 39–41): occurrences (events), endurances (societies of events), and recurrences (abstract qualities).

4. Whitehead made frequent use of emergence wherever an "organic" whole evolves from parts. But he did not make life or mind emergent, since he included rudimentary forms of them in all occasions. Also his ontology is not pluralist; all actualities are, or are composed of, actual occasions, which, while complex, are evanescent quantized entities that

must be at least as small as a quark and as short-lived as the most short-lived particle.

5. Dewey was unhappy with the term "emergence," which is surprising since he was close to Mead, who endorsed it. Dewey accepted that novelty could arise in evolution, but thought the term "emergent" was part of an attempt to surpass a dualism which he had already rejected. He disliked its "doctrinal associations," and wrote in a letter to Arthur Bentley that "'Emergent Evolution' . . . seems to be a doctrine and a rather absurd one" (Dewey 1964, 1945.04.21, no. 15429). His most substantive discussion is in *Experience and Nature*, chapter seven (Dewey 1981, pp. 206–12), but see also Dewey 2008, p. 307, Dewey 1944, and Dewey 1994. Perhaps Dewey objected to the theism and teleological notions of Alexander, with whose work he was most familiar. (I thank John Shook for his detective work.)

6. Mill's "Of the Composition of Causes" in his *System of Logic* (1943) distinguished chemistry's "heteropathic" laws and effects, which are non-mechanical, from physics' "homopathic" laws and effects, which result from the composition of forces or vector addition. Lewes first used the term "emergent" for Mill's heteropathic causes.

7. Lloyd Morgan formulated his view in 1912–15 and presented it publicly just before Samuel Alexander gave his own Gifford lectures of 1916–18, although Alexander's book was published (1920) before Lloyd Morgan's (1923). They had been corresponding since the 1880s, and Alexander cited Lloyd Morgan as his source for the term "emergent" (Blitz 1992, p. 114).

8. Another influence on Lloyd Morgan was the American "new realists" whose writings were published in 1912 in E. B. Holt et al., *The New Realism*, particularly the holism of Edward G. Spaulding and Walter Marvin, who argued for levels of reality (Holt 1925).

Chapter 3

1. While many contemporaries attribute the phrase "special sciences" to Jerry Fodor, who wrote a famous paper under that title (1974), he did not coin it. The use of the phrase to mean sciences other than physics goes back at least to Broad's 1960, published 1925, p. 76: "The advantage of Mechanism [vs. Emergence] would be that it introduces a unity and tidiness into the world which appeals very strongly to our aesthetic interests. . . . On such a view the external world has the greatest amount of unity which is conceivable. There is really only one science [mechanics], and the various 'special sciences' are just particular cases

of it." "Special sciences" also appeared earlier and repeatedly in 1912 in E. B. Holt et al. (1925), but meaning *all* non-philosophical sciences (including physics).

2. Theory-succession reductions are intra-level (at the same scale of phenomena), intransitive (reduction of theory A to a special case of B, and of B to a special case of C, does not entail reduction of A to C), and do not validate elimination of one theory by its successor. Rather, they transform an older theory into a limit case of a newer, more inclusive theory; the reduced theory remains true and useful within certain limits. Inter-level or explanatory reductions are sometimes transitive and sometimes validate replacement of one entity by a lower level set of entities (Wimsatt 2007, ch. 11).

3. The source for this often mentioned remark is the memory of P. M. S. Blackett: ". . . even a joke from Rutherford's mouth was apt to become a dogma in lesser men's minds. No very young physicist could be totally unaffected by his famous crack: 'All science is either physics or stamp collecting,' or by the often implied assumption that it only needed some further progress in physics to allow us to deduce from first and physical principles the facts and laws of the lesser sciences like chemistry. Rutherford undoubtedly should have been sympathetic to Comte's hierarchy of sciences with physics . . . at the top, and with each science deducible in principle from the next one higher up. . . . However . . . the limitations of calculation, even with the electronic machines of the future, will still leave the different sciences largely autonomous . . ." (Blackett 1962, p. 108). Rutherford also served as Broad's example of a "mathematical archangel" who could deduce complex properties from microphysics (Broad 1960, p. 70).

4. While there is a disagreement over what is the unit of natural selection, meaning what gets selected, e.g., genes, organisms, or groups, there is agreement on what does the selecting: it is the operation of the environment on the organism's phenotype. Unless we can describe the interactions of phenotype and environment as causal, we cannot explain why some genes are naturally selected for and others selected against.

5. Two quips on the priority of physics, to counter Rutherford's, come secondhand from Shimony and Wimsatt, respectively. Laszlo Tisza remarked that "the biologist should have no inferiority complex vis-à-vis the physicist, for although the latter can predict quantitatively the trajectory of a planet, the former can reliably predict that the hen's egg will hatch only into the chicken's tiny sector of an immense phase space" (Shimony 1993c, p. 224). Jack Cowan distinguished the biophysicist from the biologist: "Take an organism and homogenize it in a Waring blender. The biophysicist is interested in those properties which are invariant under that transformation" (Wimsatt 2007, p. 174–5).

6. Bickhard's 2000 notes that quantum field theory treats particles as epiphenomena of oscillating fields. He concludes process is primary: we see "process all the way down and all the way up." While I agree with his critique of componential reductionism, I will argue that reality is not all, or most fundamentally, process any more than it is structure or components.

7. There are multiple kinds of dependence including mereological (dependence of wholes on parts), predicative (dependence of the properties of X on X), and causal (effect on cause).

8. The term was first used by the emergentist Lloyd Morgan to describe the relation of a novel property to that from which it had arisen (McLoughlin and Bennett 2004).

9. As pointed out by Elizabeth Baeten (Personal communication).

10. This is a sophisticated version of Mill's notion of "homopathic" properties as mere additive sums of part-properties.

11. Wimsatt uses the term "phenomenological," but this courts conflation with Husserlian phenomenology. My use of "phenomenal" has nothing to do with phenomenal*ism*.

12. As we will see, Schrödinger made a similar point: the movement from the simple to the complex is one from the indeterminate to the determinate (Schrödinger 1992).

13. Here one could (as I will) use Colin Pittendrigh's term "teleonomic," made famous by Ernst Mayr, for Wimsatt's "teleological" (Pittendrigh 1958, Mayr 1974).

14. As Salthe points out, this is not entirely true (Salthe 1993, ch. 2). Hierarchies, hence "levels," come in two types: "specification" or "subsumptive" hierarchies, as well as "compositional" hierarchies. We will discuss this in the following chapter.

15. If subatomic levels were included, Wimsatt's dissipative wave would be more like a normal curve (an inverted U-shape), with less entification at quantum levels. Also, he does not include the larger cosmic physical scales (e.g., planets and larger) which up to a point are highly entified.

16. As a self-proclaimed "materialist" Wimsatt would not follow me here (Wimsatt 2007, p. 236).

17. To claim the objects of these sciences are physical because spatially extended is to switch definitions in mid-argument, and reap the above-mentioned problem of the spatial definition.

18. John Post writes of his sophisticated nonreductive physicalism: "like all versions of physicalism, the theory must require that any entity mentioned outside physics . . . is [token-] identical with something in the physicalist's inventory of what there is" (Post 2007, p. 163). This is definitionally required, for "*any* metaphysics is a . . . [token-token]

identity theory . . . insofar as it must claim that everything at bottom is somehow token-identical with . . . certain basic entities or processes . . ." If not, it is "not a metaphysics in the first place. . . ." He thus justifies his physicalist metaphysics—in effect, how far physicalism can be stretched in a nonreductive direction—in part by accepting that any metaphysics must make *all* of nature token-identical with some *part* of nature. I do not accept that notion of metaphysics or nature.

Chapter 4

1. Salthe's 1985 is, to my knowledge, the only natural scientific employment of Buchler.

2. By "natural kind" I mean whatever kinds are taken to be non-artificial and "fundamental" to an order. We need not clarify whether these exhibit "essences," "clustered" identities, necessity, or contingency. (See e.g., Putnam 1975, Kripke 1980, Boyd 1991.)

3. The metaphysician and math logician will note the absence of an account of mathematical objects in this book. Alas, I cannot sufficiently stretch my "incompetence in many fields at once"—mathematics is a field too far. If I were otherwise, I would suggest that mathematical and logical objects (e.g., sets) can be understood as functions on universals, hence possibilities, not unlike the "second-order" objects some project (except, in my account, not second-order *physical* objects). As will be seen in Chapter Nine, this would make them kinds of "meanings," rule-governed perspectively selected collections of possibilities of natural complexes.

4. Suppose an apple lying on the ground under an apple tree is E, and C is my having picked it and dropped it. For Lewis "C caused E" means E could not have happened without C. But the wind could have blown the apple, a deer knocked it off, etc. So "C is the cause of E" means that in any possible world that is identical in all other details to our own—no recent strong winds or hungry deer—E cannot have happened without C.

5. Compare this to Dretske's distinction between "triggering" and "structuring" causes in my Chapter 8. If we accept Salthe's language, Dretske's structuring causes would be causal but not efficient causes. I accept this, but Dretske would not (Dretske 1988).

6. A brief word about information. Claude Shannon (1948) quantified the amount of information in a message in terms of the probability of its selection from a class of possible choices. He expressed this mathematically in terms analogous to Boltzmann's "H-theorem," hence in relation to entropy. But like "complexity," "information" is variously used today,

sometimes as equivalent to the number of possible states of a system, sometimes as the amount of organization of a system. Some take information to provide an alternative dimension of nature from mass-energy, as the key to life (e.g., DNA coded information), as an observer-relative approach to reality (inspired by quantum mechanics), or even as determinative of being or reality itself (e.g., Wheeler 1990). Such expansive employment seems to rest on believing the concept somehow saves us from problems themselves dependent on entitative reductionism. Once we accept pluralism information becomes one feature of natural complexes among others, or rather, a useful way of collecting and describing subsets of those features.

7. While it is true that we only know the object through perspectival schemes, the object nevertheless puts constraints on our choice of perspective. There is no plausible perspective for which a photon is more complex than an amoeba.

8. Note that, unlike Salthe's analysis, both families of complexity imply that wholes are more complex than any subset of their parts. (Salthe defines extensional complexity as whole-part complexity.) Any whole must be extensionally more complex than any subset of its parts. It must be more intensionally complex if the whole is an individual, but not necessarily if it is an ensemble, which has low structure. The ensemble of persons walking through Times Square at this moment is more extensionally complex than its human constituents but less intensionally complex.

9. My diagram borrows Wimsatt's notion of causal thickets.

Chapter 5

1. Tian-yu Cao (Personal communication).

2. The key to this example is that from the perspective of an observer moving with the rod, the arrival of the ends of the rod at the back and front of the hole are not simultaneous.

3. This arose in solution to the "hole" problem nicely explained in Stachel 1993 and 2000.

4. A tensor is a set of component functions (in some cases, a set of vectors), one per subscript, each of which can have a number of dimensional components. Tensors with two indices are equivalent to an $i \times j$ matrix of g's (a rectangular array of numbers in i rows and j columns). In a four-dimentional space (x, y, z, and t) each index varies over 4 numbers, generating a 4×4 matrix of 16 values for g. Matrices allow us to represent, and perform operations on, many related quantities at once.

5. The fuller version of the equation is $R_{ab} - \frac{1}{2}R(g_{ab}) = -(8\pi G/c^4)\, T_{ab}$. The left side is the Einstein tensor (G_{ab}) which is equal to $R_{ab} - \frac{1}{2}R(g_{ab})$

where: R_{ab} is the Ricci tensor (which measures deformation due to tidal forces); R is the scalar Ricci Curvature, which gives the curvature of the Riemannian manifold; and g_{ab} is the metric tensor. The right side T_{ab}, or with its constant κT_{ab} (where $\kappa = 8\pi G/c^4$), is the stress-energy tensor, which gives the energy density flow over time, along with the pressure in each of three spatial directions. This assumes the representation of space-like dimensions are positive and time negative: – + + +.

6. In a vacuum the stress-energy tensor and the Ricci tensor go to zero, but the Riemann curvature and the metric tensor do not (cf. note 5 above).

7. There will be no question here of understanding QM in a deeper sense. The physicist Steven Weinberg tells the story of meeting another physicist, an old friend, during his travels. Upon inquiring about a promising graduate student he had known, the friend told Weinberg the young man had dropped out of graduate studies. Surprised and disappointed, Weinberg asked what had happened. His colleague shook his head: "He tried to *understand* quantum mechanics."

8. Thus "the outcome obtained is a matter of objective chance. . . . The chance character of the outcome . . . is not a matter of ignorance on the part of . . . observers; it is a property of the physical situation itself . . . [this] presupposes the concept of objective indefiniteness [which] . . . cannot even be formulated in classical physics" (Shimony 1989, p. 374).

9. Classical waves are also indeterminate in the sense that they spread out over a potentially infinite area. Some of the strange features of quantum phenomena are in fact simply true of any wave. The nonclassical problem is that the quantum phenomenon has *both* wave properties and quantized particle properties, so we cannot model it consistently with either particle or wave alone.

10. These operators must be quite special: they are complex numbers (hence include the imaginary number $i = \sqrt{-1}$, in the form, $x + b\ i$), and Hermitian or self-adjoint, meaning the number is equal to the transposition of the sign of its imaginary part, $(x + b\ i) = (x - b\ i)$, so the matrix of the expression can be "flipped over" at the diagonal (an $m \times n$ matrix that is equal to $n \times m$). We can express such a complex, self-adjoint linear operator through Euler's formula, $e^{i\theta} = \cos\Theta + i\sin\Theta$. According to the polar version of the classical equation for wave phases, $\Psi(x,t) = Ae^{i(kx-\omega t)}$, this makes the formula $x + iy = e^{i\theta}$. Expressed as a matrix, it is then multiplied by its opposite, in effect rotated about the origin. The point of all these steps is to just to replace the complex (containing the imaginary number i) formula with an equivalent but real one (lacking the i) in which the operators rotate and/or translate the vector in real space.

11. Kinetic energy is ½ mv^2 or in terms of momentum, $p^2/2m$. Schrödinger has replaced p by $i\hbar\nabla$, where $\nabla = (\partial/\partial x, \partial/\partial y, \partial/\partial z)$ or rate of change in a quantity in all three dimensions.

12. One may say that Everett had the right idea, that the multiple possibilities or eigenstates really do occur, but in our one world, not in "many" worlds. Their effects are just washed out.

13. Quarks, protons, neutrons, and other hadrons obey the SU(3) group, while the weak force together with electromagnetism obeys the unified SU(2) × U(1) group of rotations.

14. One last theoretical innovation was *renormalization theory*. All quantum field theories are dogged by the "divergence" problem. Fields locally interact at dimensionless points. But the ascription of energy or mass to given points of the field means the field can interact with itself. Given Heisenberg uncertainty, if Δt or Δx is infinitesimally small, then, respectively, ΔE or Δp must be infinitely large. Consequently the field's self-interaction causes unwanted infinities. Suggested by Schwinger, but made famous by Feynman, renormalization treated the divergent quantity as the result of two contributions, one from the observed quantity and one from the infinity. The infinity was manipulated so that it was part of the mass or charge at a point, and these "absorbed" infinities would then cancel each other, leaving us able simply to substitute observed mass and charge values by hand. This procedure could be physically interpreted as restricting the values to a non-zero, finite region or volume around the dimensionless point, which could be made as small as one might want. Any such theory would be "renormalizable." The remarkable thing is that it works; the resulting predictions are valid. In fact, Feynman's QED became the most accurate physical theory ever.

15. This is not mysterious mathematically; the tip of any cone has infinite curvature.

16. The Second Law also provides an argument for a past-finite universe. If the Second Law holds, an eternal universe with an infinite past would now have to be in equilibrium. Ours is not.

17. Hawking also showed that black holes emit thermal energy and eventually "evaporate."

18. As Schrödinger argued, they are reversible only if we ignore heat-exchange, which is indeed increasingly accurate as temperature approaches absolute zero or as the environment approaches a vacuum. See his *What is Life?* (1992).

19. As in the superspace model of quantum cosmology (see Chapter 11).

20. Feynman's remark violates Peirce's pragmatic theory of meaning that our notion of all our observable effects and functions of X is our notion of what X is (Peirce 1992c).

21. This does not mean uniqueness or difference from others of its kind.

Chapter 6

1. Gravity is extremely weak relative to the other fundamental forces. A one-ounce refrigerator magnet routinely defeats the gravitational pull of the Earth's 5.9×10^{24} kg.

2. In the Hertzsprung-Russell diagram temperature, the x-axis variable, goes from high to low because that corresponds to change in color from low B-V index to high. The direction of the arrow shows the direction of a star's evolution over time.

3. The principle quantum number *n* is labeled by a number or a letter (starting with K = 1, L = 2, M = 3, etc.) from the nucleus out. The azimuthal quantum number *l*, has values 0 to n – 1, and is labeled by letters: *s*(harp), *p*(rincipal), *d*(iffuse), or *f*(undamental). So a shell is the collection of electrons in one atom that occupy the same orbital; those in a shell sharing *l* as well as *n* values occupy a subshell (e.g., 1*s*, 2*p*, 3*d*, etc.) The magnetic quantum number, m_l, essentially gives the direction of the angular momentum vector within the subshell, exhibiting integral values from –l to +l. The electronic occupant of a subshell with identical *n*, *l*, and m_l can be double, generating two oppositely directed magnetic fields though the property of self-rotation or spin, m_s, the fourth quantum number, which can have the value –½ or +½.

4. This comes from Avogadro's number: a "mole" of a macroscopic substance is 6.02×10^{23} atoms or molecules.

5. Atoms are individuals but still not qualitative or unique individuals, because every atom is identical and inter-substitutable with any other individual of its natural kind.

6. Additionally, there are cases where quantum mechanical predictions do not explain, or even place, an electron configuration in its proper group (Scerri 2007, p. 242).

7. Two philosophers willing to take this step are Lombardi and Labara (2005).

Chapter 7

1. Daniel McShea has objected to the blanket claim that evolutionary change is marked by increased complexity (McShea 1991). Where it is claimed that every step in evolution exhibits increase in complexity,

McShea is certainly right. But his narrow notion of complexity as the antonym of organization (so a random pile of stones is more complex than a brick wall), operationally defined as the number of different kinds of parts of a system (so aquatic mammals are not more complex than lions, who evolved earlier, because the former have fewer distinct kinds of vertebrae), is not adequate either.

2. The modern definition of species as a closed reproductive pool applies only to sexual organisms. Since bacteria are asexual, it excludes most life on Earth! In the asexual realm biologists define species by similarity of DNA or RNA, morphology, and niche, which are secondary indicators of species in sexual organisms as well.

3. Viruses, which are far smaller than bacteria, are genes in a protein coat but generally without any cell membrane. Nor do they grow or metabolize like cells. Some take on a membrane-like fatty envelope from their host's material.

4. Other prominent elements in living cells include: calcium (Ca), potassium (K), sodium (Na), chlorine (Cl), and magnesium (Mg), with traces of boron (B), chromium (Cr), cobalt (Co), fluorine (F), iodine (I), iron (Fe), manganese (Mn), molybdenum (Mo), selenium (Se), silicon (Si), tin (Sn), vanadium (V), and zinc (Zn). Recently researchers discovered bacteria that can to some extent use arsenic (As) in place of phosphorus.

5. $G = H - TS$, where H is enthalpy (the work the system can do), T temperature, S entropy.

6. Such a device must have "universal constructor," a Turing machine that can read a set of instructions for assembling materials. The "blueprint" or instructions must be fed into the constructor. But there is a problem. If the constructor merely follows the instructions, the product will not contain a copy of the instructions; while if instructions to copy the instructions are included in the blueprint, the blueprint must include another version of itself within itself, ad infinitum. Von Neumann's (1966) solution was twofold: he attached a "supervisory unit" to the constructor to read the blueprint, and made it capable of two different tasks, following the instructions and merely copying the instructions. The supervisory unit tells the constructor to follow the blueprint as commands for copying the whole device except the blueprint, then switches into a copying mode, and inserts a copy of the blueprint into the otherwise completed replica (Poundstone 1985). Life actually accomplishes Von Neumann's prescription.

7. See the work of biosemioticians, for example Pattee (1969) and Hoffmeyer (1997). The term "proto-semiotics" is from Stillwaggon and Goldberg (2009). But I do not accept the notion that semiosis is the

essential distinction of life from non-life, nor that it implies the kind of intentionality we ascribe either to minds or a "self."

8. According to the "RNA world hypothesis" the first cells reproduced using RNA alone.

9. Fred Hoyle suggested in the early 1980s that the chances of a live bacteria forming by random combination of amino acids are like those that "a tornado sweeping through a junkyard might assemble a Boeing 747." There is an interesting history of this remark in Gert Korthop's 2009, involving Stuart Kauffman, Robert Shapiro, and Hubert Yockey.

10. Recently Craig Venter's research team has successfully replaced a bacterium's entire genome with an artificially constructed one. The new genome then took over and "remade" the living bacterium. Just how far such engineering can go, and what it will imply about the nature and origin of life, remains unclear (Wade 2010).

11. It has been thought that only during the Cambrian Explosion, when atmospheric oxygen reached a level high enough to produce sufficient ultraviolet-blocking ozone, was the land colonized by animals and plants. But evidence of land cyanobacteria has been found more than a billion years old.

12. The continental glaciations are given different but corresponding names; "Wurm" is the European, "Wisconsin" the North American name for the same glaciating period.

13. As Geist points out, from the viewpoint of the genome of a species, evolution is a "calamity just short of extinction" (Geist 1978). What, after all, is a "successful" species? It must be one that endures over time, or multiple environments, as the same species.

14. The Weismann barrier is distinct from Francis Crick's "Central Dogma" of molecular biology which holds that information passes only one way, from nucleotides to proteins.

15. Biologists critical of the neo-Darwinian synthesis point out that Darwin's own writings are theoretically more pluralistic than those of many later Darwinians. See Cobb 2007.

16. The Nidopallium Caudolaterale (NCL) seems to enable complex learning in corvids, which lack the laminated neocortex mammals have. This raises questions about the attempt to correlate particular neural structures and cognitive abilities independent of an organism's further neurology, development, and behavioral niche (Gregg DiGirolamo, Personal communication).

17. My favorite example is that, according to reports, the re-introduction of the wolf to Yellowstone in the 1990s increased the trout population. Without the wolf, the elk herds lazily fed on stream banks, stunting plant growth. The wolf keeps the elk moving, allowing plants to mature, which puts the streams in shade, which lowers the water temperature. Trout like it cold.

18. Dawkins also seems to believe that the fact that genes are the individual entities most invariant across generations makes them the proper causal agency for evolution. I do not think this premise, even if right, justifies the conclusion.

19. I thank Elizabeth Baeten for this point.

20. Namely $dN/dt = r\ N\ (1 - N/K)$, relating population size (N), rate of reproduction (r), and carrying capacity for a population (K) over time (t).

21. Maturana and Varela denied any reference to the external world. But the organism's auto-making must adequately refer to the environment which dictates survival and was instrumental in the organism's evolution. How these two processes link is the topic of evolutionary epistemology, as we will see in Chapter 10.

22. The "particularity" listed first does not gainsay symbiosis, parasitism, or the behavior of collections of organisms. Nevertheless, all life is cellular, and every cell is bounded.

23. Elizabeth Baeten (Personal communication).

Chapter 8

1. Clayton rightly calls these two "halves" of "the" hard problem (Clayton 2004, pp. 120–3).

2. Unlike myself, Horst rejects ontological pluralism.

3. Notice that an eliminative materialist or an identity theorist who claims mind is identical to brain activity, hence has no properties brain events do not have, would in principle have no reason to deny any human-like mental activity to any encephalized organisms.

4. Some consider "thinking" any mental processing of any mental content whatsoever. I will save "thinking" for the kind of discriminative mental processing humans alone seem to do, involving ideas as well as images, about which more in the following chapter.

5. I will define "mind" and "consciousness" differently, but sometimes use them interchangeably for reasons I explain below.

6. Jonas too thought emotion was "the animal translation of the fundamental drive which . . . operates in the ceaseless carrying-on of the metabolism" (Jonas 1966, p. 126).

7. I believe sensible qualia like redness are akin to feelings as the nonrepresentational but intentional stratum of perception, the qualities of images rather than bodily states.

8. A better term for image might be "schema" or even "model."

9. Velmans has summarized very interesting biofeedback studies in which humans can learn to manipulate individual neuron firings. This

has an important resonance for the causality debate, but it is not, I think, decisive (Velmans 2002).

10. See Kim, Horgan, and Cummins on Dretske, and Dretske's replies (McLoughlin 1991). Some of their criticisms of Dretske presuppose ideas I do not employ, e.g., supervenience. Unlike Dretske, I can accept that, in the human case, reasons do not just explain but can make a causal difference to subsequent behavior, since causality is for me not solely efficient and physical.

Chapter 9

1. Geist lists 29 "diagnostic features" of Homo sapiens (Geist 1978, pp. 412–13).

2. Note that we cannot assume our abilities are always most closely approximated by primates. Our capacity for vocal imitation, for example, is more similar to birds (Hauser et al. 2002).

3. Today it is common to use "hominid" for the all the great apes including humans, and "hominin" for the humans and their precursors that split from the ape lineage.

4. The remains of *Home floresiensis*, the "Hobbit," a one-meter-tall hominin more like habilis or earlier hominins than erectus, neanderthalensis, or Homo sapiens, found on the island of Flores in Indonesia, seem to be from only 12,000 ya! (Finlayson 2007, pp. 48–54).

5. As Geist writes "a horse does not look like a horse because its ancestors looked like horses; it looks like a horse because the physical and . . . social environment[s] of the modern horse . . . are not too different from those that shaped [its] ancestors . . ." (Geist 1978, p. 71–2). I would say the horse looks like a horse for both reasons, the linkage of the two being the key point.

6. The issue is where to draw the species barrier between ourselves, Neanderthals, and other hominins starting with habilis. Are we different species, or different subspecies of one species? If two or more are one species (Homo sapiens) then we who developed 200,000 years ago need an additional modifier, e.g., "Homo sapiens sapiens."

7. There is recent research on a 4.4-million-year-old non-ape precursor to Australopithecus afarensis, Ardipithecus ramidus. It was heavier than Australopithecus, had some bipedal ability, but retained an opposed big toe for climbing, and seems to have been a forest dweller.

8. Cooking with fire could have made a major difference to hominin evolution, because cooking allows a wider variety of foods to yield nutrition with less energy expenditure in eating and digestion. Richard

Wrangham believes Homo erectus must have cooked, but as of yet we have no evidence of cooking prior to Homo sapiens (Wrangham 2009).

9. "Riss" is the European, "Illinoian" the North American name for the glaciation.

10. A number of domesticated great apes have famously been taught some representational language. Washoe, a chimpanzee, was taught some American Sign Language by Allan and Beatrix Gardner, and later herself tried to teach signs to a younger chimp. Roger and Deborah Fouts worked with Washoe (Fouts 1998). The gorilla Koko was taught sign language by Francine Patterson. The bonobo Kanzi was taught by Sue Savage-Rumbaugh to press lexigrams on a keyboard that lit up pictures of referents. Kanzi could even differentiate word order, e.g., distinguish "make the doggie bite the snake" from "make the snake bite the doggie" (Anderson 2004). But prepositions, connectors, conjunctions, disjunctions, subordinating clauses, and embedded sentences seem beyond Kanzi's ability. Steven Pinker (1992, pp. 336–40) and Cheney and Seyfarth (2007) have argued that these cases are not comparable even to rudimentary human language use, spoken or sign. I cannot, and do not need, to decide this question. For present purposes it is sufficient to say at least some features of human language use are certainly unique to us. Just how many are, and why, is another matter.

11. These macaques wash their sweet potatoes before eating, unlike other troops. This case may not be so compelling—the sweet potatoes were the gift of human researchers, and it took two years for the behavior to spread throughout the troop—but there are other cases of local traditions among other mammals (Tomasello 1999, pp. 26–31).

12. In 1999 Tomasello concluded that nonhuman primates do not mind-read. A decade later he and his collaborators qualified their view after conducting studies in which chimpanzees showed evidence of the ability to attribute perceptions and goals to others, but not false beliefs (Call and Tomasello 2008, Kaminski et al. 2008). Thus for him it is not the complete absence of theory of mind that differentiates chimpanzee and human, but the kind and extent of the human mind-reading ability.

13. While Tomasello uses the language of vision ("shared gaze"), I see no reason that joint attention could not arise where touch and/or hearing are substituted for vision, just as, below, I seen no reason sign language could not substitute for spoken-heard language. Of course, whether that is true is a matter for empirical studies.

14. I will define culture differently below.

15. Philosophers since Wittgenstein have written of "seeing-as," e.g., alternately seeing an intentionally ambiguous drawing as a duck or a rabbit (Wittgenstein 1953). The current point is that there must be *joint* seeing-as.

16. There may be a connection here with D. W. Winnicott's notion of the "transitional object," the cherished thing, e.g., a blanket, that the one- to three-year old seems to take to maintain a kind of non-separation in the process of separation from the primary caregiver, which Winnicott thinks is the forerunner to cultural objects (Winnicott 1982).

17. Tomasello and Hobson cite Mead, and Tomasello calls his own approach "social-pragmatic."

18. The insight learning of orangutans and others implies rudimentary imagination, so it seems there must be imagining without language (Cheney and Seyfarth 2007, pp. 270–72).

19. Dretske would disagree because he does not accept non-efficient causality.

20. Buchler defined meaning more broadly as the allocation of function to some element of experience in subsequent human judgment (Buchler 1955 pp. 153–70).

Chapter 10

1. The task of joining one's metaphysics and epistemology has been called "closing the circle" (Shimony 1993a). Any metaphysics must be capable of cohering with an account of knowing that makes it possible, and any naturalism will inevitably constitute a (hopefully virtuous, or at least benign) circle. But we must avoid any implication of closure or completeness. I would rather say we are connecting, thereby strengthening, several discrete pieces of theory, whose mutual plausibility increases the likelihood of their joint validity.

2. When a human plays Frisbee with a dog is there no inter-special commonality among their cognitions? Is there any way to explain that commonality without reference to jointly perceived objective conditions, however distinct their cognitive styles? If the humanly cognized world is "constructed" by us, how can we explain that commonality? Is the dog a constructivist, or do I construct the Frisbee while the dog perceives it as a naïve realist but both happen magically to correspond?

3. I thank John Shook for pointing out the James reference. Buchler's theory of human judgment is also naturalistic. For him, "Man is born is a state of natural debt, being antecedently committed to the execution or the furtherance of acts that will largely determine his individual existence. . . . In the understanding of the human process, natural or animal obligation is more fundamental than . . . moral obligation" (Buchler 1955, p. 3).

4. The original thought experiment is in Borel's 1913 essay *Mécanique Statistique et Irréversibilité*. Eddington restated it to apply to the British Museum in his 1927 Gifford Lectures "The Nature of the Physical World." See also Poundstone 1985, p. 22ff.

5. There is no getting around the unique power of vision to determine a currently present object's nature at distance. However useful, olfaction of present objects is at the mercy of wind direction, and hearing alerts the hearer to events in a direction, but often not what the agent is. Even whitetail deer, with incredibly keen senses of smell and hearing, must often wait until what they smell or hear is in visual range to know whether it is a threat or not.

6. There are models of rational evaluation that can apply to any culture, and across cultures, for example, MacIntyre's "rationality-in-a-tradition" and Bartley's "comprehensive critical rationalism," which define rationality in terms of self-correction over time (MacIntyre 1989, Radnitzky and Bartley 1987).

7. Pastoral peoples share many features with segementary societies.

8. See also Chapter 12, note 2.

9. Buchler believed this implies ordinalism: "Many philosophers who would agree that when we stand before a house it is the house that we see, not an image or sense-datum or appearance of the house, would balk at the ordinal consequences. As we move away from the house . . . if what is called the 'house itself' appears smaller, it is because it gets smaller. It is in the order of vision that it gets smaller" (Buchler 1990, p. 279).

10. Putnam criticized some versions of naturalistic epistemology: "It follows that it is simply a mistake to think that evolution determines a unique correspondence . . . between referring expressions and external objects" (Putnam 1981, pp. 38–41). No unique correspondence, to be sure; but there is correspondence in the relevant medium, just as the train's filmic images have some correspondence to the train.

11. My earlier critique of Margolis (Cahoone 2002c) was, I now think, inadequate. Margolis' project to create a relativistic, or "constructivist" realism in which truth must be relative to domains of evidence is driven by his consideration of artworks and his notion of the human self as itself an artifact or cultural creation. Further, he argues that our rational reconstruction of whatever is "given" must be understood as already structured by the relation of the human agent and the world, by a subject-object symbiosis (reminiscent of James' and Dewey's notion of "primary experience"). Since the subjective contribution includes history and culture, and bivalence can't work in art, we retain at most a "construc-

tive realism," a realism allowing relativism in some areas while preserving bivalence, hence traditional realism, in science.

Chapter 11

1. I will confess that it never occurred to me in my career as a professional philosopher to make an argument for the existence of God, or that a reasonable argument could be made, until I began reading the works of current physical cosmologists.

2. This is not to accept Parmenides' conclusion that change and motion are impossible. Also my argument does not prohibit creativity, as long as we do not conceive creativity as the uncaused arising of something from Nothing, nor chance or indeterminacy, for once there is something, there can be no requirement that it be absolutely determinate (beings need only be partly determinate). Nor is my Lucretian principle a full endorsement of Leibniz's Principle of Sufficient Reason; I do not claim every being or state of affairs must have a reason. Last, I take all these principles as heuristic rules of only likely validity, not *a priori* true principles.

3. I am not assuming that every event must have a cause in the narrow sense of a discrete efficient cause. The emission of an alpha particle from an atomic nucleus in radioactive decay fails to have such a cause. But the emission is a stochastically predictable consequence of a system's properties; the alpha particle does not "come from Nothing."

4. If the Moon were removed far enough so that Earth's pull and the Moon's acceleration toward the Earth were near zero, the Moon's gravitational potential energy must be zero. But it must also be zero if the Moon were to be pulled into contact with the Earth's core. At points in between it would be moving toward Earth, so have positive kinetic energy. But potential energy plus kinetic energy must add to zero to be conserved. So the gravitational potential energy must be negative (Guth 1998, Appendix A).

5. Our real world is characterized by the four-dimensional Minkowski spacetime metric in which time is negative ($ds^2 = dx^2 + dy^2 + dz^2 - dt^2$). Using the method of Wick rotation, t can be taken to be a complex number, multiplied by the imaginary number $i = \sqrt{-1}$, so that the time factor becomes positive, yielding a Euclidean metric ($ds^2 = dx^2 + dy^2 + dz^2 + dt^2$) of three real dimensions and one imaginary dimension.

6. This can be related to the Vilenkin tunneling model. The Hartle-Hawking wavefunction retains the incoming particle-waves (A in Figure 11.1) as well as the outgoing (B), the former now interpreted as the "Big

Crunch" of a collapsing universe. The interval *a* in $-a_0 < a < +a_0$ is the "Nothing" from which a universe tunnels out.

7. Hartle-Hawking correctly implies that the universe cannot have had a beginning *in time* because time is internal to it. But we did not need QG for that; as Vilenken recognizes, St. Augustine knew that we cannot conceive the universe being created at a time (Vilenkin 2006).

8. Regarding Wittgenstein's early claim that beliefs which could not be logically said could nevertheless be shown, Frank Ramsey remarked, "If we can't say it, we can't say it, and we can't whistle it either" (Ramsey 1990, p. 146). Nothing can't be whistled either.

9. As I argued in Chapter 5 with Cao and Stachel, I resist the notion that mathematical structures by themselves can be causal in the accepted sense. In my language, they can only be formally, neither efficiently nor materially-physically, causal.

10. My answer to the question "Why is there something rather than Nothing?" is that Nothing cannot be or refer, hence is neither actual nor possible. A more substantive question is, why does our universe, the particular something we observe, exist?

11. Even Smolin's series of "naturally selected" universes must be constrained by background laws that make the cosmic evolution possible (Smolin 1997).

12. Some have been overly confident in the powers of chance. Hawking, using the well-known simian figure (see preceding chapter, note 4), imagined the likelihood of a homogeneous-isotropic universe emerging from the Big Bang: "It is a bit like the well-known horde of monkeys hammering away on typewriters . . . very occasionally by pure chance they will type out one of Shakespeare's sonnets" (Hawking 1998, p. 123). Or will they? Poundstone applied the figure to the writing of Shakespeare's *Hamlet* (Poundstone 1985, p. 22ff). Supposing 50 keys on a keyboard, and 150,000 keystrokes required for Hamlet, the likelihood of that particular series of keystrokes being produced by one primate is one in $50^{150,000}$. The age of the observable universe, we now believe, is 4.35×10^{17} seconds. If the number of monkeys were the same as the number of baryons in the universe, 10^{80}, it would still take them more than $10^{149,900}$ universe-periods as old as ours to do it. In effect, the monkeys of randomness could produce an object with the complex structure of *Hamlet*—or the universe—only if they were gods themselves.

13. That is, at least some constants must be unchanging. This does not deny that some others have evolved, e.g., through the symmetry-breaking of the basic forces.

14. See Helge Kragh's 2009 for a review of the recent history of cyclic cosmology.

15. Steinhardt is not "wedded" to there being no beginning (Personal communication).

16. I should add that my claim need not deny that there are other universe cycles, that creation has only happened once. That is beyond our knowledge. Indeed, other than a Ground of Nature, the next most plausible of the models described herein—plausible in dealing both with the origin and the improbability questions—would be an eternal version of the cyclic hypothesis, namely, that our observable universe is a phase of a past-eternal series of universes or cycles, a series whose mechanisms are finely-tuned to make universes like ours by an uncaused Ground of Nature. But currently we have no evidence of the cyclic universe, and the Steinhardt-Turok model may not be past-eternal.

Chapter 12

1. It is interesting to compare Russell's "A Free Man's Worship" to existentialism. Not only did Russell express the basic, as Jonas sees it, Gnostic idea that the human spirit is alien to nature, but also, like (and before) Albert Camus, he called the proper human attitude toward nature "revolt." And the figure that begins his essay is reminiscent of Nietzsche's eternal recurrence: God invents nature to alleviate boredom, later to snuff out the stars saying, "It was a good play. I'll have it performed again." Prominent existentialists do in fact presuppose a non-telic universe foreign to human valuation, which is to say existentialism is partly based on a popularized version of early twentieth-century physics (especially thermodynamics and the atomic theory).

2. We should note that the most "natural" natural religions are the animisms of hunter-gatherer societies. They have greatest right to the term. They constitute the most basic human religious posture and a form of religion remarkably integrated with wild nature. They are arguably ecological religions, not because their members were less environmentally destructive—although they generally were—but because theirs was a religion of the circulation of sacred energy or ecological value through the natural world and human being. In the words of the Sioux medicine man Black Elk, "nothing can live well except in a manner that is suited to the way that the sacred Power of the World lives and moves" (Neihardt 1979, p. 169). What moves is Power, or manna, or what Holmes Ralston calls value, circulating through natural kinds in an ecosystem, so that "everything an Indian does is in a circle, and that is because the Power of the World always works in circles, and everything tries to be round." Many mythical narratives metaphorically express relationships of ecologi-

cal interdependence that are objectively true by today's scientific standards. This is a key theme of Paul Shepard's work, in which, one might say, ecology recapitulates mythology (Shepard 1985, Cahoone 2006).

3. Darwin's two embedded sentences, which Popper has separated by an ellipsis, are from two separate letters to Asa Gray (Nov. 26, 1860 and Dec.11, 1861). See the Darwin Correspondence Project at http://www.darwinproject.ac.uk/entry-3342.

4. As Putnam notes, Hume's relevant passages argue both more and less than this; less, because he does not claim it to be a fallacy of logic, and more, because he does seem to think that matters of fact are by themselves value-neutral (Putnam 2004).

5. Erazim Kohák's 1987 is a marvelous phenomenological attempt to find value, even moral value, in nature. But as I have suggested, it has difficulty making the transition from natural value to values obligatory for humans (Cahoone 1999).

6. My denial is not aimed at the classical function of that doctrine in Christian theology as a guarantee of the complete dependence of the Creation on the Creator.

7. See Dombrowski 1996 for further discussion of Hartshorne's conception.

8. As noted, the evolution of laws and/or constants was suggested by Peirce and more recently by Smolin. But Smolin (like Peirce) must assume some eternal laws to govern the process of "cosmological natural selection" (Smolin 1997).

9. By "complexity" here I mean mostly, but not exclusively, organized complexity. The ambiguity is unavoidable due to the difficulty in comparing quantitative and organized complexity. See my Chapter 4. One theorist who tries to compare the two is Ulanowicz in his 2009.

10. While complex in role differentiation, bureaucratic or military order purposively constrains individuality and interaction to produce the highest level of uniformity.

11. My view implies all life, since complex, has intrinsic value, and in this sense I do agree with environmental biocentrism. However, because we are always having to trade off the destruction of some valuable things for the creation or maintenance of others, and since human individuals are more complex than the individuals of any other Earth species, my view does not take sides in the biocentrism versus anthropocentrism debate in any other sense.

12. Ilya Prigogine's widely read *Order Out of Chaos* (Prigogine and Stengers 1984, p. 313) closes with a passage from Andre Neher's discussion of the Midrashic commentary on Genesis (Neher 1975, p. 179): "Twenty-six attempts preceded the present genesis, all of which were destined to

fail . . . 'Let's hope it works.' *Halway Seyaamod*, exclaimed God as he created the World, and this hope . . . has emphasized right from the outset that this history is branded with the mark of radical uncertainty." Appealing, but Neher's rendition is questionable. The 26 attempts comes from the medieval commentator Rashi, not Genesis. "Halway seyaamod" does not appear in the passage Neher cites (*Genesis Rabbah*, vol. 1, 9:4), and the closest passage is translated by Jacob Neusner "May you [the world] always charm me." I thank Professors Neusner and Alan Avery-Peck for their help in finding and analyzing the original.

13. Joseph Schumpeter was right that some creation requires destruction. But when the destruction is greater than the creation, it is difficult to call the whole act "creative."

Bibliography

Ahl, Valerie, and T. F. H. Allen 1996. *Hierarchy Theory: A Vision, Vocabulary, and Epistemology*. New York: Columbia University Press.

Alexander, Samuel. 1920. *Space, Time, and Deity*. London: Macmillan.

Allen, T. F. H., and Thomas B. Starr. 1982. *Hierarchy: Perspectives for Ecological Complexity*. Chicago: University of Chicago.

Anaximander of Miletus. 1988. Chapter Three. *The Presocratic Philosophers*. Ed. by G. S. Kirk, J. E. Raven, and M. Schofield. Cambridge: Cambridge University.

Anderson, Philip. 1972. "More is Different." *Science* vol. 177, no. 4:393–96.

Anderson, Stephen R. 2004. "A Telling Difference." *Natural History* November. pp. 38–43.

Aquinas, St. Thomas. 1999. *On Faith and Reason*. Indianapolis: Hackett Publishing. pp.121–8.

Atkatz, David. 1994. "Quantum Cosmology for Pedestrians." *American Journal of Physics* 62 (7), July. pp. 619–27.

Austin, J. L. 1975. *How to Do Things with Words*. Cambridge, MA: Harvard University.

Baeten, Elizabeth. 1999. "Rethinking the Social Constituted Self as the Subject of Ethical Communication." *The Journal of Speculative Philosophy* vol. 13, no. 1, pp. 1–18.

Baez, John C., and Emory F. Bunn. 2006. "The Meaning of Einstein's Equation." http://math.ucr.edu/home/baez/Einstein.enstein.pdf.

Bain, Alexander. 2004. *Logic*. Whitefish, MT: Kessinger Publishing.

Barrow, John. 2002. *The Constants of Nature: From Alpha to Omega—The Numbers that Encode the Deepest Secrets of the Universe*. New York: Vintage.

Barrow, John, and Frank Tipler. 1986. *The Anthropic Cosmological Principle*. Oxford: Oxford University Press.

Bateson, Gregory. 2000. "A Theory of Play and Fantasy." In *Steps Toward an Ecology of Mind: Collected Essays in Anthropology, Psychiatry, Evolution, and Epistemology*. Chicago: University of Chicago Press.

Bechtel, William, and Robert C. Richardson. 1992. "Emergent Phenomena and Complex Systems." In Beckerman 1992b. pp. 257–88.

Beckermann, Ansgar. 1992a. "Supervenience, Emergence, and Reduction." In Beckerman 1992b. pp. 94–118.

Beckermann, Ansgar, Hans Flohr, and Jaegwon Kim. 1992b. *Emergence or Reduction? Essays on the Prospects of Nonreductive Physicalism*. New York: Walter de Gruyter.

Benfey, Theodor. 2000. "Reflections on the Philosophy of Chemistry and a Rallying Call for our Discipline." *Foundations of Chemistry* 2:195–205.

Berger, Peter L. 1970. *A Rumor of Angels: Modern Society and the Rediscovery of the Supernatural*. New York: Anchor Books.

Bergson, Henri. 1944. *Creative Evolution*. Trans. by Arthur Mitchell. New York: Modern Library.

Berkeley, George. 1906. *Three Dialogues Between Hylas and Philonous*. Chicago: Open Court.

Bertalanffy, Ludwig von. 1968. *General Systems Theory: Foundations, Development, Applications*. New York: George Braziller.

Bickhard, Mark. 2000. "Emergence." In *Downward Causation*. Ed. by P. B. Andersen, et al. Aarhus, Denmark: University of Aarhus.

Blackett, P. M. S. 1962. "Memories of Rutherford." In J. B. Birks. *Rutherford at Manchester*. London: Heywood. pp. 102–13.

Blitz, David. 1992. *Emergent Evolution: Qualitative Novelty and the Levels of Reality*. Dordrecht: Kluwer.

Borde, A., A. H. Guth, and A.Vilenkin. 2003. "Inflationary spacetimes are not past-complete." *Physical Review Letters*, 90 (15). April 18. pp. 151301-1 to 151301-4.

Boyd, Richard. 1991. "Realism, Anti-Foundationalism and the Enthusiasm for Natural Kinds." *Philosophical Studies*. 61:127–48.

Boyd, Richard, Philip Gaspar, and J. D. Trout. *The Philosophy of Science*. Cambridge, MA: MIT Press, 1991.

Brentano, Franz. 1973. *Psychology from an Empirical Standpoint*. London: Routledge, Kegan, Paul.

Broad, C. D. 1960. *The Mind and Its Place in Nature*. London: Routledge, Kegan, & Paul.

Buchler, Justus. 1951. *Toward a General Theory of Human Judgment*. New York: Dover.

———. 1955. *Nature and Judgment*. New York: Grosset and Dunlap.

———. 1990. *Metaphysics of Natural Complexes*. Ed. by Kathleen Wallace and Armen Marsoobian, with Robert S. Corrington. Albany, NY: State University of New York Press.

———. 1990a. "On the Concept of 'the World'." Appendix III to Buchler 1990.

———. 1990b. "Probing the Idea of Nature." Appendix IV to Buchler 1990.

Bunge, Mario. 1979. *A World of Systems*. Volume 4 of *Treatise on Basic Philosophy*. Dordrecht: D.Reidel.

Cahoone, Lawrence. 1999. "Whose Nature? Which Morality?: On Erazim Kohák's Moral Sense of Nature," in Robert S. Cohen and Alfred Tauber, *Philosophies of Nature: The Human Dimension*. Dordrecht: Kluwer. pp. 19–34.

———. 2002a. *The Ends of Philosophy: Pragmatism, Foundationalism, and Postmodernism*. Malden, MA: Blackwell.

———. 2002c. "Margoline Relativism," *Idealistic Studies* 32:1, 2002. pp. 27–36.

———. 2005. *Cultural Revolutions: Reason versus Culture in Philosophy, Politics, and Jihad*. University Park, PA: Pennsylvania State University Press.

———. 2006. "Our Recent Rousseau: On Paul Shepard." *Environmental Philosophy* 3.1:13–26.

———. 2008. "Emergence, Reduction, and Ordinal Physicalism." *Transactions of the Charles S. Peirce Society* 44.1. Winter. pp. 40–62.

———. 2009. "Arguments From Nothing: God and Quantum Cosmology." *Zygon: Journal of Religion and Science* 44 (4) December. pp. 777–96.

Call, Josep, and Michael Tomasello. 2008. "Does the Chimpanzee have a Theory of Mind? 30 Years Later." *Trends in Cognitive Sciences* vol.12, no.5, pp. 187–92.

Callicot, J. Baird. 1986. "On the Intrinsic Value of Nonhuman Species." Ed. by Bryan Norton. *The Preservation of Species*. Princeton: Princeton University. pp. 142–3.

Campbell, Donald. 1960. "Blind Variation and Selective Retention in Creative Thought as in Other Knowledge Processes." *Psychological Review* 67, no. 6, pp. 380–400.

———. 1985. "Pattern Matching as an Essential in Distal Knowing." In Kornblith 1985. pp. 49–70.

———. 1987. "Neurological Embodiments of Beliefs and the Gaps in the Fit of Phenomena to Noumena." In Shimony and Nails 1987. pp. 165–92.

———. 1988a. "Evolutionary Epistemology." In Campbell, *Methodology and Epistemology for Social Science: Selected Papers*. Ed. by E. Samuel Overman. Chicago: University of Chicago. pp. 393–434.

———. 1988b. "Descriptive Epistemology: Psychological, Sociological, and Evolutionary." In Campbell, *Methodology and Epistemology for Social Science: Selected Papers*. Ed. E. Samuel Overman. Chicago: University of Chicago. pp. 435–85.

———. 1990. "Levels of Organization, Downward Causation, and the Selection-Theory Approach to Evolutionary Epistemology." In *Theories of the Evolution of Knowing*. Ed. by Gary Greenberg and Ethel Tobach. Hillsdale, NJ: Erlbaum Associates.

Cao, Tian-yu. 1997. *Conceptual Developments of 20th Century Field Theories*. Cambridge, UK: Cambridge University Press.

———. 2003. "Can We Dissolve Physical Entities into Mathematical Structures?" *Synthese* 136:57–71.

———. 2004. "Ontology and Scientific Explanation." In *Explanations: Styles of Explanation in Science*. Ed. by John Cornwell. Oxford, UK: Oxford University. pp.173 96.

———. 2006. "Structural Realism and Quantum Gravity." In *Structural Foundations of Quantum Gravity*. Ed. by Dean Rickles, Steven French, and Juha Saatsi. Oxford: Oxford University Press. pp. 40–52.

Carnap, Rudolf. 1988. "Empiricism, Semantics, and Ontology." With *Meaning and Necessity: A Study in Semantics and Modal Logic*. Chicago: University of Illinois.

———. 2003. *The Logical Structure of the World and Pseudo-Problems of Philosophy*. Trans. Rolf George. Open Court.

Carter, Brendon. 1974. "Large Number Coincidences and the Anthropic Principle in Cosmology." *IAU Symposium 63: Confrontation of Cosmological Theories with Observational Data*. Dordrecht: Reidel. pp. 291–98.

Cassirer, Ernst. 1953. *Substance and Function and Einstein's Theory of Relativity*. New York: Dover.

———. 1962. An *Essay on Man: An Introduction to a Philosophy of Human Culture*. New Haven, CT: Yale University.

———. 1965. *The Philosophy of Symbolic Forms. Vol. I Language. Vol. II Mythical Thought, Vol.III The Phenomenology of Knowledge*. Trans. by Ralph Mannheim. New Haven, CT: Yale University.

Casti, John L. 1995. *Complexification: Explaining a Paradoxical World Through the Science of Surprise*. New York: Harper Perennial.

Chalmers, David J. 1995. "Facing up to the Problem of Consciousness." *Journal of Consciousness Studies* 2 (3):200–19.

Chase, Philip. 1999. "Symbolism as Reference and Symbolism and Culture." In Dunbar 1999. pp. 34–49.

Cheney, Dorothy L., and Robert M. Seyfarth. 2007. *Baboon Metaphysics: The Evolution of a Social Mind*. Chicago: University of Chicago Press.

Clayton, Philip. 1997. *God and Contemporary Science*. Grand Rapids, MI: Wm. B. Erdmans Publishing.

———. 2004. *Mind and Emergence: From Quantum to Consciousness*. Oxford: Oxford University Press.

Clayton, Philip, and Paul Davies. 2006. *The Re-Emergence of Emergence: The Emergentist Hypothesis from Science to Religion*. Oxford: Oxford University Press.

Cobb, John B. 2008. *Back to Darwin: A Richer Account of Evolution*. Grand Rapids, MI: Wm. B. Erdmans Publishing.

Collins, Robin. 2002. "God, Design, and Fine-Tuning." http://www.bilimfelsefedin.org /blog/wp-content/uploads/God,_Design,_and_Fine-Tuning_-_Robin_Collins.pdf.

Craig, William Lane, and Quentin Smith. 2003. *Theism, Atheism, and Big Bang Cosmology*. New York: Oxford University Press.

Crosby, Donald. 2002. *A Religion of Nature*. Albany: State University of New York Press.

Cummins, Robert. 1991. "The Role of Mental Meaning in Psychological Explanation." In McLoughlin 1991.

Curie, Pierre. 1894. "*Sur la symétrie dans les phénomènes physiques. Symétrie d' un champ électrique et d'un champ magnétique.*" *Journal de Physique* 3rd series, vol. 3, pp. 393–415.

D'Espagnat, Bernard. 2006. *On Physics and Philosophy*. Princeton: Princeton University.

Damasio, Antonio. 1994. *Descartes' Error: Emotion, Reason, and the Human Brain*. New York: HarperCollins.

———. 2000. *The Feeling of What Happens: Body and Emotion in the Making of Consciousness*. San Diego, CA: Harcourt.

———. 2003. *Looking for Spinoza: Joy, Sorrow, and the Feeling Brain*. Orlando, FL: Harcourt.

Darwin, Charles. 1860. *Darwin Correspondence Project*. Jim Secord director, November 26, 1860 and Dec.11.1861 letters to Asa Gray. http://www.darwinproject.ac.uk/entry-3342.

Davidson, Donald. 1980. "Mental Events." In *Actions and Events*. Oxford: Clarendon. pp. 207–27.

Davies, Paul C. W. 1982. *The Accidental Universe*. Cambridge, UK: Cambridge University Press.

———. 1984. *God and the New Physics*. New York: Simon and Schuster.

———. 1989. *The New Physics*. Cambridge, UK: Cambridge University Press.

Dawkins, Richard. 1982. *The Extended Phenotype*. Oxford: Oxford University.

———. 1989. *The Selfish Gene*. Oxford: Oxford University.

Del Re, Giuseppe. 1998. "Ontological Status of Molecular Structure." *HYLE: International Journal for Philosophy of Chemistry* vol. 4, no. 2, pp. 81–103.

Deleuze, Giles, and Felix Guattari. 1987. *A Thousand Plateaus: Capitalism and Schizophrenia.* Trans. by Brian Massumi. Minneapolis: University of Minnesota.

DeMarco, Wes. 2006. "A Zero-Method for the 'Big Why'." Unpublished manuscript.

Depew, David J., and Bruce H. Weber. 1996. *Darwinism Evolving: System Dynamics and the Genealogy of Natural Selection.* Cambridge, MA: MIT Press.

Dewey, John. 1896. "The Reflex Arc Concept in Psychology." *Psychological Review* 3:357–370.

———. 1944. "Anti-naturalism in Extremis." In *Naturalism and the Human Spirit.* Ed. by Yervant H. Krikorian. New York: Columbia University. pp. 1–16.

———. 1958. *Experience and Nature.* Chicago: Open Court.

———. 1964. *John Dewey and Arthur F. Bentley: A Philosophical Correspondence, 1932–1951.* Ed. by Sidney Ratner and Jules Altman. New Brunswick: Rutgers Univ. Press.

———. 1981. *John Dewey: The Later Works, 1925–53.* Ed. by JoAnne Boydston. Vol 1. Carbondale, IL: Southern Illinois University.

———. 2008. *John Dewey: The Later Works, 1925–53.* Ed. by JoAnne Boydston. Vol 6. Carbondale, IL: Southern Illinois University.

Dewey, John, Sidney Hook, and Ernest Nagel. 1994. "Are Naturalists Materialists?" In *American Philosophic Naturalism in the Twentieth Century.* Ed. by John Ryder. Amherst, NY: Prometheus Books. pp.102–20.

Diamond, Jared. 1999. *Guns, Germs and Steel: The Fates of Human Societies.* New York: W. W. Norton & Company.

Dicke, R. H. 1961. "Dirac's Cosmology and Mach's Principle." *Nature* 192:440–41.

Dombrowski, Daniel. 1996. *Analytic Theism, Hartshorne, and the Concept of God.* Albany, NY: State University of New York Press.

Dowell, J. L. 2006. "Formulating the Thesis of Physicalism: An Introduction." *Philosophical Studies* 131:1–23.

Dretske, Fred. 1988. *Explaining Behavior: Reasons in a World of Causes.* Cambridge, MA: MIT.

———. 1991. "Dretske's Replies." In *Dretske and His Critics.* Ed. by Brian P. McLaughlin. Cambridge: Blackwell. pp. 180–221.

Dunbar, Robin, Chris Knight, and Camilla Power, eds. 1999. *The Evolution of Culture: An Historical and Scientific Overview.* New Brunswick, NJ: Rutgers University Press.

Dupré, John. 1993. *The Disorder of Things: Metaphysical Foundations of the Disunity of Science*. Cambridge, MA: Harvard University.

Eddington, Arthur S. 1933. *The Nature of the Physical World*. Cambridge University Press: Cambridge UK.

Edmunds, Bruce. 1999. *Syntactic Measures of Complexity*. PhD thesis. University of Manchester, UK.

Einstein, Albert. 1921. *Relativity: The Special and General Theory*. Trans. by Robert Lawson. New York: Holt and Co.

Einstein, Albert, and David Infeld. 1966. *The Evolution of Physics*. New York: Simon & Schuster.

Einstein, Albert, B. Podolsky, and N. Rosen. 1935. "Can Quantum-Mechanical Description of Reality be Considered Complete?" *Physical Review* 47:777–80.

Eisenstadt, S.N. 2000. *Fundamentalism, Sectarianism, and Revolution: The Jacobin Dimension of Modernity*. Cambridge, UK: Cambridge University.

Eliade, Mircea. 2005. *The Myth of the Eternal Return: Cosmos and History*. Princeton, NJ: Princeton University.

Feibleman, James K. 1954. "Theory of Integrative Levels." *British Journal for the Philosophy of Science* 5:59–66.

Feynman, Richard. 1994. *The Character of Physical Law*. Cambridge, MA: MIT Press.

Feynman, Richard, with Robert Leighton, and Matthew Sands. 1970. *Lectures On Physics. Vol. I*. Reading, MA: Addison, Wesley, Longman.

Fichte, Gottlob. 1970. *Science of Knowledge (Wissenschafslehre) with First and Second Introductions*. Ed and trans by Peter Heath and John Lachs. New York: Appleton-Century-Crofts.

Fine, Arthur. 1991. "The Natural Ontological Attitude." In Richard Boyd, Philip Gasper, and J. D.Trout, *The Philosophy of Science*. Cambridge, MA: MIT Press. pp. 261–78.

Finlayson, Clive. 2010. *The Humans Who Went Extinct: Why Neanderthals Died Out and We Survived*. Oxford: Oxford University.

Flannery, Tim. 2001. *The Eternal Frontier: An Ecological History of North America and its Peoples*. New York: Grove.

Fodor, Jerry. 1974. "Special Sciences (Or: The Disunity of Science as a Working Hypothesis)." *Synthese* 28:97–115.

Fouts, Roger, Deborah Fouts, and Thomas Van Cantfort. 1989. "The Infant Loulis Learns Signs from Cross-fostered Chimpanzees." In *Teaching Sign Language to Chimpanzees*. Ed. by R. A. Gardiner, B. T. Gardiner, and Thomas Van Cantfort. Albany, NY: State University of New York Press. pp. 280–92.

French, Steven and James Ladyman. 2003. "Remodeling Structural Realism: Quantum Physics and the Metaphysics of Structure." *Synthese* 136:31–56.

Gadamer, Hans-Georg. 1994. *Truth and Method*. Trans. by Joel Weinsheimer and Donald Marshall. New York: Continuum.

Galileo. 1957. "Excerpts from The Assayer." In *Discoveries and Opinions of Galileo*. Trans. by Stillman Drake. New York: Anchor Books. pp. 229–280.

Geist, Valerius. 1978. *Life Strategies, Human Evolution, Environmental Design: Toward a Biological Theory of Health*. New York: Springer Verlag.

———. 1994. "Culture and its Biological Origins: A View from Ethology, Epigenesis, and Design." In R. A. Gardner, B. Chiarelli, and F. X. Plooij, eds. *The Ethological Roots of Culture*. Dordrecht: Kluwer. pp.441– 60.

———. 2001. *Whitetail Tracks: The Deer's History and Impact in North America*. Iola, WI: Krause Publications.

Gellner, Ernest. 1988. *Plow, Sword, and Book: The Structure of Human History*. Chicago: University of Chicago.

Genesis Rabbah, The Judaic Commentary to the Book of Genesis: A New American Translation, Vol.I: Parashiyyot One through Thirty-Three on Genesis 1:1 to 8:14. 1985. Trans. by Jacob Neusner. Providence, RI: Brown University Press.

Gibson, J. J. 1986. "The Theory of Affordances." In *The Ecological Approach to Visual Perception*. Hillsdate, NJ: Lawrence Erlbaum Associates. pp. 127–43.

Gilbert, Scott, and David Epel. 2009. *Ecological Developmental Biology: Integrating Epigenetics, Medicine, and Evolution*. Sunderland, MA: Sinauer Associates Inc.

Globus, Gordon, G. Maxwell, and I. Savodnik. 1976. *Consciousness and the Brain: A Scientific and Philosophical Inquiry*. New York: Basic Books.

Goodenough, Ursula. 1998. *The Sacred Depths of Nature*. Oxford: Oxford University.

Gray, Michael, et al. 2010. "Irremediable Complexity." *Science* 12, vol. 330, no. 6006. November. pp. 920–21.

Griffin, Donald R. 2001. *Animal Minds: Beyond Cognition to Consciousness*. Chicago: University of Chicago Press.

Guth, Alan. 1998. *The Inflationary Universe: The Quest for a New Theory of Cosmic Origins*. New York: Perseus.

———. 2007. "Eternal Inflation and its Implications." *Journal of Physics* A40:6811–6826.

Hartee, Douglas. 1957. *The Calculation of Atomic Structures*. New York: John Wiley & Sons.

Hartle, J. B., Hawking, S. W. 1983. "Wave function of the Universe." *Physical Review D*. 28 (12):2960–75.

Hartmann, Nicolai. 1952. *New Ways of Ontology*. Trans. by Reinhard C. Kuhn. Westport, CT: Greenwood Press.

Hartshorne, Charles. 1984. *Omnipotence and Other Theological Mistakes*. Albany: State University of New York Press.

Hauser, Marc, Noam Chomsky, and W. Tecumseh Fitch. 2002. "The Faculty of Language: What is it, Who Has it, and How Did it Evolve?" *Science* vol. 298. November 22. pp. 1569–79.

Hawking, Stephen. 1988. *A Brief History of Time*. New York: Bantam.

Hegel, G. W. F. 1979. *Hegel's Phenomenology of Spirit*. Trans. by A. V. Miller. Oxford: Oxford University Press.

Hellman, Paul, and Frank Wilson Thompson. 1975. "Physicalism: Ontology, Determination, and Reduction." *Journal of Philosophy* 72, 17:551–564.

Hempel, Carl. 1969. "Reduction: Ontological and Linguistic Facets." In Sidney Morgenbesser, et al., eds., *Essays in Honor of Ernest Nagel*. New York: St Martin's Press.

Hobson, R. Peter. 2004. *The Cradle of Thought: Exploring the Origins of Thinking*. London: Pan Macmillan.

Hodgson, Marshall. 1974. *The Venture of Islam: Conscience and History in a World Civilization*. Two Volumes. Chicago: University of Chicago.

Hoffmeyer, Jesper. 1997. *Signs of Meaning in the Universe*. Bloomington: Indiana University.

Holt, B. Edwin, Walter T. Marvin, William Pepperrell Montague, Ralph Barton Perry, Walter B. Pitkin, and Edward Gleason Spaulding. 1925. *The New Realism: Cooperative Studies in Philosophy*. New York: Macmillan.

Horgan, Terence. 1991. "Actions, Reasons, and the Explanatory role of Content." In McLoughlin 1991. pp. 73–101.

Horst, Stephen. 2007. *Beyond Reduction: Philosophy of Mind and Post-Reductionist Philosopohy of Science*. Oxford: Oxford University Press.

Hume, David. 1910. "Of Providence and a Future State." In *An Enquiry into Human Understanding*. Section XI. New York: Collier.

———. 1998. *Dialogues Concerning Natural Religion*. Indianapolis: Hackett.

Humphreys, Nicholas. 2006. *Seeing Red: A Study in Consciousness*. Cambridge, MA: Harvard University.

Husserl, Edmund. 1982. *Ideas Pertaining to a Pure Phenomenology and to a Phenomenological Philosophy: First Book: General Introduction to a Pure Phenomenology*. Trans. by Fred Kersten. The Hague: Nijhoff.

Illich, Ivan. 1990. *Gender*. Berkeley, CA: Heyday Books.

Isham, Christopher. 1993. "Quantum Theories of the Creation of the Universe." In Russell et al. 1993. pp. 49–89.

Jablonka, Eva, and Marion J. Lamb. 2004. *Evolution in Four Dimensions: Genetic, Epigenetic, Behavioral, and Symbolic Variation in the History of Life*. Cambridge, MA: MIT Press.

James, William. 1950. *The Principles of Psychology*. Two Volumes. New York: Dover.

———. 1981. *Pragmatism*. Indianapolis: Hackett.

Jaspers, Karl. 1965. *The Origin and Goal of History*. Trans. by Michael Bullock. New Haven: Yale University Press.

Jonas, Hans. 1966. *The Phenomenon of Life: Toward a Philosophical Biology*. Chicago: University of Chicago.

Kaminski, Juliane, Josep Call, and Michael Tomasello. 2008. "Chimpanzees know what others know, but not what they believe." *Cognition* 109:224–34.

Kant, Immanuel. 1998. *Critique of Pure Reason*. Trans. by Paul Guyer and Allan Wood. Cambridge, UK: Cambridge University.

Kauffman, Stuart. 1993. *The Origins of Order: Self-Organization and Selection in Evolution*. Oxford: Oxford University Press.

Kim, Jaegwon. 1991. "Dretske on How Reasons Explain Behavior." In McLoughlin 1991. pp. 285–308.

———. 1993. *Supervenience and Mind*. Cambridge: Cambridge University Press.

———. 1998. *Philosophy of Mind*. Boulder, CO: Westview Press.

———. 2000. *Mind in a Physical World: An Essay on the Mind-Body Problem and Mental Causation*. Cambridge, MA: MIT.

———. 2005. *Physicalism, or Something Near Enough*. Princeton: Princeton University Press.

Kirsch, Janina A., Onur Güntürkün, and Jonas Rose. 2008. "Insight without Cortex: Lessons from the Avian Brain." *Consciousness and Cognition* 17:475–483.

Kirschner, Robert P. 2004. *The Extravagant Universe: Exploding Stars, Dark Energy, and the Accelerating Cosmos*. Princeton: Princeton University Press.

Kitcher, Philip. 2003. "Function and Design." *In Mendel's Mirror: Philosophical Reflections on Biology*. Oxford: Oxford University. pp. 159–76.

Knight, Chris. 1999. "Sex and Language as Pretend-Play." In Dunbar et al. 1999. pp. 228–47.

Knight, Chris, Robin Dunbar, and Camilla Power. 1999. "An Evolutionary Approach to Human Culture." In Dunbar et al., 1999. pp.1–14.

Kohák, Erazim.1987. *The Embers and the Stars: A Philosophical Inquiry into the Moral Sense of Nature*. Chicago: University of Chicago.

Kornblith, Hillary. 1985. *Naturalizing Epistemology*. Cambridge, MA: MIT.

———. 2002. *Knowledge and its Place in Nature*. Oxford: Clarendon.

Korthop, Gert. 2009. "A Memorable Misunderstanding: Fred Hoyle's Boeing Story in the Evolution/Creation Literature." http://home.wxs.nl/~gkorthof/kortho46a.htm.

Kragh, Helge. 2009. "Continual Fascination: The Oscillating Universe in Modern Cosmology." *Science in Context* 22(4):587–612.

Krech, Shepard III. 2000. *The Ecological Indian: Myth and History*. New York: W. W. Norton & Co.

Krieger, Leonard. 1957. *The German Idea of Freedom: History of a Political Tradition*. Boston: Beacon Press.

Krikorian, Yervanth. 1944. *Naturalism and the Human Spirit*. New York: Columbia University Press.

Kripke, Saul. 1980. *Naming and Necessity*. Oxford: Basil Blackwell.

Leggett, Anthony. 1987. *The Problems of Physics*. Oxford: Oxford University.

Lewes, G. H. 1875. *Problems of Life and Mind*. London: Kegan Paul, Trench, and Turbner.

Lewis, David. 1973. "Causation." *The Journal of Philosophy* vol. 70, no. 17, pp. 556–67.

———. 2001. *On the Plurality of Worlds*. Cambridge, UK: Wiley-Blackwell.

Linde, Andre. 1994. "The Self-Reproducing Inflationary Universe." *Scientific American*. November. pp. 48–55.

———. 2007. "Inflationary Cosmology." *Lecture Notes in Physics*. 738, May:1–54

Lloyd, Elisabeth. 2001. "Units and Levels of Selection: An Anatomy of the Units of Selection Debates." In *Thinking about Evolution, Volume Two: Historical, Philosophical, and Political Perspectives*. Ed. by R. S. Singh, et al. Cambridge: Cambridge University. pp. 267–91.

Lloyd Morgan, Conwy. 1926. *Emergent Evolution*. New York: Henry Holt and Co.

Lombardi, Olimpia, and Martín Labarca. 2005. "The Ontological Autonomy of the Chemical World." *Foundations of Chemistry* 7:125–48.

Lorenz, Konrad. 1941. "Kant's Doctrine of the A Priori in the Light of Contemporary Biology." In *Konrad Lorenz: The Man and His Ideas*. Ed. by Richard I. Evans. New York and London: Harcourt Brace Janovich. pp. 129–217.

———. 1973. *Behind the Mirror: A Search for a Natural History of Human Knowledge*. New York and London: Harcourt Brace Janovich.

Lucretius. 1921. *On the Nature of Things*. Trans. by William Ellery Leonard. New York: E. P. Dutton.

Luhmann, Niklas. 1990. "The Cognitive Program of Constructivism and a Reality that Remains Unknown." In *Self-organization. Portrait of a scientific revolution*. Ed. by W. Krohn. Dordrecht: Kluwer. pp. 64–85.

———. 1996. *Social Systems*. Trans. by John Bednarz, Jr. with Dirk Baecker. Stanford, CA: Stanford University Press.

Luisi, Pier Luigi. 2002. "Emergence in Chemistry: Chemistry as the Embodiment of Emergence." *Foundations of Chemistry* 4:183–200.

Lyotard, Jean-François. 1984. *The Postmodern Condition: A Report on Knowledge*. Trans. by Brian Massumi and Geoff Bennington. Minneapolis: University of Minnesota.

MacArthur, Robert, and E. O. Wilson. 2001. *The Theory of Island Biogeography*. Princeton, NJ: Princeton University Press.

MacIntyre, Alasdair. 1988. *Whose Justice? Which Rationality?* Notre Dame, IN: University of Notre Dame.

Magee, Glenn. 2001. *Hegel and the Hermetic Tradition*. Ithaca, NY: Cornell University.

Mandlebrot, Benoit. 1967. "How Long is the Coast of Britain? Statistical Self-Similarity and Fractional Dimension." *Science* 156, no. 2775. pp. 636–638.

Margolis, Joseph. 1995. *Historied Thought, Constructed World: A Conceptual Primer for the Turn of the Millennium*. Berkeley: University of California.

———. 2001. *Selves and Other Texts: The Case for Cultural Realism*. State College, PA: Penn State Press.

———. 2010. *Pragmatism's Advantage: American and European Philosophy at the End of the Twentieth Century*. Stanford, CA: Stanford University Press.

Margulis, Lynn, and Dorion Sagan. 1999. "The Role of Symbiogenesis in Evolution." In *Back to Darwin: A Richer Account of Evolution*. Ed. by John Cobb. Grand Rapids, MI: Wm. B. Eerdmans. pp. 176–84.

Marsoobian, Armen, Robert S. Corrington, and Kathleen Wallace. 1990. *Nature's Perspectives: Prospects for an Ordinal Metaphysics*. Albany: State University of New York.

Marsoobian, Armen, and John Ryder. *The Blackwell Guide to American Philosophy*. Cambridge, MA: Wiley-Blackwell, 2004.

Maturana, Humberto, and Francisco Varela. 1991. *Autopoiesis and Cognition: The Realization of the Living*. Springer.

Mayr, Ernst. 1965. "Cause and Effect in Biology." In D. Lerner. *Cause and Effect*. New York: Free Press. pp. 33–50.

———. 1974. "Teleological and Teleonomic: A New Analysis." *Boston Studies in the Philosophy of Science*. Vol.14, pp. 91–117.

———. 1982. *The Growth of Biological Thought: Diversity, Evolution, and Inheritance*. Cambridge, MA: Belknap, 1982.

———. 1992. "The Idea of Teleology." *Journal of the History of Ideas* 53, no. 1, pp. 117–35.

———. 1997. "The Objects of Selection." *Proceedings of the National Academy of Sciences of the United States of America*. Vol. 94, no. 6, pp. 2091–94.

Mayr, Ernst, and John Brockman. 2001. "Interview with Ernst Mayr." Edge Foundation. www.edge.org/3rd_culture/mayr/mayr_print.html.

McCrea, W. H., and M. J. Rees. 1983. *The Constants of Physics*. Great Neck, NY: Scholium International.

McKee, Jeffrey K., Frank E. Poirier, and W. Scott McGraw. 2005. *Understanding Human Evolution*. Upper Saddle River, NJ: Pearson, Prentice Hall.

McLoughlin, Brian. 1991. *Dretske and His Critics*. Cambridge, MA: Basil Blackwell.

———. 1992. "The Rise and Fall of British Emergentism." In Beckerman et al., 1992b. pp. 49–93.

McLoughlin, Brian, and Karen Bennett. 2004. "Supervenience." *Stanford Encyclopedia of Philosophy*. http://plato.stanford.edu/entries/supervenience.

McShea, Daniel. 1991. "Complexity and Evolution: What Everybody Knows." *Biology and Philosophy* 6:303–24.

Mead, George Herbert. 1932. *The Philosophy of the Present*. Chicago: University of Chicago.

———. 1938. *The Philosophy of the Act*. Chicago: University of Chicago.

———. 1962. *Mind, Self, and Society: From the Standpoint of a Social Behaviorist. Vol. I*. Ed. by Charles Morris. Chicago: University of Chicago.

Melnyk, Andrew. 2003. *A Physicalist Manifesto: Thoroughly Modern Materialism*. Cambridge, UK: Cambridge University.

Merleau-Ponty, Maurice. 1962. *Phenomenology of Perception*. Trans. by Colin Smith. New York: Humanities Press.

———. 1968. "The Intertwining—The Chiasm." In *The Visible and the Invisible, Followed by Working Notes*. Trans. by Alphonso Lingis. Evanston: Northwestern University Press. pp. 130–55

Mill, John Stuart. 1843. *System of Logic*. London: Longmans, Green, Reader, and Dyer.

Mitchell, Melanie.2009. *Complexity: A Guided Tour*. Oxford: Oxford University Press.

Monod, Jacques. 1971. *Chance and Necessity: An Essay on the Natural Philosophy of Modern Biology*. Trans. by Austryn Wainhouse. New York: Vintage Books.

Nagel, Ernest. 1979. *The Structure of Science: Problems in the Logic of Scientific Explanation*. Indianapolis: Hackett Publishing.

Nagel, Thomas. 1989. "The View from Nowhere." *In The View from Nowhere*. Oxford: Oxford University Press.

Neher, Andre. 1975. "*Visions du temps et l'histoire dans la culture juive.*" In *Les cultures et le temps*. Ed. by Paul Ricoeur. Paris: Payot. pp. 171–91.

Neihardt, John. 1979. *Black Elk Speaks: Being the Life Story of a Holy Man of the Oglala Sioux*. Lincoln: University of Nebraska.

Neville, Robert C. 1968. *God the Creator: On the Transcendence and Presence of God* Chicago: University of Chicago.

———. 1989. *Recovery of the Measure: Interpretation and Nature*. Albany, NY: State University of New York Press.

Newell, A., J. C. Shaw, and H. A. Simon. 1958. "Elements of a Theory of Human Problem Solving." *Psychological Review* 65, no. 3. pp. 151–66.

Nicolis, Gregoire. 1989. "Physics of Far-From-Equilibrium Systems and Self-Organisation." In Paul Davies 1989. pp. 316–47.

Nussbaum, Martha. 1990. "Aristotelian Social Democracy." In *Liberalism and the Good*. Ed. by R. Bruce Douglass et al. New York: Routledge.

Oakeshott, Michael. 1991. *On Human Conduct*. Oxford: Clarendon.

Oppenheim, Paul, and Hilary Putnam. 1958. "The Unity of Science as a Working Hypothesis." *Minnesota Studies in Philosophy of Science*. Vol. 2, pp. 3–36.

Pattee, H. H. 1969. "How Does a Molecule Become a Message?" *Developmental Biology Supplement* 3. pp. 1–16.

———. 1973. *Hierarchy Theory: The Challenge of Complex Systems*. New York: G. Baziller.

Peacocke, Arthur. 1993. *Theology for a Scientific Age*. Minneapolis, MN: Fortress Press.

Peirce, Charles Sanders. 1931. *The Collected Papers of Charles Sanders Peirce*. Vol. 5. Ed. by Charles Hartshorne and Paul Weiss. Cambridge, MA: Harvard University.

———. 1992a. "Questions Concerning Certain Faculties Claimed for Man." *The Essential Peirce: Selected Philosophical Writings, Vol. I. 1867–1893*. Ed. by Nathan Houser and Christian Kloesel. Bloomington, IN: Indiana University.

———. 1992b."Some Consequences of Four Incapacities." In *Essential Peirce. Vol. I.*

———. 1992c. "Grounds of Validity of the Laws of Logic" In *Essential Peirce. Vol. I.*

———. 1992d. "How to Make our Ideas Clear." In *Essential Peirce. Vol. I.*

———. 1992e. "The Order of Nature." In *Essential Peirce. Vol. I.*

Penrose, Roger. 2005. *The Road to Reality: A Complete Guide to the Laws of the Universe*. New York: Knopf.

Pinker, Steven. 1994. *The Language Instinct: How the Mind Creates Language*. New York: William Morrow and Company.

Pittendrigh, Colin. 1958. "Adaptation, natural selection, and behavior." In *Behavior and Evolution*. Ed. by A. Roe and G. G. Simpson. New Haven: Yale University Press. pp. 390–416.

Plato. 1999. *Meno*. In *Great Dialogues of Plato*. Trans W.H.D. Rouse. New York: Signet.

Poland, Jeffrey. 1994. *Physicalism: The Philosophical Foundations*. Oxford: Clarendon.

Polanyi, Michael. 1962. *Personal Knowledge: Towards a Post-Critical Philosophy*. Chicago: University of Chicago.

Polkinghorne, John. 1984. *The Quantum World*. Princeton: Princeton University.

Popper, Karl. 1963. *Conjectures and Refutations: The Growth of Scientific Knowledge*. New York: Harper & Row.

———. 1972. "Of Clouds and Clocks: An Approach to the Problem of Rationality and the Freedom of Man." In *Objective Knowledge: An Evolutionary Approach*. Oxford: Clarendon. pp. 206–55.

———. 1987. "Natural Selection and the Emergence of Mind." In Radnitzky and Bartley. 1987. pp. 139–56.

———. 1996. "The Myth of the Framework." In *The Myth of the Framework: A Defense of Science and Rationality*. New York: Routledge. pp. 33–69.

———. 2002. *The Logic of Scientific Discovery*. New York: Routledge.

Post, John. 2007. *The Faces of Existence: An Essay in Nonreductive Metaphysics*. Ithaca, NY: Cornel University.

Poundstone, William. 1985. *The Recursive Universe: Cosmic Complexity and the limits of Scientific Knowledge*. Chicago, IL: Contemporary Books.

Premack, David, and Guy Woodruff. 1978. "Does the Chimpanzee have a theory of mind?" *Behavioral and Brain Sciences*. 4:515–26.

Prigogine, Ilya. 1977. "Time, Structure, and Fluctuations." Nobel lecture. December 8. http://nobelprize.org/nobel_prizes/chemistry/laureates/1977/prigogine-lecture.pdf.

———, and Gregoire Nicolas. 1989. *Exploring Complexity: An Introduction*. New York: Freeman and Co.

———, and Isabelle Stengers. 1984. *Order out of Chaos: Man's New Dialogue with Nature*. New York: Bantam.

Primas, Hans. 1998. "Emergence in Exact Natural Sciences." *Acta Polytechnica Scandinavica* Ma 91:83–98.

Putnam, Hillary. 1975. *Mind, Language and Reality: Philosophical Papers vol. 2*. Cambridge, UK: Cambridge University Press.

———. 1981. *Reason, Truth, and History*. Cambridge, UK: Cambridge University.

———. 1994. "Sense, Nonsense, and the Senses: An Inquiry into the Powers of the Human Mind." *The Journal of Philosophy* XCI, no.9. September. pp. 445–517.

———. 2004. *Collapse of the Fact-Value Dichotomy and Other Essays*. Cambridge, MA: Harvard University.

Quine, W. V. O. 1962. "Translation and Meaning." In *Word and Object*. Cambridge, MA: MIT Press. pp. 26–79.

———. 1968a. "Epistemology Naturalized." In *Ontological Relativity and Other Essays*. New York: Columbia University. pp. 69–90.

———. 1968b. "Natural Kinds." In *Ontological Relativity and Other Essays*. pp. 114–38.

———. 1980. *From a Logical Point of View*. Cambridge, MA: Harvard University Press.

Radnitzky, Gerard, and W. W. Bartley III. 1987. *Evolutionary Epistemology, Rationality, and the Sociology of Knowledge*. Chicago and La Salle, IL: Open Court.

Ramsey, Frank. 1990. "Law and Causality B: General Propositions and Causality." *Philosophical Papers*. Ed. by D. H. Mellor. Cambridge: Cambridge University Press. pp. 145–63.

Ramsey, Jeffrey. 1997. "Molecular Shape, Reduction, Explanation and Approximate Concepts." *Synthese* 111:233–51.

Randall, John Herman. 1958. *Nature and Historical Experience*. New York: Columbia University Press.

Regan, Thomas. 1985. *The Case for Animal Rights*. Berkeley: University of California.

Reichenbach, Hans. 1961. *The Rise of Scientific Philosophy*. Berkeley, CA: University of California.

Reynolds, Andrew. 2002. *Peirce's Scientific Metaphysics: The Philosophy of Chance, Law, and Evolution*. Nashville, TN: Vanderbilt University.

Rockwell, W. Teed. 2005. *Neither Brain Nor Ghost: A Nondualist Alternative to the Mind-Brain Identity Theory*. Cambridge, MA: MIT Press.

Rolston III, Holmes. 1988. *Environmental Ethics: Duties to and Values in the Natural World*. Philadelphia: Temple University.

———. 1989. "The River of Life: Past, Present, and Future." In *Philosophy Gone Wild: Environmental Ethics*. Amherst, NY: Prometheus Books. pp. 61–71.

———. 2010. *Three Big Bangs*. New York: Columbia University.

Rorty, Richard. 1989. *Contingency, Irony, and Solidarity*. Cambridge, UK: Cambridge University.

———. 1991. *Objectivity, Relativism, and Truth*. Cambridge, UK: Cambridge University.

Royce, Josiah. 1908. *The Philosophy of Loyalty*. New York: Macmillan.

Ruse, Michael. 1995. "The View from Somewhere." In *Evolutionary Naturalism*. London: Routledge.

Russell, Robert John. 1993. "Finite Creation Without a Beginning: The Doctrine of Creation in Relation to Big Bang and Quantum Cosmologies." In *Quantum cosmology and the Laws of Nature: Scientific Perspectives on Divine Action*. Ed. by Russell, Nancy Murphy, and C. J. Isham. Vatican City: Vatican Observatory. pp. 293–329.

———. 1995. "Quantum Physics in Philosophical and theological Perspective." In *Physics, Philosophy, and Theology: A Common Quest for Understanding*. Ed. by Robert J. Russell, William Stoeger, and George V. Coyne. Vatican City: Vatican Observatory. pp. 343–74.

Ryder, John. 1994. *American Philosophic Naturalism in the Twentieth Century*. Amherst, NY: Prometheus Books.

Salthe, Stanley. 1985. *Evolving Hierarchical Systems*. New York: Columbia.

———. 1993. *Development and Evolution: Complexity and Change in Biology*. Cambridge, MA: MIT Press.

Sarkar, Sahotra. 1998. *Genetics and Reductionism*. Cambridge, UK: Cambridge University.

Scerri, Eric R. 1991. "The Electronic Configuration Model, Quantum Mechanics and Reduction." *British Journal of the Philosophy of Science* 42:309–325.

———. 2007 *The Periodic Table: Its Story and Significance*. Oxford: Oxford University.

Scerri, Eric R., and Lee McIntyre. 1997. "The Case for the Philosophy of Chemistry." *Synthese* 111:213–32.

Schelling, Friedrich Wilhelm Joseph von. 1936. *Philosophical Inquiries into the Nature of Human Freedom*. Trans. James Guttman. LaSalle, IL: Open Court.

———. 1988. *Ideas for Philosophy of Nature*. Trans. by Errol Harris and Peter Heath. Cambridge, UK: Cambridge University.

Schmidt, James, ed. 1996. *What is Enlightenment? Eighteenth-Century Answers and Twentieth-Century Questions*. Berkeley: University of California.

Schneider, Eric, and Dorion Sagan. 2005. *Into the Cool: Energy Flow, Thermodynamics and Life*. Chicago: University of Chicago Press.

Schröder, Jürgen. 1998. "Emergence: Non-Deducibility or Downwards Causation." *The Philosophical Quarterly* 48:193.

Schrödinger, Erwin. 1992. *What is Life? with Mind and Matter and Autobiographical Sketches*. Cambridge, UK: Cambridge University.

Searle, John. 1980. "Minds, Brains, and Program." *Behavioral and Brain Sciences* 3. pp. 450–6.

Sellars, Roy Wood. 1922. *Evolutionary Naturalism*. Chicago: Open Court.

Shannon, Claude. 1948. "A Mathematical Theory of Communication." *The Bell System Technical Journal* 27, July, October: 379–423, 625–56.

Shepard, Paul. 1973. *The Tender Carnivore and the Sacred Game*. Athens, GA: University of Georgia Press.

———. 1982. *Nature and Madness*. San Francisco: Sierra Club Books.

———. 1998. *Coming Home to the Pleistocene*. Ed. by Florence Shepard. Washington, DC: Island Press/Shearwater.

Shepard, Paul, with Barry Sanders. 1985. *The Sacred Paw: The Bear in Nature, Myth, and Literature*. New York: Viking Press.

Shimony, Abner. 1989. "Conceptual Foundations of Quantum Mechanics." In Davies 1989. pp. 373–95.

———. 1993a. "Reality, Causality, and Closing the Circle." In *The Search for a Naturalistic Worldview, Vol. I: Scientific Method and Epistemology*. Cambridge: Cambridge University. pp. 21–61.

———. 1993b. "Search for a Worldview which can Accommodate our Knowledge of Microphysics." *The Search for a Naturalistic Worldview, Vol. I: Scientific Method and Epistemology*. pp. 62–76.

———. 1993c. "Some Proposals Concerning Parts and Wholes." *The Search for a Naturalistic Worldview, Vol. II: Natural Science and Metaphysics*. Cambridge: Cambridge University. pp. 218–27.

Shimony, Abner, and Debra Nails. 1987. *Naturalistic Epistemology: A Symposium of Two Decades*. Dordrecht: Reidel.

Simon, Herbert A. 1962. "The Architecture of Complex Systems." In *Proceedings of the American Philosophical Society*. pp. 467–82.

———. 1973. "The Organization of Complex Systems." In Pattee 1973. pp.1–28.

———. 2003. "Can there be a Science of Complex Systems." In *Unifying Themes in Complex Systems. Vol. One. Proceedings of the First International Conference on Complex Systems*. Ed. by Yaneer Bar-Yam. Boulder, CO: Westview Press. pp. 1–12.

Simpson, Lorenzo. 2001. *The Unfinished Project: Toward a Postmetaphysical Humanism*. New York: Routledge.

Singer, Peter. 2002. *Animal Liberation*. New York: HarperCollins.

Smith, Quentin. 1998. "Why Steven Hawking's Cosmology Precludes a Creator." *Philo: A Journal of Philosophy* vol. 1, no. 1, pp. 75–94.

Smolin, Lee. 1997. *The Life of the Cosmos*. Oxford: Oxford University Press.

———. 2001. *Three Roads to Quantum Gravity*. New York: Basic Books.

Smuts, Jan Christian. 2006. *Holism and Evolution*. Whitefish, MT: Kessinger Publishers.

Snow, C. P., and Stefan Collini. 1993. *The Two Cultures*. Cambridge, UK: Cambridge University Press.

Snow, Dale. 1996. *Schelling and the End of Idealism*. Albany: State University of New York.

Spaulding, Edward Gleason. "A Defense of Analysis." In Holt 1925. pp. 155–302.

Sperry, R. W. 1976. "Mental Phenomena as Causal Determinants in Brain Function." In Globus, et al. 1976. pp. 163–77.

Stachel, John. 1993. "The Meaning of General Covariance: The Hole Story." In *Philosophical Problems of the Internal and External Worlds: Essays on the Philosophy of Adolf Grünbaum*. Ed by Jean Earman, All Janis, Gerald Massey, and Nicholas Rescher. Pittsburgh: University of Pittsburgh. pp. 129–60.

———. 2002. "Einstein's Search for General Covariance, 1912–1915." In John Stachel. *Einstein from 'B' to 'Z.'* Boston: Birkhäuser. pp. 301–37.

———. 2006. Stachel, John. "Structure, Individuality and Quantum Gravity." In *Structural Foundations of Quantum Gravity*. Ed. by Dean Rickles, Steven French, and Juha Saatsi. Oxford: Oxford University Press. pp. 53–82.

Stein, Howard. 1989. "Yes, but . . . Some Skeptical Remarks on Realism and Anti-Realism." *Dialectica* vol. 43, no. 1, pp. 47–65.

Steinhardt, Paul, and Neil Turok. 2006. "Why the Cosmological Constant is Small and Positive." *Science* vol. 312, no. 5777. May 26:1180–83.

———. 2007. *Endless Universe: Beyond the Big Bang*. New York: Doubleday.

Stenger, Victor. 2006. "Why is there Something rather than Nothing?" *Skeptical Briefs Newsletter*. Committee for Skeptical Inquiry. June. Vol. 16. http://www.csicop.org/ sb/2006-06/reality-check.html.

Stillwaggon, Liz, and Louis J. Goldberg. 2009. "Biosymbols: Symbols in Life and Mind." *Biosemiotics* 3. 1:17–31.

Stone, Jerome. 2008. *Religious Naturalism Today: The Rebirth of a Forgotten Alternative*. Albany: State University of New York.

Strawson, P. F. 1990. *Individuals*. New York: Routledge.

Stueber, Karston. 2006. *Rediscovering Empathy: Agency, Folk Psychology, and the Human Sciences*. Cambridge, MA: MIT Press.

Tegmark, Max. 2003. "Parallel Universes." *Scientific American* May. pp. 41–51.

Teller, Paul 1992. "A Contemporary Look at Emergence." In Beckerman et al., 1992b. pp. 139–55.

Thoreau, Henry David. 1937. "Walking." Ed. by Brooks Atkinson, in *Walden and Other Writings of Henry David Thoreau*. New York: The Modern Library. pp. 597–632.

Tillich, Paul. 1952. *The Courage to Be*. New Haven: Yale University Press.

Tomasello, Michael. 1999. *The Cultural Origins of Human Cognition*. Cambridge, MA: Harvard University.

Tomasello, Michael, et al. 2005. "Understanding and Sharing Intentions: The Origins of Cultural Cognition." *Behavioral and Brain Sciences* 28:675–735.

Tomasello, Michael, and Josep Call. 1997. *Primate Cognition*. Oxford: Oxford University.

Tryon, Edward. 1973. "Is the Universe a Vacuum Fluctuation?" *Nature* 246, pp. 346–7.

Ulanowicz, Robert. 2009. *A Third Window: Natural Life beyond Newton and Darwin*. West Conshohocken, PA: Templeton Foundation.

Van Brakel, J. 1997. "Chemistry as the Science of the Transformation of Substances." *Synthese* 111:253–82.

Van Lawick-Goodall, Jane. 1971. *In the Shadow of Man*. Boston: Houghton Mifflin.

Velmans, Max. 2002. "How Could Conscious Experience Affect Brains? *Journal of Consciousness Studies* 9(11). pp. 3–29.

Vilenkin, Alexander. 1982. "Creation of Universes from Nothing." *Physical Letters* 117B:25–28.

———. 1984. "Quantum Creation of Universes." *Physical Review D*. 30 (2). July 15. pp. 509–11.

———. 2006. *Many Worlds in One: The Search for Other Universes*. New York: Hill and Wang.

Von Neumann, John. 1966. *Theory of Self-Reproducing Automata*. Ed. by Arthur W. Burks. University of Illinois: Urbana and Champaign, IL.

Von Uexküll, Jacob. 1926. *Theoretical Biology*. Trans. by D. L. MacKinnon. New York: Harcourt, Brace.

Waddington, Conrad. 1957. *The Strategy of the Genes: A Discussion of Some Aspects of Theoretical Biology*. London: Allen & Unwin.

Wade, Nicholas. 2010. "Researchers Say They Created a 'Synthetic Cell'." *New York Times*. May 20.

Wallace, Kathleen. 1990. "Ordinal Possibility: A Metaphysical Concept." In Marsoobian et al. 1990. pp. 171–88.

Walzer, Michael.1994. *Thick and Thin: Moral Argument at Home and Abroad*. Notre Dame: University of Notre Dame.

Warnock, Mary. 1978. *Imagination*. Berkeley: University of California.

Watts, Ian. 1999. "The Origin of Symbolic Culture." In Dunbar 1999. pp. 113–46.

Weber, Bruce. 2007. "Fact, Phenomenon, and Theory in the Darwinian Research Tradition." *Biological Theory* 2:168–78

Weber, Max. 1972. "Science as a Vocation." In *From Max Weber: Essays in Sociology*. Trans. by H. H. Gerth and C. Wright Mills. New York: Oxford University. pp. 129–56.

Weinberg, Steven. 1992. *Dreams of a Final Theory: The Scientist's Search for the Ultimate Laws of Nature*. New York: Vintage.

———. 1993. *The First three Minutes. A Modern View of the Origin of the Universe*. New York: Basic Books.

Weiss, Phil. 1990. "Possibility: Three Recent Ontologies." Marsoobian et al. 1990. pp. 145–70.

Wheeler, John A. 1990. "Information, Physics, Quantum: The Search for Links." In *Complexity, Entropy, and the Physics of Information*. Ed. by W. H. Zurek. Reading, MA: Addison-Wesley. pp. 3–28.

Wheeler, William Morton. 1928. *Emergent Evolution and the Development of Societies*. New York: W. W. Norton.

Whitehead, A. N. 1925. *Science and the Modern World*. New York: Mentor Press.

———. 1957. *The Concept of Nature*. Ann Arbor, MI: University of Michigan Press

———. 1978. *Process and Reality*. New York: Free Press.

———. 1982. *An Enquiry Concerning the Principles of Natural Knowledge*. New York: Dover Press.

———. 2005. *Principle of Relativity*. New York: Barnes and Noble.

Wiener, Norbert. 1948. *Cybernetics: Or Control and Communication in the Animal and the Machine*. Cambridge, MA: MIT Press.

Wilczek, Frank, and Betsy Divine. 1989. *Longing for the Harmonies: Themes and Variations from Modern Physics*. New York: Norton.

Wilson, Jessica. 2006. "On Characterizing the Physical." *Philosophical Studies* 131:61–99.

Wilson, E. O. 1983. "An Introduction." *Nature Conservancy News* 33, no. 6. November-December.

Wimsatt, William C. 1972. "Teleology and the Logical Structure of Function Statements," *Studies in the History and Philosophy of Science* 3, no.1. pp.1–80.

———. 1974. "Complexity and Organization." In Wimsatt 2007. pp. 179–89.

———. 1976a. "Reductive Explanation: A Functional Account." In Wimsatt 2007. pp. 241–73.

———. 1976b. "Reductionism, Levels of Organization, and the Mind-Body Problem." In G. G. Globus 1976. pp. 199–267.

———. 1986. "Forms of Aggregativity." In *Human Nature and Natural Knowledge: Essays Presented to Marjorie Grene on Occasion of Her Seventy-Fifth Birthday*. Ed. by Alan Donagan, Anthony Perovich, and Michael Wedin. Dordrecht: Reidel. pp. 259–91.

———. 1994. "The Ontology of Complex Systems: Levels of Organization, Perspectives, and Causal Thickets." In Wimsatt 2007. pp. 193–240.

———. 2000. "Emergence as Non-Aggregativity and the Biases of Reductionisms." In Wimsatt 2007. pp. 274–312.

———. 2007. *Re-Engineering Philosophy for Limited Beings: Piecewise Approximations to Reality*. Cambridge: Harvard University.

Winnicott, D. W. 1982. *Playing and Reality.* New York: Routledge.

Witmer, D. Gene. 2006. "How to Be a (Sort of) A Priori Physicalist." *Philosophical Studies.* 131:185–225.

Wittgenstein, Ludwig. 1953. *Philosophical Investigations.* Trans. by G. E. M. Anscombe. Oxford University Press.

Woolley, R. G. 1978. "Must a Molecule Have a Shape?" *American Chemical Society* 100:4. pp. 1073–78.

Wrangham, Richard. 2009. *Catching Fire: How Cooking Made Us Human.* New York: Basic Books.

Zurek, Wojciech H. 2002. "Decoherence and the Transition from quantum to Classical—Revisited." *Los Alamos Science* 27:2–25.

Zurek, Wojciech H, and Juan Pablo Paz. 1994. "Decoherence, Chaos, and the Second Law." *Physical Review Letters* 72. 16:2508–2512.

Index